Olena Slavinska · Viktor Danchuk ·
Olga Kunytska · Oksana Hulchak
Editors

Intelligent Transport Systems: Ecology, Safety, Quality, Comfort

Proceedings of ITS ESQC 2024, Volume 1

Springer

Editors
Olena Slavinska
National Transport University
Kyiv, Ukraine

Olga Kunytska
National Transport University
Kyiv, Ukraine

Viktor Danchuk
National Transport University
Kyiv, Ukraine

Oksana Hulchak
National Transport University
Kyiv, Ukraine

ISSN 2367-3370 ISSN 2367-3389 (electronic)
Lecture Notes in Networks and Systems
ISBN 978-3-031-87375-1 ISBN 978-3-031-87376-8 (eBook)
https://doi.org/10.1007/978-3-031-87376-8

This Springer imprint is published by the registered company Springer Nature Switzerland AG
The registered company address is: Gewerbestrasse 11, 6330 Cham, Switzerland

Preface

The 2nd International Scientific Conference "Intelligent Transport Systems: Ecology, Safety, Quality, Comfort" (ITS ESQC 2024) was held on November 26–27, 2024. The National Transport University organized the conference with the Ministry of Education and Science of Ukraine. It brought together leading academic scientists, researchers, and industry representatives to exchange experiences, best practices, and research results in ITS.

The event was conducted in a hybrid format: offline in Kyiv and with the possibility of online participation, ensuring safety and accessibility for all participants under the conditions of the ongoing full-scale aggression against Ukraine. We extend our sincere gratitude to the Armed Forces of Ukraine for making it possible to host this conference. The conference program included plenary presentations, panel discussions, and special sessions on topics such as "Smart Cities and Sustainable Development," "Intelligent Transport Technologies and Smart Logistics," and "Smart Mobility and Traffic Safety." Keynote speakers included international experts such as Adnan Rahman, Luca Persia, Antonio Comi, Bruno Dalla Chiara, Torben Heinemann, Stanisław Iwan and others.

This volume of proceedings contains selected papers covering pressing issues such as the ecological sustainability of transport systems, road safety, implementation of energy-efficient technologies, and the integration of smart solutions into urban infrastructure. All contributions offer valuable insights and foster experience exchange among scientists in the field of transport engineering. The organizing committee of the conference received 119 papers through the Microsoft CMT platform, of which 76 were accepted. The papers were submitted by 44 universities from about 10 countries (Ukraine, Poland, the Philippines, England, Italy, Brazil, Spain, and Lithuania). All submitted papers were evaluated by the editors for compliance with the requirements presented on the conference website https://www.itsesqc.ntu.edu.ua and reviewed by the reviewers, which included scientists from Europe and Ukraine.

The organizers of ITS ESQC 2024 made significant efforts to ensure the success of the conference. We express our gratitude to all members of the Program and Organizing Committees, the reviewers for their work in evaluating and selecting the papers, and all the authors for their contributions to creating a shared intellectual environment for solving relevant scientific challenges.

Special thanks go to Springer, particularly Thomas Ditzinger and Janusz Kacprzyk, for their support in publishing these selected works in the " Intelligent Transport Systems: Ecology, Safety, Quality, Comfort" series.

Kyiv, Ukraine

Olena Slavinska
Viktor Danchuk
Olga Kunytska
Oksana Hulchak

Organization

Organizers

Ministry of Education and Science of Ukraine

National Transport University

Steering Committee

Honorary Chair

Oleksandr Gryshchuk — Candinate of Technical Sciences, Full Professor, Rector of the National Transport University, Kyiv, Ukraine

Executive Chair

Olena Slavinska — Dr. Tech. Sciences, Full Professor, Vice-rector for scientific work of the National Transport University, Kyiv, Ukraine

Program Committee Chair

Viktor Danchuk — Dr. Phys.-Math. Sciences, Prof., Full Professor, Dean of the Faculty of Transport and Information Technologies of the National Transport University, Kyiv, Ukraine

Publication Chair

Olga KunytskaProfessor of the Department of Transport Systems and Road Safety, Ph.D., National Transport University, Kyiv, Ukraine

International Program Committee

Oleksandr Gryshchuk	Rector of the National Transport University, Candinate of Technical Sciences, Full Professor, head of the Program Committee, Ukraine
Olena Slavinska	Vice-rector for scientific work of the National Transport University, Dr. Tech. Sciences, Full Professor, deputy chairman of the Program Committee, Ukraine
Viktor Danchuk	Dean of the Faculty of Transport and Information Technologies of the National Transport University, Dr. Phys.-Math. Sciences, Prof., Full Professor, Deputy Chairman of the Program Committee, Ukraine
Luca Persia	Ph.D., President of Sapienza Startup CTL-up, Professor of Sustainable Transport Planning at La Sapienza University of Rome, Italy
Bruno Dalla Chiara	Ph.D.; M.Sc. Mech Eng., Full Professor of Deptartment DIATI – Transport systems, Politecnico di Torino, Italy
Stanisław Iwan	Ph.D., Professor, Dean of Faculty of Economics and Transport Engineering, Maritime University of Szczecin, Poland
Norbert Gruenwald	Dr. rer. nat., Professor, Rector a.D. Hochschule Wismar University of Applied Sciences Technology, Business and Design, Faculty of Engineering, Director Robert-Schmidt-Institute, Germany
Antonio Comi	Ph.D., Professor, Department of Enterprise Engineering University of Rome Tor Vergata, Italy
Dmytro Bespalov	Director of "Pro Mobility" LLC, lecturer of Urban Planning department, KNUCA, Ukraine
Ilie Bricicaru	Ph.D., road safety auditor and trainer, director of the Road Infrastructure Safety Observatory within the Technical University of Moldova, Moldova

Serhii Turpak	Dr. Sc., Head of the Department of Transport Technology, Full Professor, National University Zaporizhzhia Polytechnic, Ukraine
Andrii Vozniuk	Ph.D., Head of the digitalization development department, State Enterprise "National Institute for Development Infrastructure", candidate technical Sciences, Ukraine
Volodymyr Kyselov	Dr. Sc., Full Professor, Director of the Educational and Scientific Institute of Municipal Management and Urban Economy at V.I. Vernadsky TNU, Ukraine
Elena Nechita	Ph.D., Vice-rector, Ph.D., Vasile Alecsandri University of Bacau, Romania
Christian Weiß	Dr., Full Professor,Dean of Faculty 4, Director of the Natural Sciences, Hochschule Ruhr West University of Applied Science, Germany
Anastasiia Makarenko	Executive Director of U-Cycle (Civic Organization "Kyiv Cyclists' Association"), Ukraine
Karl-Georg Steffens	Dr., Professor, Institute for Post-Fossil Logistics, Germany
Vitaliy Snytyuk	Dr. Tech. Sciences, Full Professor, Dean of Faculty of Information Technology, Taras Shevchenko National University of Kyiv, Ukraine
Alla Novikova	Doctor of Economics, Chef of Research Center of Complex Transport Problems, Ukraine
Francisco Alonso	Dr., Director of INTRAS, Full Professor, University of Valencia, Spain
Serhii Hnatiuk	Dr. Tech. Sciences, Full Professor, Dean of Faculty of Computer Sciences and Technologies, National Aviation University, Ukraine
Małgorzata Zakrzewska-Grünwald	Dr, Institute of Economics and Finance, University of Szczecin, Poland
Oleksandr Chemerys	Dr. Tech. Sciences, Deputy Director for Scientific Work, Senior Research Officer, G.E. Pukhov Institute for Modelling in Energy Engineering National Academy of Sciences of Ukraine
Viktoriia Nykonchuk	Dr. Economics Sciences, Full Professor, Head of Department of Transport Technologies and Technical Service, National University of Water and Environmental Engineering, Ukraine

Volodymyr Kaskiv	Ph.D., Deputy Director on Scientific Work, State Enterprise "National Institute for Development Infrastructure", Ukraine
Olena Bakulich	Ph.D., Dean of the Faculty of Management, Logistics and Tourism, Full Professor of the Department of Management, National Transport University, Ukraine
Andrii Bubela	Dr. Tech. Sciences, Dean of the Faculty of Transport Construction, Full Professor of the Department of Transport Construction and Property Management, National Transport University, Ukraine
Ihor Taran	Dr. Tech. Sciences, Head of the Department of Transport Management, Full Professor, National Technical University "Dniprovska Polytechnic", Ukraine
Petro Horbachov	Dr. Tech. Sciences, Head of the Department of Transport Systems and Logistics, Full Professor, Kharkiv National Automobile and Highway University, Ukraine
Olexandr Mischenko	Candidate of Technical sciences, Director of the Traffic Management Center of Kyiv, Ukraine
Volodymyr Sakgno	Dr. Tech. Sciences, Full Professor, Head of the Department of Automobiles the National Transport University, Ukraine
Yuriy Gutarevich	Dr. Tech. Sciences, Prof. department of engines and heat engineering, Full Professor, National Transport University, Ukraine
Ali Al-Ammouri	Dr. Tech. Sciences, Full Professor, Head of the Department of Information Analysis and Information Security, National Transport University, Ukraine
Myroslav Oliskevych	Dr. Tech. Sciences, Full Professor of agricultural engineering and technical service department, L'viv National University of Nature Management, Ukraine
Volodymyr Polishchuk	Dr. Tech. Sciences, Full Professor, Head of Department of Transport Systems and Road Safety, National Transport University, Ukraine
Georgii Prokudin	Dr. Tech. Sciences, Full Professor, Head of the Department of International Transportation and Customs Control, National Transport University, Ukraine

Vasyl Mateychyk	Dr. Tech. Sciences, Full Professor of the Department of Ecology and Environmental Protection Technologies, National Transport University, Ukraine
Ramazan Khabutdinov	Dr. Tech. Sciences, Full Professor, Head of the Department of Transport Technologies, National Transport University, Ukraine
Viktoriia Khrutba	Dr. Tech. Sciences, Full Professor, Head of the Department of Ecology and Environmental Protection Technologies, National Transport University, Ukraine
Andriy Galkin	Dr. Tech. Sciences, Professor, Professor University of Antwerp MSCA Postdoctoral fellow, Belgium. Kharkiv National University of Urban Economy named after O.M. Beketova, Ukraine
Oleksandr Dobrovolskyi	Ph.D., Professor, Dean of the Faculty of Automotive Mechanics, National Transport University, Ukraine
Viacheslav Savenko	Dr. Tech. Sciences, Full Professor, Head of the Department of Transport Construction and Property Management, National Transport University, Ukraine
Mykolay Moroz	Dr. Tech. Sciences, Full Professor, Head of the Department of Transport Technologies, Kremenchuk Mykhailo Ostrohradskyi National University, Ukraine
Valeriy Havrylenko	Dr. Phys.-Math. Sciences, Full Professor, Head of the Department of Information Systems and Technologies, National Transport University, Ukraine
Yurii Papirovyi	Head of the "Transport Safety" NGO, Ukraine

Reviewers

lexander Chemeris	Pukhov Institute for Modeling in Energy Engineering NASU, Ukraine
Ali Al-Ammouri	National Transport University, Ukraine
Andrii Galkin	O. M. Beketov National University of Urban Econ-omy in Kharkiv, Ukraine
Andrii Vozniuk	National Institute for Development of Infrastructure, Ukraine
Andriy Goncharenko	National Aviation University, Ukraine

Andriy Musienko	National Technical University of Ukraine "Igor Si-korsky Kyiv Polytechnic Institute", Ukraine
Antonio Comi	University of Rome Tor Vergata, Italy
Arsen Klochan	National Transport University, Ukraine
Bruno Dalla Chiara	Politecnico di Torino (University), Dept. DIATI—Transport systems, Italy
Dmytro Bespalov	Kyiv National University of Construction and Archi-tecture, Ukraine
Dmytro Burko	O. M. Beketov National University of Urban Economy in Kharkiv, Ukraine
Dmytro Shevchuk	Nation aviation university, Ukraine
Edgar Sokolovskij	Vilnius Gediminas Technical University (VILNIUS TECH), Lithuania
Elena Nechita	Vasile Alecsandri University of Bacau, Romania
Elzbieta Macioszek	Silesian University of Technology, Poland
Francisco Alonso	University of Valencia, Spaine
Georgii Prokudin	National Transport University, Ukraine
Ievgen Medvediev	Gdansk University of Technology, Poland
Ihor Kozarchuk	National Transport University, Ukraine
Ihor Taran Dnipro	University of Technology, Ukraine
Ilie Bricicaru	Technical University of Moldova, Moldova
Kateryna Vakulenko	O.M. Beketov National University of Urban Econo-my in Kharkiv, Ukraine
Liudmyla Abramova	Kharkiv National Automobile and Highway Univer-sity, Ukraine
Maksym Afonin	Lviv Polytechnic National University, Ukraine
Marco Petrelli	Roma Tre University, Italy
Maria Cieśla	Silesian University of Technology, Poland
Mykola Moroz	Kremenchuk Mykhailo Ostrohradskyi National Uni-versity, Ukraine
Mykola Tsiuman	National Transport University, Ukraine
Myroslav Oliskevych	L'viv National University of Nature Management, Ukraine
Natalya Shramenko	Karlsruhe University of Applied Sciences, Germany
Norbert Grünwald	Wismar University, RSI, Germany
Oksana Hulchak	National Transport University, Ukraine
Olena Bakulich	Nationale Transport Universite, Ukraine
Olena Slavinska	National Transport University, Ukraine
Olexander Krasnoshtan	National transport university, Ukraine
Olga Kunytska	National Transport University, Ukraine
Olha Ushchenko	National Aviation University, Ukraine

Paulo Figueiras	Instituto de Desenvolvimento de Novas Tecnologias UNINOVA, Portugal
Petro Horbachov	Kharkiv national automobile and highway universi-ty, Ukraine
Serhii Taraban	State Enterprise "State Road Transport Research Institute", Ukraine
Serhii Turpak	National University Zaporizhzhia Polytechnic, Ukraine
Sevket Oguz Kagan Capkin	Sapienza University of Rome, Italy
Stanisław Iwan	Maritime University of Szczecin, Poland
Taras Postranskyy	Lviv Polytechnic National University, Ukraine
Umberto Crisalli	University of Rome Tor Vergata, Italy
Viktor Danchuk	National Transport University, Ukraine
Viktoriia Khrutba	National Transport University, Ukraine
Viktoriia Budnyk	State University of Infrastructure and Technologies, Ukraine
Vitalii Svatko	National Transport University, Ukraine
Vitaliy Zaytsev	Alfred Nobel University, Ukraine
Vitaliy Snytyuk	Taras Shevchenko National University of Kyiv, Ukraine
Vjacheslav Trushevski	National University "Zaporizhzhia Polytechnic", Ukraine
Volodymyr Kaskiv	State enterprise "National Institute for Development Infrastructure", Ukraine
Volodymyr Kyselov	V.I.Vernadsky Taurida National Yniversity, Ukraine
Yevhen Kush	O. M. Beketov National University of Urban Econ-omy in Kharkiv, Ukraine
Yuriy Royko	Lviv Polytechnic National University, Ukraine
Andriy Lyamzin	National Aviation University, Ukraine
Viktoria Nikonchuk	National University of Water Management and En-vironmental Management, Ukraine
Lyudmila Nagrebelna	State Enterprise "National Institute for Development Infrastructure" , Ukraine
Yury Davidych	O. M. Beketov National University of Urban Econ-omy in Kharkiv, Ukraine

Organizing Team

(National Transport University, Kyiv, Ukraine)
Olga Kunytska
Oksana Hulchak

Marina Dekhtyar
Viktoriia Lebid
Inna Vyhovska
Inesa Halona
Alina Korchevska
Arsen Klochan
Natalia Katrushenko
Stanislav Popov
Artem Oharkov
Nataliia Luzhanska
Alona Shpyh
Alina Sevostyanova
Oksana Semenchenko
Yevheniia Shapenko
Vitaliy Svatko
Nataliia Zozulia
Maksym Roi
Serhii Yanishevskyi

Contents

Smart Cities and Sustainable Development

Integral Approach to Assessing the Sustainable Urban Development of Kyiv

Olena Tymoshenko, Andriy Tymoshenko, Serhii Yahodzinskyi, and Arthur Amber(✉)

Private Higher Educational Establishment "European University", 16v. Vernadskoho Akademika Blvd, Kyiv 03115, Ukraine
arthur.amber@e-u.edu.ua

Abstract. This article analyzes theoretical and methodological approaches to assessing sustainable urban development. It identifies comprehensive indicator groups for evaluating the social, environmental, and economic aspects of urban development. Integrating these indicators have been developed as an integral measure of the Sustainable Urban Development Index (SUDI) for Kyiv. This index facilitates both quantitative and qualitative diagnostics of the city's development parameters in alignment with sustainable development principles.

According to the developed assessment scale for Kyiv's sustainable urban development level, from 2016 to 2018, the city's Sustainable Development Index was at a basic level, ranging from 0.4767 to 0.4849. From 2019 to 2021, the city demonstrated an average level of sustainable development, with values ranging from 0.5109 to 0.5293.

The analysis of correlation coefficients among the components of the city's sustainable urban development reveals significant interdependence between different aspects. Notably, a strong negative correlation between economic and environmental components (-0.823) suggests that increases in economic indicators are often accompanied by decreases in environmental indicators. Conversely, the positive correlation between economic and social components (0.810) underscores the importance of economic growth for enhancing social well-being.

Keywords: Sustainable Urban Development · Sustainable City · Sustainability Assessment · Smart City · Sustainability Indicators · Quality of Life · Environmental Safety · Urban Economics

O. Slavinska et al. (Eds.): ITSESQC 2024, LNNS 1335, pp. 3–13, 2025.
https://doi.org/10.1007/978-3-031-87376-8_1

1 Introduction

In the context of global urbanization, the concept of smart cities represents an innovative response to the numerous challenges facing urban agglomerations. Smart cities embody the integration of advanced information technologies and communication systems into urban infrastructure to enhance resource management efficiency, improve the quality of life for residents, and reduce environmental impact. They function as experimental platforms for implementing sustainable development principles, ensuring a balance between economic growth, social equity, and environmental sustainability.

One of the main goals of smart cities is to ensure sustainable development. The World Commission on Environment and Development defined sustainable development as "development that meets the needs of the present without compromising the ability of future generations to meet their own needs" [1]. Recently, this principle has become increasingly relevant in smart city research, highlighting its critical importance in shaping effective and resilient urban systems [2, 3].

The issue addressed in this research is the necessity for a comprehensive metric of urban sustainability that encompasses economic, environmental, and social dimensions. Current methodologies frequently concentrate on individual aspects of sustainability, lacking a unified index that represents the overall performance of urban development.

The primary objective of this research is to develop a Sustainable Urban Development Index for Kyiv. The index is designed to provide a holistic evaluation of the city's sustainability performance by integrating various dimensions of urban development.

To achieve this, the study aims to identify comprehensive indicator groups by establishing a set of indicators that thoroughly cover the social, environmental, and economic aspects of urban sustainability, integrate these indicators into a single SUDI that reflects the overall sustainability of Kyiv, and analyze correlations to examine the interactions and impacts of different sustainability components on the city's development.

2 The Bibliographic Review

Most scientific studies assert that smart cities have the potential to enhance resilience [4, 5]. At the same time, other research points to possible negative consequences of implementing smart technologies, noting that an excessive focus on economic benefits and complex technologies can both create new and exacerbate existing social and environmental issues [6].

D. Michalina, P. Mederly, H. Diefenbacher and B. Held reviewed global and European urban sustainability indicator frameworks, highlighting commonalities and differences in sustainability dimensions and thematic categories. They identified key indicators of urban sustainability and pointed out inconsistencies in assessing and measuring sustainable development [7].

Tomadon, E. Couto, W. Vries and Y. Moretto delved into the pivotal role that indicators play in designing, assessing, and guiding policies for sustainable urban development. They highlighted the importance of comprehensive indicators and the growing interest in smart city initiatives [8].

B. Mengiste, W. Shi and M. Wong conducted a systematic review of empirical studies on urban landscape development effects (ULDE), categorizing them into energy consumption impacts, transport and mobility, climate change, land use change, and livability and resilience. They emphasized the need for holistic assessment methods and integrated urban spatial growth strategies [9].

Thus, while most studies focus on the economic and environmental aspects of sustainability, the social dimension remains largely underestimated [10].

Despite the substantial number of scientific works on the topic of urban sustainability assessment, the methodology for evaluating the sustainability of its infrastructure remains unapproved.

In recent years, methodologies for comparative analysis that use a range of indicators to assess various aspects of urban sustainability have become widely applied. These methods were developed for assessing urban sustainability and sometimes used for evaluating the sustainability of municipal infrastructures, due to the lack of a universally method.

According to C. Baloloy, the Cost and Benefits Analysis (CBA) method is a suitable tool for measuring sustainability and the resilience of infrastructure systems to future climate uncertainties [11].

Among Ukrainian researchers, it is worth noting the studies of S. Yahodzinskyi, who examined the role of public transportation as a phenomenon in the context of sociocultural relations in the modern urban space [12, 13]. A. Hrechko dedicated her research to methodological approaches for assessing the sustainable development of productive forces in Ukraine's regions [14]. S. Leonova and K. Barvinska analyzed the prospects for sustainable development of Ukraine's road transportation system [15]. M. Tymoshenko researched methodological principles for assessing sustainable development in rural areas [16].

3 Research Methodology

To obtain an objective conclusion regarding the level of sustainable urban development in the city of Kyiv, we propose a methodology for integral assessment. The first stage of the research involves defining the purpose, object, subject, and tasks of the assessment. The next stage involves creating an information base for the research, which based on complete and reliable data. It is important to note that due to the full-scale invasion of the Russian Federation into Ukraine, the State Statistics Service of Ukraine up to 2021, which complicates the analytical stage of the research, mostly provides certain categories of statistical information.

The analytical stage includes calculating the level of sustainable development based on functional components. This involves forming a set of indicators to determine the state of functional components of sustainable development and their characteristics. When forming the set of indicators, it is important to ensure the informational consistency of the indicators. To achieve this, indicators are divided into stimulators and dissimulators.

The selection of indicators for assessing the level of sustainable urban development of a city should follow the principles of comprehensiveness (adequate representation of all aspects), measurability (realistic and calculable), informational simplicity (based on official documents, reports, and expert assessments), uniqueness (indicators should not reflect the same characteristics of sustainable development subsystems), and normalization (the ability to bring selected indicators to a uniform format and units of measurement) (Table 1).

Table 1. Indicators for the integral assessment of the sustainable urban development level of Kyiv by functional components

Indicator	2016	2017	2018	2019	2020	2021	Weighting coefficient
Share of profitable enterprises, %	65.5	66.1	68.2	69.4	65.5	68.5	0.25
Sales volume relative to 2015, %	129.5	162.8	199.7	208.4	215.8	285.5	0.20
Labor costs relative to 2015, %	141.6	186.5	235.2	288.0	312.8	375.0	0.15
Labor productivity relative to 2015, %	111.9	137.0	163.4	162.4	168.0	220.0	0.20
Share of labor costs in the cost of goods sold relative to 2015, %	5.25	5.51	5.66	6.64	6.97	6.31	0.10
Consumer price indices, %	114.0	113.4	108.8	103.9	105.8	109.7	0.10
CO_2 emissions relative to 2015, %	111.8	105.9	105.9	104.0	90.2	76.5	0.25
Volume of untreated wastewater discharged into surface water bodies relative to 2015, %	94.1	90.1	87.8	85.9	76.3	76.3	0.20
Volume of recycled and reused water relative to 2015, %	202.4	215.5	256.0	307.8	349.3	148.8	0.15
Share of waste recycled relative to 2015, %	32.8	65.0	70.7	80.7	82.5	96.0	0.10
Share of disturbed land, %	3.60	3.60	3.60	3.60	3.60	3.60	0.10
Share of forests and other forested areas, %	41.7	41.7	41.7	41.7	41.7	41.7	0.15
Current expenditures on environmental protection relative to 2015, %	148.4	149.0	217.8	245.6	250.1	243.0	0.25

(continued)

Table 1. (*continued*)

Indicator	2016	2017	2018	2019	2020	2021	Weighting coefficient
Employment rate of the working-age population relative to the total population of the same age, %	68.8	68.3	69.5	70.9	69.5	69.7	0.20
Unemployment rate of the working-age population relative to the total population of the same age, %	7.1	7.4	6.6	6.2	7.3	7.5	0.10
Share of reported criminal offenses relative to the total number in Ukraine, %	13.0	13.5	12.3	12.1	11.0	12.0	0.15
Share of the population that reported using internet services, %	72.2	72.2	76.4	81.4	83.5	86.9	0.10
Share of urban housing stock relative to the total in Ukraine, %	10.7	10.6	10.5	10.0	9.4	9.0	0.10
Share of higher education students relative to the total in Ukraine, %	26.0	26.5	26.0	25.6	22.7	24.3	0.15
Share of doctors relative to the total population in Ukraine, %	13.1	13.3	13.6	13.6	14.0	9.9	0.20

Source: State Statistics Service of Ukraine

Normalization of the stimulator indicators carried out according to the corresponding formula [17]:

$$P_i = \frac{x_{i\,\max} - x_i}{x_{i\,\max} - x_{i\,\min}}, \tag{1}$$

Destimulator indicators are normalized according to the formula:

$$P_i = \frac{x_i - x_{i\,\min}}{x_{i\,\max} - x_{i\,\min}}, \tag{2}$$

where:

P_i- is the value of the normalized i-th indicator;

x_i- is the value of the i-th sustainability indicator;
$x_{i\,\max}$- is the upper threshold of the i-th sustainability indicator;
$x_{i\,\min}$- is the lower threshold of the i-th sustainability indicator.

These normalized values allow for a comprehensive assessment of the sustainable development level, facilitating a more accurate comparison and analysis of the functional components across different periods and scenarios.

4 Results and Discussion

To ensure a quality assessment of the process, it is necessary to conduct a gradation of the levels of sustainable urban development of Kyiv, which defined based on the use of normative values of indicators and the method of expert assessments (Table 2).

This comprehensive approach allows stakeholders to pinpoint key areas for improvement and track the effectiveness of various strategies.

Table 2. Characteristics of sustainable urban development levels

Sustainable Development Level	Sustainable Development Level Indicator	Level Characteristics
Advanced	0.8 –1.0	A smart city demonstrates effective integration of smart technologies to optimize resource use, reduce carbon emissions, and enhance residents' quality of life. Such cities characterized by advanced infrastructure, a high level of public engagement, and robust policies that ensure long-term sustainability
Moderate	0.5 –0.8	The city demonstrates positive trends toward sustainable development, although certain aspects require further improvement. The achieved balance between economic, environmental, and social components is moderate, indicating the need for continued optimization of processes to enhance sustainability indicators

(*continued*)

Table 2. (*continued*)

Sustainable Development Level	Sustainable Development Level Indicator	Level Characteristics
Basic	0.3 –0.5	The city is in the early stages of integrating sustainable development into its urban development strategies. It has begun implementing smart technologies and sustainable practices, but these efforts not realized yet. There may be inconsistencies or gaps in policy implementation, as well as limited public participation in sustainable initiatives
Low	0 –0.3	Cities with a low level of sustainable development face challenges in meeting basic sustainability criteria. They may experience significant environmental degradation, social inequality, and economic instability. In these cities, the use of smart technologies and sustainable practices is minimal or ineffective, necessitating substantial improvements and support to achieve sustainability goals

Source: authors' own development

To determine the weighting coefficients of the indicators for assessing the level of sustainable urban development by functional subsystems, experts among the scientists and stakeholders identified for each functional component. Through a survey, respondents asked about their preferences regarding the importance of the proposed indicators and their prioritization using a pairwise comparison scale developed by the Analytic Hierarchy Process (AHP) method. The results of the calculations indicate that the highest weighting value among the functional subsystems of sustainable development attributed to the economic subsystem, with a weight of 0.5 (50%). The social subsystem has a weight of 0.3 (30%), and the environmental subsystem has a weight of 0.2 (20%). Based on the use of reports, primary official documents, and expert evaluations, the assessment of the level of sustainable urban development of Kyiv for the years 2016–2022 was conducted (Table 3).

Based on the data analysis, it could be concluded that the level of sustainable urban development in Kyiv shows an upward trend during the period 2016–2021 (Fig. 1).

Table 3. Integral Assessment of the Sustainable Urban Development Level of Kyiv for the Period 2016–2021

Years	Economic	Environmental	Social	Sustainability
2016	0.2653	0.0894	0.1220	0.4767
2017	0.2784	0.0824	0.1227	0.4835
2018	0.2852	0.0753	0.1244	0.4849
2019	0.3123	0.0705	0.1281	0.5109
2020	0.3213	0.0659	0.1331	0.5203
2021	0.3093	0.0815	0.1386	0.5293

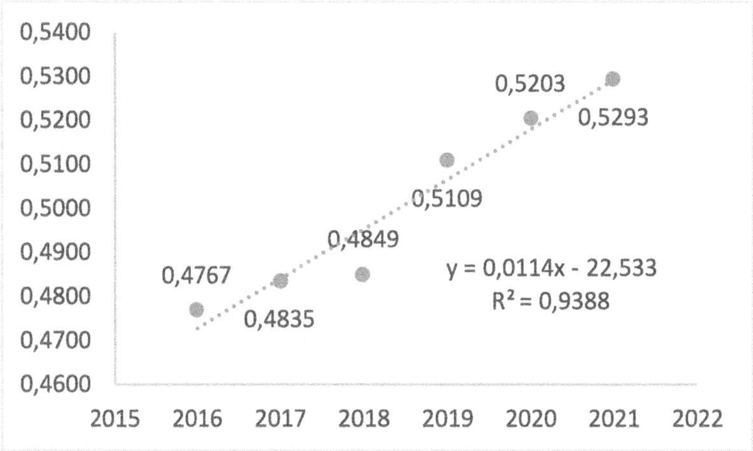

Fig. 1. Dynamics of the Sustainable Urban Development Index in Kyiv for the Period 2016–2021.

Period between 2016 and 2018, the level of sustainable development could be classified as basic, with indicators ranging from 0.4767 to 0.4849. From 2019 to 2021, the city experienced a moderate level, with indicators ranging from 0.5109 to 0.5293. Therefore, it could be asserted that the achieved balance between economic, environmental, and social components is moderate, indicating the need for further optimization of processes to improve sustainability indicators.

The detailed analysis of correlation coefficients reveals a significant interdependence between the subsystems of sustainable development. As shown in Table 4, these correlations highlight the intricate connections and mutual influences among various subsystems. This high degree of correlation underscores the complex and interconnected nature of sustainable development, indicating that changes in one subsystem can significantly impact others. Understanding these relationships is crucial for developing effective strategies and policies aimed at promoting sustainability.

A negative correlation between the economic and environmental components, with a correlation coefficient of -0.823, indicates that economic growth may be associated

Table 4. Matrix of Pairwise Correlation Coefficients of Sustainable Development Level by Functional Components

Functional Components	Economic Subsystem	Environmental Subsystem	Social Subsystem
Economic Subsystem	1		
Environmental Subsystem	−0.823	1	
Social Subsystem	0.810	−0.381	1

Source: authors' own development

with negative environmental consequences. This highlights the need to integrate environmental aspects into economic development strategies to prevent potential negative environmental effects.

A positive correlation between the economic and social components (correlation coefficient of 0.810) demonstrates that improvements in economic conditions generally accompanied by enhancements in social indicators. This underscores the importance of economic growth for improving social well-being and supporting economic strategies within the framework of overall sustainable development policy.

A moderately negative correlation between the environmental and social components (correlation coefficient of −0.381) indicates some contradictions between environmental initiatives and social goals, though this relationship is less pronounced. This suggests possible challenges in achieving a balance between environmental sustainability and social well-being, which requires further consideration within the context of sustainable development policy.

Thus, the analysis results confirm the need for a comprehensive approach to sustainable development that takes into account the interrelationships between economic, environmental, and social aspects to ensure harmonious and sustainable growth.

5 Conclusion

The analysis of Kyiv's sustainable urban development from 2016 to 2021 reveals a positive trend, with the index rising from 0.4767 to 0.5293. This indicates a gradual improvement in balancing economic, environmental, and social components.

Correlation analysis shows a significant negative correlation (-0.823) between economic and environmental components, highlighting the need to integrate environmental considerations into economic strategies. Conversely, the positive correlation (0.810) between economic and social components suggests that economic growth generally enhances social conditions. The moderate negative correlation (-0.381) between environmental and social components indicates some existing conflicts that need addressing.

To promote balanced sustainable development, the following recommendations are:

- Integrate Environmental Aspects
- Support Social Initiatives

- Balance Goals
- Monitor and Evaluate
- Engage the Public

By implementing these recommendations, Kyiv can advance towards a more balanced and effective approach to sustainable development, ensuring long-term progress across economic, environmental, and social dimensions.

References

1. Kazmi, A.: Sustainability (World Commission on Environment and Development Definition). Encyclopedia of Sustainable Management. Springer, Cham (2023). https://doi.org/10.1007/978-3-031-25984-5_84
2. Komninos, N., Kakderi, C., Panori, A., Tsarchopoulos, P.: Smart city planning from an evolutionary perspective. J. Urban Technol. **26**(2), 3–20 (2019). https://doi.org/10.1080/10630732.2018.1485368
3. Martínez-Bravo, M., Martínez-del-Río, J., Antolín-López, R.: Trade-offs among urban sustainability, pollution and livability in European cities. J. Clean. Prod. **224**, 651–660 (2019). https://doi.org/10.1016/j.jclepro.2019.03.110
4. Bolívar, M.P.R.: Influence of smart cities sustainability on citizen's quality of life. Handbook of Smart Cities, 1–27. Springer International Publishing, Cham (2020). https://doi.org/10.1007/978-3-030-15145-4_72-1
5. Shin, S., Kim, D., Chun, S.A.: Digital divide in advanced smart city innovations. Sustainability **13**(7), 4076 (2021). https://doi.org/10.3390/su13074076
6. Martin, C.J., Evans, J., Karvonen, A.: Smart and sustainable? Five tensions in the visions and practices of the smart-sustainable city in Europe and North America. Technol. Forecast. Soc. Chang. **133**, 269–278 (2018). https://doi.org/10.1016/j.techfore.2018.01.005
7. Michalina, D., Mederly, P., Diefenbacher, H., Held, B.: Sustainable Urban Development: A Review of Urban Sustainability Indicator Frameworks. Sustainability **13**, 9348 (2021). https://doi.org/10.3390/su13169348
8. da Silva Tomadon, L., do Couto, E.V., de Vries, W.T., et al.: Smart city and sustainability indicators: a bibliometric literature review. Discov. Sustain. **5**, 143 (2024). https://doi.org/10.1007/s43621-024-00328-w
9. Mengiste, B.M., Shi, W., Wong, M.S.: Urban landscape and development effects on city sustainability: a systematic review of empirical studies. Environmental Development and Sustainability (2024). https://doi.org/10.1007/s10668-024-05405-8
10. Križaj, D., Bratec, M., Kopić, P., Rogelja, T.: A technology-based innovation adoption and implementation analysis of European smart tourism projects: Towards a smart actionable classification model (SACM). Sustainability **13**(18), 10279 (2021). https://doi.org/10.3390/su131810279
11. Baloloy, C.: Benefit cost analysis as a tool for sustainability and resiliency of transport infrastructure system to future climate uncertainty. NTNU (2018). https://ntnuopen.ntnu.no/ntnu-xmlui/handle/11250/2570003
12. Stryhula, M., Khomerikia, O., Mykhailycha, O., Yahodzinskyi, S., Romanenko, Y., Perelyhina, T.: Functioning of public transport in the social space of the city. Transportation Research Procedia **54**, 610–616 (2021). https://doi.org/10.1016/j.trpro.2021.02.113
13. Drotianko, L., Yahodzinskyi, S.: International science conference SPbWORCE – 2016 'SMART City' (November 15-17, 2016). MATEC Web of Conferences **106**, 01006 (2017). https://doi.org/10.1051/matecconf/201710601006

14. Hrechko, A. V.: Methodological approaches to evaluating the sustainability of the development of productive forces in the regions of Ukraine. Effective Economy, 8 (2018). http://www.economy.nayka.com.ua/pdf/8_2018/56.pdf
15. Leonova, S. V., Barvinska, Kh. V.: Identification and evaluation of the prospects for sustainable development of the road transport system of Ukraine. Market Infrastructure **40**, 103–109 (2020). http://www.market-infr.od.ua/journals/2020/40_2020_ukr/19.pdf
16. Tymoshenko, M. M.: Methodical bases of estimation of sustainable development of rural territories: algorithm, structural scheme and research tools. Mykolaiv National University named after V.O. Sukhomlynskyi **2**, 214–220 (2018). http://global-national.in.ua/archive/21-2018/44.pdf
17. Sydorchuk, O.: Integral assessment of the level of social security in Ukraine. Scientific Bulletin of Kherson State University **38**, 62–68 (2020). https://doi.org/10.32999/ksu2307-8030/2020-38-11

The Realisation of Sustainable Development Goals Through the Implementation of International Environmental Projects with the Implementation of Transport Infrastructure Solutions

Olena Barabash(✉), Olga Khrutba, Tetiana Pokshevnytska, and Mykola Tarasiuk

National Transport University, Mykhaila Omelianovycha-Pavlenka Str.1, Kyiv 01001, Ukraine
barabashelena29@gmail.com

Abstract. The objective of this article is to examine the experience of international environmental projects designed to conserve and restore biodiversity. In order to achieve the goals of sustainable development, the article will identify and generalise the characteristics of effective management approaches that have been applied in such projects. The authors examine the experience of implementing joint international environmental projects with the objective of achieving sustainable development goals. The role of donors and the direction of international project implementation for large protected areas, such as national parks, biosphere reserves or nature reserves, is considered. This study analyses a selection of projects supported by the Global Conservation organisation. The Danube Transnational Programme (Interreg Danube) is examined in detail, with particular attention paid to the principal objectives and strategic orientations of the various projects. The principal focus of the Horizon Europe programme has been established. This analysis considers international projects involving Ukraine that aim to conserve the biodiversity of flora and fauna in the territories and objects of the nature reserve fund. The primary vector of international environmental projects is then identified based on the data obtained. The analysis of the directions and implementation of international environmental projects has led to the formulation of recommendations for the improvement of the management of such projects at the national level.

Keywords: International Environmental Project · Transport Infrastructure Solutions · Biodiversity · Sustainable Development · Protected Areas

1 Introduction

In the context of contemporary globalisation, the development of environmentally-friendly investments to ensure the sustainable development of countries and initiatives to minimise and prevent the negative effects of climate change are becoming increasingly prevalent. Consequently, there is a growing number of projects aimed at reducing

O. Slavinska et al. (Eds.): ITSESQC 2024, LNNS 1335, pp. 14–28, 2025.
https://doi.org/10.1007/978-3-031-87376-8_2

the impact of human activities on the environment. The practical implementation of projects serves as a conduit for the implementation of novel solutions and the preservation and restoration of Ukraine's territories and environmental sites. The implementation of project-based environmental protection activities necessitates the collation of a substantial corpus of data pertaining to the objectives and outcomes of international projects, in addition to the data concerning the results achieved as a consequence of the environmental project. A comprehensive understanding of the nuances of international project implementation enables the timely and constructive identification of potential risks and outcomes associated with international environmental projects. The acquisition of knowledge regarding the potential objectives and outcomes of such projects will facilitate the timely implementation of the requisite management approaches to enhance the efficacy of environmental project management.

2 Analysis of Recent Research and Publications

The necessity for the development of international conservation projects can be attributed to the fact that scientists from disparate countries engage in communication and achieve results during the implementation phase of a project, which in numerous instances culminates in enhanced management functions of protected areas. The interaction of scientists, joint research, and new ideas that arise during communication facilitate the resolution of problems and the attainment of project objectives.

Presently, a significant challenge exists whereby states, invoking national security concerns, are pursuing initiatives to construct infrastructure for border protection, energy production, and transportation. Projects pertaining to climate change and the conservation and restoration of biodiversity in protected areas are becoming less prominent [1].

A significant aspect of the relationship between governmental entities and environmental non-governmental organizations (ENGOs) is the potential for mutual benefit through the utilization of each entity's respective strengths, ultimately leading to the attainment of sustainable development goals [2].

The primary objective of the CLICTOUR project, entitled "Climate Resilient Tourism in Protected Areas of Northern Portugal," was to ascertain the reasons why tourists do not repeat their previous hiking route. It was determined that the implementation of communication strategies to obtain feedback from tourists and to improve routes is essential for the sustainable development of natural areas [3].

The ethical approaches to biodiversity conservation vary considerably between conservation projects. The Pisavaara Nature Reserve (New Zealand), Ostvaardersplassen, and the Great Green Wall project have all yielded insights that have informed the creation of a heuristic tool in the form of a biological conservation map for each reserve [4].

The project to conserve the biodiversity and natural heritage of the Egyptian Nile Islands allowed the islands to be classified and a database of endemic plant and animal species to be created [5].

The evaluation of the efficacy of nature reserves as exemplars for biodiversity and ecosystem service conservation is an emerging area of research. A project designed to assess the efficacy of implemented environmental protection measures at the source of rivers on the Tibetan Plateau permitted an evaluation of the current environmental situation in these areas [6].

In order to achieve comprehensive biodiversity protection, it is necessary to develop and calculate a range of diversity indices. The Galiformes Conservation Project in China has facilitated the identification of priority areas for the conservation of species, phylogenetic and functional diversity of Galiformes, as well as the spatial dynamics of species distribution in response to climate change [7].

The mangrove protection project in the Futian Mangrove Reserve in Shenzhen resulted in the reserve being classified as a Class A area, which does not require carbon adjustment. [8].

The analysis of nature reserve policies enables the assessment of the efficacy of nature reserve management. It was determined that nearly all of the territories within one of China's regions underwent a change in status to that of protected areas following the implementation of the "nature reserve policy" [9].

Conservation projects permit the quantification of the extent of environmental degradation affecting the existence of flora and fauna species, as well as the assessment of biodiversity losses, which in the majority of cases are proportional to urban expansion [10].

The development of multifactorial models of the state of biosystems and the graphical interpretation of modeling results permit the prediction of the level of environmental safety of cities and regions. The application of mathematical modeling in biodiversity conservation projects facilitates the processing of a substantial quantity of data and the visualization of the resulting information [11].

The utilization of management strategies to guarantee the efficacy of international initiatives entails the formulation of a distinctive procedure for the selection of approaches at each phase of project implementation [12].

The sustainable development of transport infrastructure is persistently confronted with the dearth of efficacious financial mechanisms and management methods. The Metro Rail project in Manila, Philippines, has demonstrated that the most efficacious tools for sustainable development are a combination of vertical and horizontal project and process management. This guarantees the cohesion of stakeholders in the construction of infrastructure facilities and the conservation of protected areas [13].

The absence of a unified set of sustainability criteria for the development and operation of transport infrastructure necessitates their investigation in international projects. The authors developed a set of infrastructure sustainability criteria, comprising 43 relevant indicators. To assess the adequacy of the criteria, it was proposed that field research be conducted. It is advisable to assess the adequacy of infrastructure sustainability criteria in both urban and protected areas for biodiversity monitoring [14].

Consequently, international environmental projects that involve the participation of multiple countries and aim to conserve biodiversity and achieve sustainable development goals require the implementation of innovative transport and infrastructure solutions.

3 Highlighting Previously Unresolved Parts of the General Problem

A review of the literature revealed the most prevalent goals, objectives, and areas of focus within international conservation projects. The potential for the development of protected areas and the conservation of biodiversity during the implementation of conservation projects has been identified. Nevertheless, the stages of project implementation, the processes of selecting management approaches to achieve project goals, and the use of transport infrastructure solutions to achieve project goals remain insufficiently studied.

4 Formulation of the Purpose of the Article and Statement of Tasks (Purpose)

The objective of this article is to examine the experience and synthesize the principal approaches to the implementation and practical realization of international environmental projects. These projects are designed to preserve and restore biodiversity in protected areas through the use of effective transport infrastructure solutions and management approaches that facilitate the achievement of sustainable development goals.

In order to address the main objectives of the article, the following tasks were set: to analyze the experience of implementing international conservation projects; to consider donors and areas of conservation projects; to analyze the Danube Transnational Program (Interreg Danube) and Global Conservation projects; to consider the vector of Horizon Europe's activities; to propose an algorithm for choosing management approaches to achieve project goals; and to structure project stages through the use of effective transport infrastructure solutions for their successful implementation.

5 Methods

To address the research questions, the study was conducted in three phases. In the initial phase, the scientific problem was formulated and an analysis was conducted of the existing scientific literature. The second stage comprised a synthesis of the findings from international projects involving Ukraine that were supported by international organizations and programs. In the final stage of the research, an algorithm for selecting management approaches and a structural phased model through the use of effective transport infrastructure solutions for achieving project goals were proposed.

6 Results

The conservation of critical ecosystems in developing countries is a significant challenge, and the establishment of UNESCO World Heritage Sites and national parks represents a crucial step in addressing this issue. However, the success of these initiatives hinges on the provision of international support. Consequently, the most pressing objective is to establish a framework that enables the integration of contemporary environmental technologies with communication systems for the purpose of safeguarding biodiversity.

After all, reserve administrations and rangers lack the skills to use technology, communications, and equipment. An important element for continuous monitoring in protected areas, for controlling, protecting and restoring protected areas is the use of the right transport infrastructure solutions. Transportation and infrastructure solutions are solutions that lead to the use of the only possible means of transportation (aircraft, ropeway, underground or overhead tunnel) at a certain time and in strictly defined protected areas of the protected area.

The international conservation organization Global Conservation (GC) was established with the objective of protecting the most significant and vulnerable areas. The organization's primary objective is to provide direct financial assistance to conservation entities with the aim of preserving national parks and UNESCO World Heritage Sites in developing countries. The objective of GC is to introduce the Global Park Defense program by 2025 and implement the SMART system for monitoring territories in 25 national parks. The implementation of the Global Park Defense program is projected to occur within a two-to-three-year timeframe. It is anticipated that the program will enhance the efficiency of ranger patrols by approximately 80%, with the assistance of transport infrastructure solutions that will facilitate the conservation of wildlife species in critical ecosystems. A number of projects have been implemented under the GC to ensure effective management in national parks. Two such projects were implemented for El Mirador National Park (Guatemala, Central America), a UNESCO World Heritage Site. El Mirador is the largest virgin primary forest in Central America. The El Mirador Park Management System Implementation project facilitated the establishment of a park management system and a professional ranger service. The project was implemented in collaboration with the National Council for the Defense of Territories of Guatemala, the Guatemalan Institute of Tourism, and local communities. The second project, entitled "Implementation of SMART Ranger Patrols for El Mirador," was implemented for El Mirador with the objective of providing SMART anti-poaching ranger patrols with radar and transport infrastructure e equipment for surveillance in dense forest cover, identification, and documentation of inspection data [15]. Training rangers in the Mirador National Park is a crucial component of combating safeguarding biodiversity. During the 2022–2023 period, a series of training programs were conducted, including the SMART Patrol system was introduced to the rangers, who were trained in its effective use for the planning and execution of patrols. The SMART system (Spatial Monitoring and Reporting Tool) was also demonstrated to the rangers, who were informed of its potential for data collection on protected areas, thus increasing their effectiveness. The training also encompassed the management of protected areas, equipping rangers with an understanding of optimal practices in this domain. This included insights into strategies for the conservation and monitoring of ecosystems. The rangers were instructed in the methodology of monitoring flora and fauna, including the utilisation of photo traps for the assessment of wildlife populations. Training Abroad: Some rangers were afforded the opportunity to attend a two-week park management training program at the University of Colorado, which served to further enhance their skills. The purchase of new vehicles was also a priority. The provision of modern vehicles to rangers constitutes an essential component of their responsibilities. During the 2022–2023 period, Global Conservation and FundaEco initiated a series of pivotal initiatives in this domain. The purchase of new equipment, including trucks, motorcycles, ATVs, and trailers, has significantly enhanced the rangers' mobility, enabling a more rapid response to threats such as illegal logging and poaching. The thransport snfrastructure solutions have facilitated the implementation of a more comprehensive patrolling strategy, enabling rangers to access and monitor remote areas of the park that were previously inaccessible due to the lack of suitable transportation infrastructure.

Two projects have been implemented within the GC framework for the Loizer Ecosystem National Park (Sumatra, Indonesia). Gunung Leuser National Park is a significant heritage site within the rainforest ecosystem of Sumatra, encompassing an area of 2.6 million hectares and supporting over 500 species of birds and mammals. The objective of the projects was to enhance the rangers' capacity to oversee and monitor the entire park through the utilisation of unmanned aerial vehicles, satellite communications, and the IoT for remote monitoring with relay support [16].

The Species Habitat and Wildlife Area Protection Project in Loizer has enhanced ecological connectivity and ecosystem integrity. As a result of the efforts of the Global Park Defense program, Murchison Falls National Park has been linked to its most remote regions. A network of cellular trail cameras and long-range thermal cameras has been installed with the objective of improving the management and control of the areas in real time. The data obtained enabled the organization of data on the status of wildlife populations [17].

A project was initiated in the Sharski Mountains National Park (Balkan Peninsula) with the objective of training rangers in the utilisation IoT technology for the purpose of monitoring [18].

The Carpathian Biosphere Reserve is a protected area designated for the conservation of species identified as endangered by the IUCN and included in the Red Data Book of Ukraine. The Global Park Defense initiative enabled the establishment of a command center for real-time monitoring of the territory, which had a beneficial impact on the restoration of vulnerable species of animals and plants [19].

In summary, the objective of Global Conservation projects is to enhance the level of control and monitoring of extensive areas for the purpose of biodiversity conservation. The main activity of the Canadian Social Sciences and Humanities Research Council (SSHRC) is to fund research in the humanities and social sciences [20].

The largest SSHRC project is the TRANSdisciplinary Education Collaboration for Transformations in Sustainability (TRANSECTS). The project's objective is to develop innovative and flexible intercultural educational programs that will facilitate the implementation of the concept of sustainable development (Fig. 1).

The project yielded the creation of the TRANSECTS Resource Platform, which serves as a training, sharing, and archiving tool for project activities and results. It is utilized as a comprehensive package of standardized internal communication flows.

The Danube Transnational Programme represents a European Territorial Cooperation (ETC) funding instrument, the objective of which is to facilitate the sustainable development of the Danube Region through the integration of socio-economic and environmental sectors. (Table 1).

The implementation of several projects under the Danube Transnational Programme has facilitated the integration of international expertise in nature conservation with

Fig. 1. Partners of the project TRANSECTS

national achievements, thereby positively influencing the conservation of extensive areas of pristine natural habitat in Ukraine. One of the primary objectives of the Danube Transnational Programme is the organisation of high-quality transport infrastructure that is environmentally sustainable and efficient. As a result of the SOLEZ project, measures to enhance the quality of transport infrastructure with an emphasis on low-carbon mobility were implemented in eight Ukrainian cities.

Horizon Europe represents the European Union's principal program for the funding of research and innovation activities until 2027. The program's principal objective is to combat climate change by addressing global challenges. The program facilitates the attainment of the UN Sustainable Development Goals and fosters collaboration and enhances the impact of research to develop, support, and implement EU policies. Horizon Europe supports projects that create jobs, stimulate economic growth, promote industrial competitiveness, and optimize investment impact in the European Research Area. The Horizon Europe program has become a driving force for reducing the effects of climate change in the Carpathian Biosphere Reserve. The use of artificial intelligence-based software has made it possible to identify flood zones and implement preventive measures.

The Zoologische Gesellschaft Frankfurt (ZGF) is an international conservation organization with its headquarters located in Germany. The ZGF implements conservation projects in a variety of natural areas, including those located in Africa, South America, Southeast Asia, and Europe. In Ukraine, ZGF-supported projects are primarily concerned with the conservation of pristine forests and wetlands of global significance. Priority support is extended to protected areas in two regions: the Carpathians and Polissia. This includes the Spatial Monitoring and Reporting Tool program, which has been adapted for use in protected areas in the Carpathians since 2021.

Table 1. Implemented projects of the Danube Transnational Programme

Project Title	Purpose of the Project	Project implementation	Trends and commonalities
GE_NOW: Climate & Nature Activity	Creative, playful research on climate change and biodiversity	Six regions of the border region of Austria and Switzerland have encouraged the public to develop environmental measures to protect climate and biodiversity	Community involvement in environmental protection activities; emphasis on innovative approaches to education
Ecosystem services	Preservation of ecosystem services and resilience of Nagelfluhkette Nature Park landscapes in the face of climate change and increased awareness of the performance of small-scale mosaic landscapes	Cross-border conservation of ecosystem services and management of sustainability of high-quality cultural landscapes and habitats in Nagelfluhkette Nature Park	Resilience of landscapes and ecosystems; focus on climate change adaptation
Nature for biodiversity conservation	Development of tourist offers for the study of nature	Tourist routes have been developed to familiarize and inspect the territories of the reserve fund	Development of ecological tourism; raising awareness of biodiversity
Open borders for wildlife in the Carpathians	Development of a system of accounting and communication for the control and observation of predatory mammals	Measures implemented to ensure monitoring, accounting, and protection of predatory mammals	International cooperation in species conservation; emphasis on scientific monitoring

(*continued*)

Table 1. *(continued)*

Project Title	Purpose of the Project	Project implementation	Trends and commonalities
Open borders for bears in the Romanian and Ukrainian Carpathians	Development of a system of accounting and communication for the control and observation of predatory mammals	Measures implemented to ensure monitoring, recording, and protection of bears in the Romanian and Ukrainian Carpathians	Joint initiatives for species protection; integration of efforts between countries
Cross-border cooperation program Hungary-Slovakia-Romania-Ukraine	Reducing conflict between humans and wildlife	Measures were taken to maintain and improve forest habitats for large carnivores and connectors between them, taking into account climate change	Reducing conflicts between people and nature; conservation of large carnivores as indicators of ecosystem health

One of the largest organizations providing financial support for environmental projects that involve international student and researcher exchange is the German Academic Exchange Service (DAAD).

The Eberswalde University of Sustainable Development in Germany is implementing a number of projects, including the "Biosphere Forests and Rivers - Building Support Systems and Linkages for Conflict Transformation" program and the "Areas of Conflict - Discourse on Socio-Ecological Forest Restoration during War and Multiple Crises" project. The projects are part of the East-West Dialogue Program, which encompasses academic exchange and research cooperation for the advancement of security, collaboration, and civil society development in Europe. The projects are designed to address the complex challenges inherent to forest ecosystem restoration, particularly in the context of armed conflict, where the prevalence of uncertainty impedes the formulation of effective management strategies. The project offers scientists the opportunity to engage in exchange and internship programs, thereby facilitating geographic expansion of their research activities.

The Center for Economics and Ecosystem Management of the Eberswalde University of Sustainable Development is currently engaged in a number of international projects. These include collaborative initiatives with partner universities in Greece, Montenegro,

Kosovo, and North Macedonia. Additionally, the Center is implementing a series of summer schools with the objective of identifying pivotal aspects and efficacious strategies for the stewardship of transboundary ecosystems. This endeavor is designed to advance ecosystem development and facilitate strategic planning for biodiversity conservation.

It was thus demonstrated that international environmental projects contribute to the achievement of sustainable development goals in a number of ways. Ukrainian scientists and the administrators of environmental institutions are pursuing an expansion of the geographical scope of international projects. Participation in projects enables the resolution of pivotal aspects of the activities of environmental institutions and universities. It is important to note that a lack of knowledge regarding the structuring of pre-project activities, project management approaches, and the selection of optimal management approaches makes it challenging to participate in international projects and achieve the project's ultimate goal. In light of the aforementioned procedure for forming a project strategy, we put forth an algorithm for selecting management approaches and a structural, step-by-step model for achieving project goals through the use of effective transport infrastructure solutions.

The principal stages of formulating a project strategy, establishing its internal context, and populating it with goals, objectives, research methods, and anticipated outcomes are the phased implementation of each of its components, commencing with preliminary project analysis and concluding with the monitoring of goal achievement (Fig. 2).

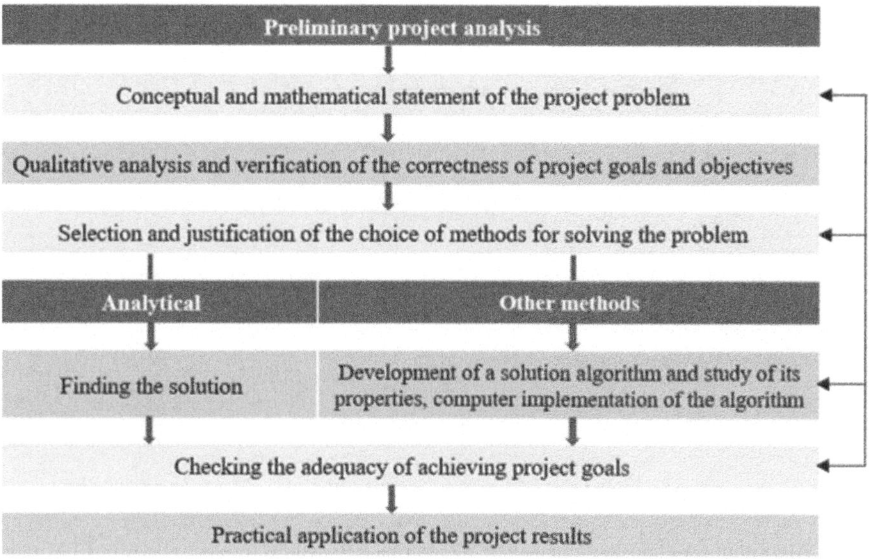

Fig. 2. Generalized stages of project strategy development

The collection of data on ecosystems and the range of relationships within ecosystems provides a comprehensive understanding of the ecological status. The next crucial step is to define the specific objectives of the project, which may include the conservation of particular species, the restoration of specific ecosystems, or the implementation of improved management practices. A variety of models are employed to construct models that accurately reflect the actual processes occurring in natural systems, including those based on mathematics, computers, and statistics. The veracity and dependability of models are evaluated through a process of comparison with empirical data and outcomes.

Given the necessity of a multifaceted approach to biodiversity conservation, encompassing a range of technological, environmental, strategic management, and transport and infrastructure considerations, we put forth the integration of an overarching methodology for the governance of international environmental initiatives. This approach, illustrated by the El Mirador National Park project, is designed to facilitate the realization of project goals and objectives through the selection of optimal management approaches and strategies (Fig. 3).

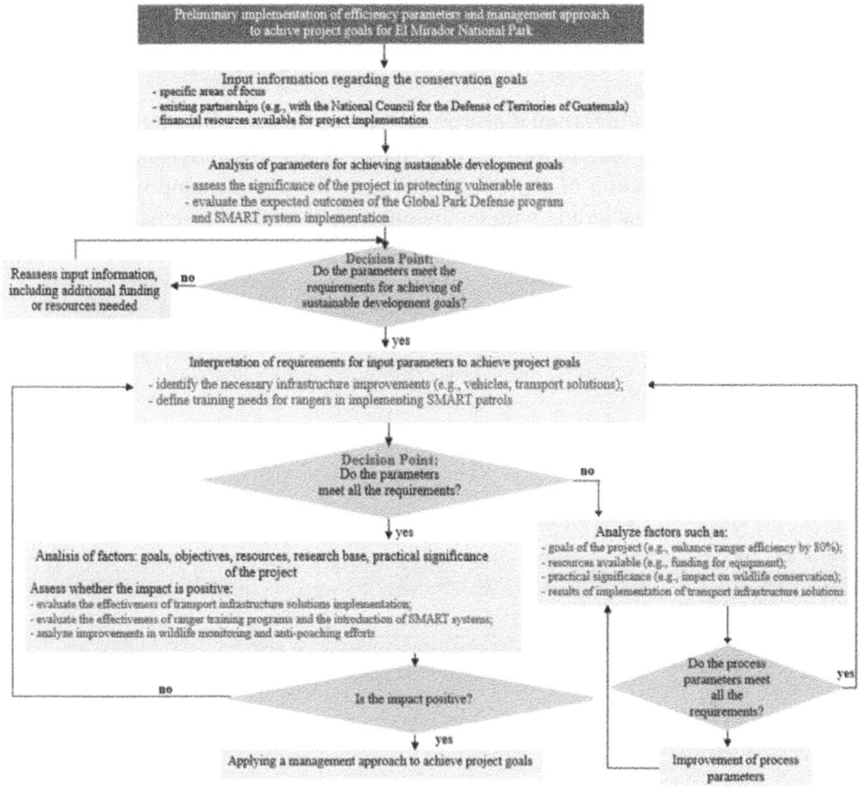

Fig. 3. Algorithm for choosing a management approach to achieve project goals for El Mirador National Park

The implementation of projects in accordance with the algorithm for selecting a management approach enables the requisite adjustments to be made to management decisions, thus ensuring the attainment of goals and objectives at each stage of the project.

In order to implement and achieve the goals of environmental projects, it is proposed that a structural phased model be used to decipher the stages of the project, from the inception of the project idea to the realization of its goals and objectives (Fig. 4).

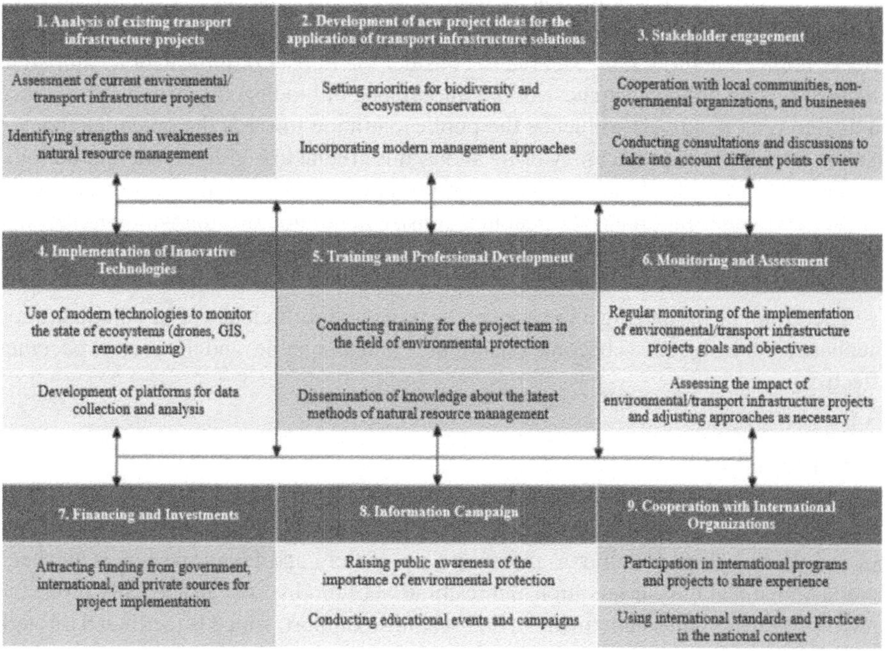

Fig. 4. A detailed, step-by-step structural model for the implementation and achievement of project goals

The preliminary phase of the process entails a comprehensive examination of the existing projects and the strategies employed to oversee their management. At this juncture, it is advisable to convene brainstorming sessions with the objective of developing novel ideas and strategies that will satisfy the identified needs. In order to make optimal decisions at the stage of sharing experiences and opinions, it is essential to engage all relevant stakeholders, including local communities, representatives of the transportation industry, businesses, academic institutions, and non-governmental organizations.

In selecting methodologies for attaining the project's goals and objectives, it is essential to consider the advancement of transport infrastructure technologies and neural networks, which facilitate the effective realization of the project's goals. Evolutionary programming, a branch of artificial intelligence, is an effective method for modeling the processes of viability and reproduction of biodiversity. The use of innovative technologies, particularly modern transport and infrastructure solutions during project implementation, will facilitate a rapid response to changes in ecosystems, allowing for the adjustment and restoration of their elements. At the subsequent stage, it is crucial to provide training and professional development for project team members, which is inextricably linked to the study of contemporary technologies. At the stage of monitoring and evaluating the effectiveness of achieving project goals, it is essential to strike a balance between implementing necessary adjustments and adaptations to new conditions. It is essential to implement transparent indicators that will provide dependable mechanisms for gauging the progress of goal attainment. The financing stage of international environmental projects necessitates the attraction of both public and private investment, in addition to the pursuit of grants from international organizations. In order to ensure effective project implementation, it is essential to consider the impact of information campaigns aimed at raising public awareness of the importance of biodiversity conservation. Such campaigns can positively influence the public's attitude towards the potential changes that the implemented project may bring about. It is crucial to highlight that international conservation projects can be applied to a multitude of aspects within a protected area. It is of paramount importance to maintain consistent communication with international partners and gradually cultivate these relationships throughout the entirety of project implementation.

Therefore, by organizing all phases of the project, the efficiency of its administration is enhanced, the objectives become quantifiable and attainable, and the project becomes effective.

7 Conclusions

In the context of persistent anthropogenic influence on ecosystems, biodiversity is undergoing a decline that is leading to the extinction of flora and fauna species. It has been demonstrated that the conservation and restoration of biodiversity in specially protected areas necessitates consistent financial and scientific support, which is facilitated through the involvement of organizations in international environmental projects. Therefore, an algorithm for selecting management approaches and a structural phased model for achieving project goals using transport infrastructure solutions were proposed to systematize pre-project activities and achieve goals while participating in international environmental projects.

The practical application of the phased model, which considers the utilisation of transport infrastructure solutions, will facilitate the scaling of research in protected areas. The implementation of an algorithm for the selection of management approaches will influence the velocity and calibre of decision-making by the project team.

In light of the accelerated advancements in technology and environmental science, future research will concentrate on assessing the efficacy of transport and infrastructure monitoring solutions deployed during project implementation in comparison to the utilisation of cutting-edge artificial intelligence-based monitoring tools that are currently being deployed on a large scale in protected areas.

References

1. Nowak, K., Bear, D., Dutta, A., Traphagen, M., Żmihorski, M., Jaroszewicz, B.: Threats to conservation from national security interests. Conserv. Biol. **38** (2024). https://doi.org/10.1111/cobi.14193
2. Zhao, J., Jin, T., Zhang, P., et al.: Political embeddedness in public–private partnership for nature conservation: A land trust reserve case from China. Ambio **53**, 324–338 (2024). https://doi.org/10.1007/s13280-023-01936-y
3. Silva, S., Marques, T., Novais, J., Folharini, S., Vieira, A.: After the Hiking Tour: Walkers' Behaviour. Management, Tourism and Smart Technologies. ICMTT 2023. Lecture Notes in Networks and Systems. 774. Springer, Cham (2024). https://doi.org/10.1007/978-3-031-43733-5_25
4. Wienhues, A., Luuppala, L., Deplazes-Zemp, A.: The moral landscape of biological conservation: Understanding conceptual and normative foundations. Biol. Conserv. **288** (2023). https://doi.org/10.1016/j.biocon.2023.110350
5. Esmail, E., Yousry, A., Qutb, S., Abdelrahman, A.: Sustainability Assessment Scale for the Nile Islands (SASNI) as a Guideline to Determine Sustainable Urban Development Policies Received. Environ. Res., Eng. Manag. **79**(3), 7–23 (2023). https://doi.org/10.5755/j01.erem.79.3.33207
6. Zhao, C., et al.: Effectiveness of protected areas in the Three-river Source Region of the Tibetan Plateau for biodiversity and ecosystem services. Ecol. Indic. **154** (2023). https://doi.org/10.1016/j.ecolind.2023.110861
7. Wang, B., et al.: Climate change affects Galliformes taxonomic, phylogenetic and functional diversity indexes, shifting conservation priority areas in China. Divers. Distrib. **29**(3), 335–461 (2022). https://doi.org/10.1111/ddi.13667
8. Lu, R., et al.: Carbon credit measurement method for mangrove conservation carbon sink project: a Case Study of Futian Mangrove Nature Reserve in Shenzhen. Scientia Silvae Sinicae **59**(3), 44–53 (2023). https://doi.org/10.11707/j.1001-7488.LYKX20220624
9. Wang, Q., Zhou, D., An, P., Jiang, G.: Impacts of nature reserve policy on regional ecological environment quality: A case study of Sanjiangyuan region. Chin. J. Appl. Ecol. **34**(5), 1349–1359 (2023).https://doi.org/10.13287/j.1001-9332.202305.023
10. Ren, Q., He, C., Huang, Q., Shi, P., Zhang, D., Güneralp, B.: Impacts of urban expansion on natural habitats in global drylands. Nature Sustainability. 5 (10), 869–878 (2022). https://doi.org/10.1038/s41893-022-00930-8/
11. Barabash, O., Weigang, G.: Mathematical Modeling of the Summarizing Index for the Biosystems Status as a Tool to Control the Functioning of the Environmental Management System at Business Entities. Mathematical Modeling and Simulation of Systems (MODS'2020) **1265**, 56−66 (2021). https://doi.org/10.1007/978-3-030-58124-4_6
12. Barabash, O., Weigang, G., Dychko, A., Belokon, K., Zhelnovach, G.: Modeling a set of management approaches for the effective operation of the environmental management system at the business entities. Ecol. Eng. & Environ. Technol. **22**(6), 1–10 (2021). https://doi.org/10.12912/27197050/141895
13. Endo, K., Edelenbos, J., Gianoli, A.: Finance arrangements and governance modes toward sustainable infrastructure: the case of urban railway projects in Manila, the Philippines. Discover Sustainability **5**, 235 (2024). https://doi.org/10.1007/s43621-024-00405-0
14. Kustova, I., Hudenko, J., Lace, N.: A Systematic review of sustainability criteria in infrastructure development. Sustainability **16**(11), 4564 (2024). https://doi.org/10.3390/su16114564
15. LaRocque, A., Leblon, B., Ek, J., Folan, W.J.: Detection of maya sacbeob (sacbes) using optical and sar imagery in northern petén, Mexico: Int. Arch. Photogramm. Remote Sens.

Spatial Inf. Sci., XLIII-B3, 633–639 (2022). https://doi.org/10.5194/isprs-archives-XLIII-B3-2022-633-2022
16. Sun, Y.: Distributed transmission and optimization of relay-assisted space-air-ground IoT systems. EURASIP J. Adv. Signal Process. **27** (2024). https://doi.org/10.1186/s13634-024-01123-5
17. Pomeroy, D., Matovu, B., Kibuule, M.: Habituation of large mammals to passing vehicles. Afr. J. Ecol. **62**(1) (2024). https://doi.org/10.1111/aje.13243
18. Yang, L.: Efficient identity security authentication method based on improved R-LWE algorithm in IoT environment. EURASIP J. on Info. Security **4** (2024). https://doi.org/10.1186/s13635-024-00153-8
19. Roopashree, S., Anitha, J., Challa, S., et al.: Mapping of soil suitability for medicinal plants using machine learning methods. Sci. Rep. **14**, 3741 (2024). https://doi.org/10.1038/s41598-024-54465-3
20. Holloway, S.M., Xu, S., Ma, S.: Chinese and Canadian preservice teachers in face-to-face dialogues. Situating teaching in cultural practices for West-East Reciprocal Learning: Teaching and Teacher Education **122** (2023). https://doi.org/10.1016/j.tate.2022.103930

Management of an Automated Air Quality Monitoring System to Achieve Sustainable Development Goals

Serhii Kolomiiets[1](✉), Viktoriia Khrutba[1], Oksana Kobzysta[1], Anna Kolomiiets[2], and Valentyna Shevchenko[3]

[1] National Transport University, Mykhaila Omelianovycha-Pavlenka Str.1, Kyiv 01001, Ukraine
s.kolomiiets@ntu.edu.ua

[2] Taras Shevchenko National University of Kyiv, 64/13, Volodymyrska Str, Kyiv 01601, Ukraine

[3] Ukrainian State Dragomanov University, 9, Pirogova Str, Kyiv 01601, Ukraine

Abstract. The paper discusses the algorithms for developing a network for automated monitoring of the city's atmospheric air quality, which provides for independent monitoring to fulfill the eleventh sustainable development indicator for Ukraine. The study considers the implementation of automated monitoring of atmospheric conditions through real-time sensors that provide indicative measurements of various air pollution parameters. The possibility of using the "Oxygen" air quality monitoring station is considered. This station allows for monitoring and recording key atmospheric indicators, such as humidity, temperature, dust concentrations (PM10 and PM2.5), as well as levels of CO, NO2, and NH3. The functionality of the "Oxygen" station not only provides real-time air quality data but also enables the archiving of large datasets for forecasting and assessing pollution indices and health risks. This can be managed through a personal account on the EcoCity website.

Experimental research was conducted by measuring the peak concentrations of CO, NH3, PM10, and PM2.5 in the air using an air quality monitoring station. During the study, the TLV were not exceeded, although there was a periodic excess of NO_2 concentration. The data show a harmful effect, mostly of transport vehicles, on the environmental and public healthy in the studied area.

Keywords: Automatic Monitoring Station · Air Quality Management · Air Quality Index · Pollutants · Sustainable Development · Public Awareness

1 Introduction

The automated environmental monitoring system in Ukraine consists of a collection of technological resources and software developed to collect, process, analyze, and transmit data on the state of the environment on a systematic basis. The primary purpose of the automated environmental monitoring system is to ensure operational control over environmental quality and timely responses to threatening environmental situations as one of the components of sustainable indicators.

O. Slavinska et al. (Eds.): ITSESQC 2024, LNNS 1335, pp. 29–41, 2025.
https://doi.org/10.1007/978-3-031-87376-8_3

As per EU Directive 2008/50/EC, the rationale for setting up automated monitoring stations should rely on predicting air pollution levels in the areas under study. This involves using data from mathematical models that simulate the spatial dispersion of pollutants in the lower atmosphere, enabling the assessment of inhalation risks to public health from pollution sources such as industrial facilities and vehicles across the city. Such a "chaotic" method is appropriate for the realities of our country, as it will save public funds, which will help organize the existing observation network and increase the relevance of management solutions to improve air quality [11].

The quality of the city's air environment is important as a component of the habitat, and therefore has an impact on the health of its citizens. As outlined in the CMU resolution on state air pollution monitoring [3], and following a comparison of monitoring practices between Ukraine and the European Union, several shortcomings have been identified. These include the absence of automated analysis for air pollution, especially concerning particulate matter, and a lack of public awareness and engagement regarding air quality in urban areas. An essential task is to create a network for monitoring the state of urban air pollution, which will allow for effective air pollution regulation and management. Thus, automatic monitoring can become not only a tool for engaging the community in the city's current problems, but also eliminate a number of drawbacks of the current air monitoring system and help in monitoring the sustainable development goals in Ukraine.

2 Analysis of Previous Research and Problem Setting

Studies on modeling the distribution of pollutants in the atmospheric air are mainly focused on certain aspects within a specific stationary source of air pollution, which allows solving environmental monitoring tasks for a single source or a specific territory.

At present, there are no valid models (or their adaptations) that take into account the total emissions from all sources located in a certain territory: enterprises, vehicles, emissions from fires, accidents or emergencies, while taking into account the separation of the emission plume (span structures, buildings of complex shape, etc.), and are focused on a large-scale grid of cities.

Despite the results of solving a range of basic gas-dynamic and atmospheric physics problems obtained by such large centers as the International Institute for Systems Analysis in Austria, the National Research Center for Information Technology in Germany, the US Meteorological Union, the Main Voyeikov Geophysical Observatory, and others, there are no methods that allow for the formation of comprehensive models covering the scale of a city[12].

Today, a number of automated air quality monitoring systems have been created and are operating around our planet.

Among the main requirements for such systems is to ensure high reliability of the measurement results of the concentration of ingredients and related atmospheric parameters. The use of reliable information on the level of air pollution in solving a range of environmental tasks would minimize the likelihood of a second kind of error, which leads to irreparable economic damage.

The term "information reliability" can be interpreted in two ways. First, it can be considered in terms of measurement and data transmission errors. Alternatively, it may have the sense of representativeness of readings at a control point for some of its neighborhood. This is the sense in which the concept of reliability of measured ingredient concentrations will be interpreted in the following. The relevance of the values of an ingredient at a control point can be assessed by examining the nature of spatial statistical relationships in the concentration fields.

The analysis of the data on the concentration of impurities is carried out using critical information control.

To assess the accuracy of meteorological information coming from the measurement network, methods based on the comparison of the measured value at a certain point to the results received by interpolating of meteorological areas are commonly used. To solve this problem, polynomial interpolation can be used, the essence of which is to approximate a two-dimensional domain by an algebraic polynomial. The disadvantages of this method are, on the one hand, its somewhat formal nature, and, on the one other hand, that the interpretation errors increase significantly with distance from the center of the considered area. Therefore, optimal interpolation is preferable [14].

The optimal interpolation procedure is based on the data on the statistical structure of the studied meteorological fields and consists of three stages: searching for influencing stations, calculating interpolation weights, and interpolation itself. This method has a great benefit when it comes to meteorological fields. It lies in its uniformity and relatively low sensitivity to the density of the station network and measurement accuracy. However, this technique cannot be used to solve the problem of critical control of measured ingredient concentrations due to the small radius of their spatial correlation.

The next group of techniques on which critical control algorithms can be based involves the use of statistical correlation of time series. If the controlled processes have stationarity, then for the purposes of critical control, the methods of optimal statistical extrapolation, formulated for the first time in [15], can be used. They make it possible to build an optimal filter that predicts the value of the time series at a moment in time, with which the concentration of an impurity measured at that moment in time is compared. A significant obstacle to the use of these methods is the uncertainty conditions that characterize the value range within in which the true value of a variable with a specified level of confidence is located.

According to the "Recommendations on Uncertainty Assessment in the System of Monitoring, Reporting and Validation of Greenhouse Gas Emissions", in order to conduct continuous measurements of atmospheric air quality in an automated mode, the operator must provide a list of all measuring instruments installed at the measurement points in the monitoring plan, indicating their operating range, uncertainty and measurement frequency. The "Procedure for Monitoring and Reporting Greenhouse Gas Emissions" [16] does not provide for any simplifications to assess uncertainty when applying a methodology based on continuous measurements, as opposed to a methodology based on calculations. This procedure requires that all measurements be made using methods based on the standard DSTU EN 14181:2014 "Stationary source emissions. Ensuring the quality of automated measurement systems".

If there is no relevant national standard, the operator has the right to use relevant international or EU standards, draft standards, established industry practices or other scientifically recognized methods that can reduce distortions during sampling or measurement.

For example, the standard DSTU EN 14181:2014 contains information on quality assurance levels (QAL 2 and QAL 3) that should minimize uncertainty, as well as instructions for assessing uncertainty itself. Recommendations for the QAL 1 quality assurance level can be found in DSTU EN ISO 14956:2008 Air quality.

There are currently no national standards for measuring greenhouse gas emissions that set requirements for measurement sections and locations. Therefore, methods based on the European standard EN 15259 "Air Quality - Measuring stationary source of emissions - Demands for measurement sections and sites and for the measuring objective, plan and reporting" or its adapted version for Ukraine "Ukrainian Air Quality Index" (UAQI) should be used.

Clause 45 of the Procedure for Monitoring and Reporting on Greenhouse Gas Emissions also states: "The operator is obliged to take into account all aspects of the continuous measurement system, including equipment placement, calibration, measurement technology, quality assurance and control."

That is precisely why the technologies for creating a city's automated air pollution monitoring network, which provides for independent monitoring of the city's air pollution, including indicators of the Sustainable Development Goals in Ukraine, are an extremely relevant issue.

3 Using the Ukrainian Air Quality Index to Assess the Quality of Atmospheric Air

The Ukrainian Air Quality Index (UAQI) was used to assess the quality of atmospheric air, which is a scale for qualitative and quantitative assessment of atmospheric air quality. UAQI as a communication tool allows stakeholders [6] to.

- better understand the air pollution where they are living, working, or traveling;
- make an informed decision on safety actions in case of air quality deterioration;
- to conduct current and retrospective air quality analysis and justify appropriate air quality management decisions in Ukraine, in individual zones, agglomerations or locations;
- to distribute up-to-date information on air quality in graphic and digital form, adapted for perception by citizens without special professional training.

The UAQI consists of four groups (Table 1) of the main pollutants, as well as physical parameters that affect air quality and safety.

UAQI provides an integrated approach to a comprehensive assessment of air quality and safety in three stages: a) individual assessment of the air quality category for each of the monitoring parameters included in the air quality monitoring program at a particular location or territory; b) determination of the air quality category for each group of parameters; c) formation of a general conclusion about the air quality and safety in a particular location or community [9].

This approach protects the user from distorted, false, or incomplete air quality information. The technical capabilities of each individual public air quality monitoring station

Table 1. Groups of monitoring parameters that form the air quality and safety of the Ukrainian Air Quality Index [9]

Group A	The content of aerosols and other finely dispersed solid particles in the air
Group B	Air content of gaseous pollutants of the priority list defined by WHO
Group C	The content of other pollutants in the air from the priority list, which is used in the practice of protecting atmospheric air from pollution in Ukraine
Group B	Parameters of physical factors that affect air quality and safety

may not provide the ability to assess the quality of air for all pollutants and physical characteristics.

Considering the capacities and development practices of the Ukrainian Eco City network of citizen air quality monitoring, Table 2 provides a detailed breakdown of the UAQI used in the Ukrainian citizen science model of air monitoring in the country.

The UAQI establishes 10 categories of ambient air quality (Table 3) - six basic, two additional and two special.

Six main categories provide the user with information on ambient air quality and safety according to EEA EU9 practice [7]. Two additional categories are used to indicate cases where air quality is not good for various reasons. Special ambient air quality categories indicate special cases of abnormal monitoring results with a lower level of confidence or special conditions of use. These categories are not used to alert the public, but are considered in the course of a detailed analysis of retrospective monitoring results. For a clear and unambiguous designation of each air quality category, the following are used:

- Numerical designation of the category, Roman numerals from I to X.
- Name of the quality category that characterizes the level of risk and hazard (key message).
- Color designation that helps to intuitively understand the risk and hazard level.
- Range of values (scale) for each monitoring parameter.

Determination of the air quality category (both for each individual monitoring parameter and for a particular location or territory) is carried out by simply comparing the monitoring result with the value ranges (scale) of each air quality parameter [9].

Table 2. Detailed structure of the UAQI, which is used by the network of public air quality monitoring Eco City [6, 9]

Group of parameters of air quality		Individualized air quality parameters		
Designations	Title	Designation in communication	Designation in databases	Title
Group (A)	The amount of aerosols and other fine particulates in the atmosphere	TSP	A1	Total content of undifferentiated dust (aerosol)
		PM_{10}	A2	Fine dust, solid particles with a fraction of up to 10 microns
		$PM_{2,5}$	A3	Fine dust, solid particles with a fraction of up to 2.5 microns
		PM_1	A4	Fine dust, solid particles with a fraction of up to 1 micron
		PM_{100}	A5	Total dust content, solid particles differentiated by size, fraction up to 100 microns
		others	A7-A99	Others
Group (B) & **Group** (C)	Air content of gaseous pollutants of the WHO priority list Air content of other pollutants from the priority list	NO_2	B1	Nitrogen dioxide
		CO	B2	Carbon monoxide (carbon monoxide)
		O_3	B3	Ozone (ground-level ozone)

(*continued*)

Table 2. (*continued*)

Group of parameters of air quality		Individualized air quality parameters		
Designations	Title	Designation in communication	Designation in databases	Title
		SO_2	B4	Sulfur dioxide (sulfur dioxide)
		others VOC (H_2CO)	B7-B99 C1	Other volatile organic compounds[1]
		NH_3	C2	Ammonia
		H_2S	C3	Hydrogen sulfide
		CI_2	C4	Chlorine
		PH_3	C5	Phosphine
		DNO	C6	Equivalent odor index (odors)
		others	C7-C99	Others
Group (D)	Physical factors influencing air quality and safety	T	D1	Atmospheric air temperature[2]
		W	D2	Relative humidity of atmospheric air 3
		p	D3	Atmospheric pressure [4]
		RAD (mc/hr)	D4	Exposure dose rate in ambient air
		RAD (nSv/hr)	D5	Equivalent dose rate/effective dose rate in ambient air
		others	D6-D99	Others

Notes to the table:
1) Including volatile organic compounds in terms of formaldehyde
2) In degrees Celsius, °C
3) Percentage of the ratio of absolute humidity to its maximum value at a given temperature, %
4) In mmHg without additional calculations to bring the indicator to the sea level of geographical latitude 45°)

4 Experimental Studies of Atmospheric Air Quality Determination Using an Automated Monitoring System

Automated indicative systems of monitoring of air quality, similar to those that are effectively operating in Europe, are designed to continuously monitor the content of pollutants in the air. These systems (Table 4) should address the issue of continuous

Table 3. Categories of ambient air quality and safety UAQI [9]

Ukrainian air quality index, uaqi		
	Category	
Main categories	I	Good air quality
	II	Satisfactory air quality
	III	Ambient air quality is unfavorable for air pollution-sensitive groups
	IV	Poor air quality
	V	Very poor air quality
	VI	Extremely poor air quality
Additional categories	VII	Ambient air quality temporarily unknown, technical reasons
	VIII	Ambient air quality is unknown, no monitoring is carried out
Special categories	IX	Ambient air quality needs to be clarified, abnormal results
	X	Ambient air quality needs to be clarified, the result is outside the sensitivity range of the monitoring tool

measurements of pollutant content and the issue of wide territorial coverage of these measurements, issues that cannot be resolved by stationary monitoring of air quality stations [4, 5].

This is a not-for-profit project that has quickly gained popularity in Ukraine and has become the largest application of environmental data on pollution, polluters, and environmental protection tools. It partially replaces state functions by collecting, storing and disseminating online all environmental information on pollution, pollutants and environmental protection tools available in Ukraine. To find out the level of pollution (quality of atmosphere) in Kyiv, you should select the corresponding monitoring stations on the map [6].

In 2022, the Kyiv Digital app was updated to include the option to sign up for alerts about high levels of air contamination in the capital. Users will receive notifications when the air quality index for the entire metropolitan area is above 75, i.e., "high pollution." This service will be especially useful for parents with small children, people with poor health, and the elderly. If you receive a message about the deterioration of urban air pollution in Kyiv, you should postpone long walks and refrain from physical activity.

Therefore, the authors analyzed the data obtained using the Oxygen air monitoring system, which allows real-time monitoring and recording of air quality, namely temperature, humidity, PM_{10} and $PM_{2.5}$ dust concentrations, CO (carbon monoxide), NO_2 (nitrogen oxides), and NH_3 (ammonia).

The experimental research was conducted as part of the international program "Clean Air for Ukraine", which National Transport University joined in March 2023 [13].

Table 4. Comparison of public monitoring stations

Characteristics of the station	Name of the station
1. The most accessible station for Ukrainian residents, it also measures the main indicators of air contamination and weather data - temperature, moisture and air pressure 2. The station is equipped with a dust meter for three fractions of fine dust PM_1, $PM_{2.5}$, PM_{10} with an automatic thermal compensation module for foggy weather 3. The station has a simplified installation for users in any accessible place	AirFreshMax
A device that allows real-time monitoring and recording of air conditions, such as temperature, humidity, PM_{10} and $PM_{2.5}$ dust concentrations, CO (carbon monoxide), NO_2 (nitrogen oxides), and NH_3 (ammonia)	Oxygen
Measures dust level, air temperature, humidity, and atmospheric pressure. It has a heating chamber, power converter, automatic update, and UV printing	SaveEcoSensor 3.0
A new type of station from EcoCity. First of all, it is designed for mobile monitoring. The station is portable and operates on rechargeable batteries that can provide power to the station for more than 48 h. It collects data from more than a dozen sensors every 5 s, then displays all the readings on a TFT display and saves each individual measurement for memory with the exact time. The sensors installed on the Sapphire32 measure 13 indicators of air quality plus an optional sound level meter. It is possible to connect to Wi-Fi to display data on the eco-city.org.ua website. It is equipped with a GPS module, so each measurement is accompanied by the coordinates where it was made, as well as the time of the satellite	Sapphire 32

As a part of the research, the Department of Ecology and Environmental Protection Technologies laboratory (audience number 357a) in the central building of the university at Omelyanovycha-Pavlenka Street, 1 (Fig. 1).

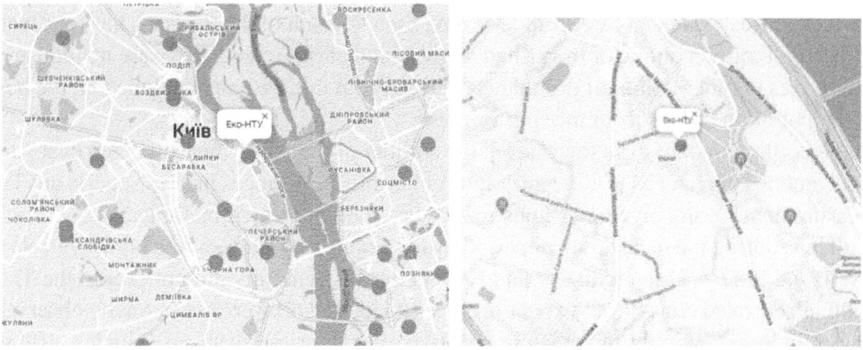

Fig. 1. Location map of the monitoring station

Universal air monitoring station for air quality "Oxygen" (Fig. 2) allows monitoring and recording of air conditions, namely temperature, humidity, PM_{10} and $PM_{2.5}$ dust concentrations, CO (Carbon oxide), NO_2 (Nitrogen dioxide), NH_3 (Ammonia) in real time. These pollutants are included in the "A" priority list, for which air quality assessment is carried out in accordance with the Procedure for Conducting State Air Protection Monitoring [2].

Fig. 2. Universal air quality monitoring station «Oxygen»

The study of pollutant indicators analyzed the dynamics of changes in the concentrations of harmful substances (NO_2, CO, NH_3, dust) n the ambient air were measured and evaluated three times a day for two weeks, which allowed us to analyze the state of air quality in the area where the monitoring station was installed (Fig. 3).

The university is located at the intersection of three busy highways (Omelyanovycha-Pavlenka Street, Lavrska Street, and Mazepa Street). Each of these streets has four lanes with a total traffic volume of more than 40 thousand cars per day. There are no industrial enterprises or other major air pollutants nearby. From this we can conclude that the bulk of the pollution comes from moving vehicles.

According to data from the station, the authors of the paper measured and evaluated the dynamics of changes in concentrations of harmful substances in atmospheric air. The measurement results (average values for 14 days) are presented in Table 5.

Thus, during the measurement period by air monitoring stations, the highest levels of PM10 and PM2.5, along with CO and NH3 in the air, generally did not exceed the TLV, although elevated concentrations of nitrogen dioxide (NO2) were occasionally observed. These findings highlight the existing negative effects, primarily from vehicles [10], on public health and the environment in the studied area. Future research aims to evaluate the health risks for residents of the Pechersk district in Kyiv following the guidelines

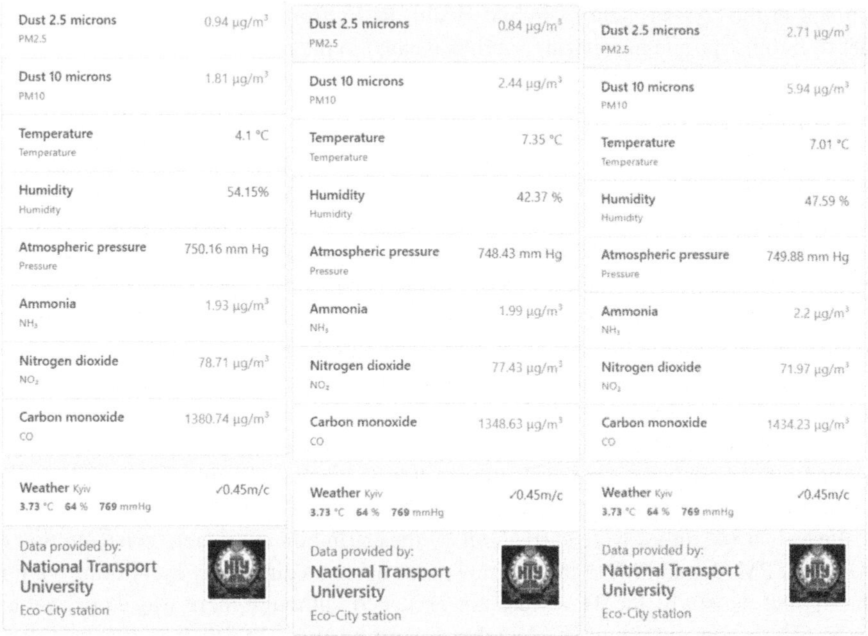

Fig. 3. Visualization of the data obtained in the personal account on the website eco-city.org.ua [1]

Table 5. Average values of stationary station measurements

Title	**Average value per day, µg/m³**							**TLV, µg/m³**
Dust 2.5 microns	7.19	9.07	18.63	21.04	5.61	4.63	5.89	25
Dust 10 microns	9.8	12.92	22.66	26.98	10.62	7.23	9.05	50
Ammonia	3.77	4.436	4.13	4	2.87	2.91	2.59	40
Nitrogen dioxide	39.3	36.02	35.55	36.67	49.21	48.95	54.85	40
Carbon monoxide	1214	1265	1218	1217	1113	1150	1093	3000
Date	**27.08**	**28.08**	**29.08**	**30.08**	**31.08**	**01.09**	**02.09**	
Title	**Average value per day, µg/m³**							**TLV, µg/m³**
Dust 2.5 microns	6.38	7.37	2.91	4.83	4.93	3.08	2.33	25
Dust 10 microns	8.78	9.79	5.37	8.4	7.2	6.68	5.57	50
Ammonia	2.85	2.87	2.48	2.84	2.87	2.42	2.46	40
Nitrogen dioxide	49.55	49.03	54.91	49.51	47.46	56.84	54.99	40
Carbon monoxide	1092	1100	1076	1156	1109	1115	1074	3000
Date	**03.09**	**04.09**	**05.09**	**06.09**	**07.09**	**08.09**	**09.09**	

outlined in the "Assessment of Public Health Risks from Air Pollution," considering data from other public monitoring stations located in the region.

5 Conclusion

Thus, the study explored the algorithms for establishing a system of automated air quality monitoring in the city, aimed at independent assessment to meet Ukraine's eleventh sustainable development indicator. It is substantiated that the introduction of automated monitoring of the quality of atmospheric air based on indicative measurements of several pollution indicators through real-time sensors is necessary. The potential use of the Oxygen air quality monitoring station, which enables the tracking and recording of key atmospheric air quality indicators, such as humidity, temperature, PM10 and PM2.5 dust concentrations, as well as CO, NO2, and NH3 levels, is examined. The Oxygen station's features make it possible to gather real-time air quality data and store large volumes of information for predicting and evaluating pollution levels and public health risks, all of which can be accessed through a personal account on the EcoCity website.

Investigative studies were carried out by measuring the maximum concentration of CO, NH3, PM10, and PM2.5 in the atmosphere using an air quality monitoring station. Throughout the study, the TLV were not exceeded, although there was an occasional excess of NO_2 concentration. The data show a harmful effect, mostly from transportation, on the environment and public health in the studied area.

References

1. EcoCity Homepage, https://eco-city.org.ua
2. Risk Assessment for the Health of the Population from Air Pollution: Methodological Recommendations MR 2.2.12-142-2007, Approved by the Order of the Ministry of Health of Ukraine. 13.04.2007 No.184. (2007)
3. Some issues of state monitoring in the field of atmospheric air protection. Resolution of the Cabinet of Ministers of Ukraine. 14.09.2019 No. 827. (2019)
4. World Health Organization. WHO global air quality guidelines: particulate matter (PM2.5 and PM10), ozone, nitrogen dioxide, sulfur dioxide and carbon monoxide. World Health Organization. https://apps.who.int/iris/handle/10665/345329 (2021)
5. SaveEcoBot Homepage, https://www.saveecobot.com
6. Van den Elshout, S., Bartelds, H., Heich, H., Léger, K.: CITE AirII, Common Information to European Air, CAQI Air Quality Index Comparing Urban Air Quality across Borders–2012. (2012)
7. European Air Quality Index Homepage, https://airindex.eea.europa.eu
8. On Approval of the List of the Most Common and Hazardous Pollutants, Emissions of which into the Atmospheric Air are Subject to Regulation. Resolution of the Cabinet of Ministers of Ukraine. 29.11.2001 No.1598. (2001)
9. Soroka, M.L.: Radiation and Smog Alarm. Guidelines and principles of public notification of air quality, radiation and chemical hazards. Arnika, Ivano-Frankivsk (2022)
10. Kolomiiets, S., Kolomiiets, A.: Integral Criterion of Environmental Safety as an Indicator of the Effectiveness of a Motor Transport Enterprise Management. Lecture Notes in Computational Intelligence and Decision Making. Advances in Intelligent Systems and Computing, 1246. Springer, Cham (2021)

11. Basic Plan of Adaptation of Environmental Legislation of Ukraine to the Legislation of the European Union (Basic Plan of Approximation): approved by the Order of the Ministry of Ecology and Natural Resources of Ukraine. 17.12.2012 No.659. (2012)
12. Kovalov, O., Sobyna, V., Sokolov, D., Harbuz, S., Vasyliev, S., Kokhanenko, V.: Method of organization of atmospheric air monitoring. Technogenic and Ecological Safety **9**(1/2021), 16–25 (2021)
13. Yesypova A., Kobzysta O., Kolomiets S., Mykhailova M.: Assessment of air quality based on data from an automatic monitoring station. Automobile Roads and Road Construction: Scientific journal. 113 (2023)
14. Loieva I.D., Vladimirova O.G., Verlan V.A.: Assessment of the state of atmospheric air pollution in a large city (methods of analysis, forecasting, regulation). KNT, Kyiv (2008)
15. Jain, L.C., Lim, C.P.: Advances in Intelligent Methodologies and Techniques. Intelligent Systems and Technologies. SCI. 217, 3–28. Springer, Heidelberg (2009)
16. The Procedure for Monitoring and Reporting on Greenhouse Gas Emissions, approved by the Resolution of the Cabinet of Ministers of Ukraine 23.09.2020. No. 960. (2020)

Intelligent Technologies for Management of Transport Highways: Safety and Environmental Aspects

Volodymyr Kyselov, Igor Yeremeyev, Alina Dychko(✉), and Yuliia Minaieva

Taurida National V. I. Vernadsky University, John McCain St., 33, Kyiv 01042, Ukraine
aodi@ukr.net

Abstract. The paper is devoted to the research of the bottleneck problems of transport highways and their solving with the use of the methods of the theory of fuzzy sets and the theory of possibilities. The aim of the research is to develop the algorithm to choose the extreme parameters of highway bottlenecks for traffic and environmental safety ensuring. The effective model of the intellectual technology of management of regional transport safety - the model of conflicting possibilities is presented. Conditions for the optimal functioning of highway in real time with the problem of bottlenecks arise are considered. The scientific novelty of the presented approach is the designed methodology of bottleneck indicators definition under conditions of uncertain data. The proposed method of the optimal capacity of the highway determination based on the criterion of road accident minimization should be used in practice of the transport highways safety and environmental management.

Keywords: Transport Highway · Bottlenecks · Theory of Fuzzy Sets · Theory of Possibilities

1 Introduction

There is a Latin winged expression: "*Dignus vindice nodus*", i.e. "Knot that requires God's intervention", in other words, a problem that requires a special, intellectual approach. Generally speaking, a node is any link in a complex system. And if the effective functioning of the entire system depends on the functioning of this link, then such a node is called the "bottleneck" of the system. A bottleneck can also be considered a link of the system, or a connection that (which) limits the freedom of choice. One of the most common "carriers" of potential bottlenecks are urban transport highways.

The problems of highway safety, parking, providing optimal conditions for solving the problems of transporting goods and passengers from the point of origin to the destination have recently received a lot of attention and achieved significant success [1–6], but as about the actual overcoming of the bottlenecks on transport highways, the matter has not yet gained the necessary emphasis.

O. Slavinska et al. (Eds.): ITSESQC 2024, LNNS 1335, pp. 42–51, 2025.
https://doi.org/10.1007/978-3-031-87376-8_4

In order to identify problems and ways to solve them regarding transport highways in particular and road traffic in general, the Strategy for Improving the Level of Road Traffic Safety and the State Program for the Improving the Level of Road Traffic Safety have been developed in Ukraine [7, 8]. The main objectives of the documents are to increase the level of road safety and reduce social and economic losses and damages from road accidents. The Law of Ukraine "On making changes to Certain Laws of Ukraine Regarding Highway Safety Management" is also adopted, in which it is proposed to carry out an audit of the safety of transport highways and roads [9]. "The State Target Economic Program for the Development of Public Highways of State Importance" defines directions for improving the safety and economy of highways and roads, taking into account the requirements for the integration of Ukraine into the European transport system [10, 11].

In addition to ensuring road safety, the issue of reducing environmental pollution from road transport remains important [12]. Since a significant number of motor vehicles are located in cities, highly toxic exhaust gases from car engines are concentrated in areas of residential construction with a low level of ventilation of blocks, in the surface layers of the atmosphere, due to which dangerous emissions enter the human body. Moreover, such pollutants, which make up more than 200 toxic chemical compounds, can remain in the atmosphere for up to six months under such conditions [13].

The aim of the work is to develop the algorithm to choose the extreme parameters of highway bottlenecks for traffic and environmental safety ensuring.

The methodology of the research is based on the use of the methods of the theory of fuzzy sets and the theory of possibilities.

2 Transport Highways Bottleneck Problems Decision

Urban transport highways are characterized, first of all, by the capacity Q and carrying capacity T. It is these indicators that can create the conditions for the emergence of the bottleneck under certain circumstances. Capacity characterizes the property of a transport highway under certain conditions to pass one or another (maximum) number of vehicles per unit of time in one direction. This property depends on the average speed at which vehicles can move on the highway V, on its width S (more precisely, from the number of rows of vehicles that can simultaneously move in one direction) and on the vehicles concentration S (the number of vehicles per 1 km of highway moving in one direction), i.e.

$$Q = F(V, S, K) \tag{1}$$

Since the average speed depends on the traffic conditions, first of all, on the state of the road surface, visibility, the number of intersections at which vehicles must reduce speed or stop to allow traffic in the orthogonal direction, as well as on the organization of traffic at intersections, the essential increasing capacity can be achieved by creating transport solutions at different levels in order to avoid crossings and related problems. However, during peak hours, the formation of "traffic jams" is likely under certain conditions. In order to eliminate this type of bottleneck, it is possible to foresee in advance the possibility of organizing alternative traffic routes to relieve the main transport highway, and in some cases, even to the transition of the selected highway to one-way traffic (which virtually doubles the highway and, thus, increases the throughput) with simultaneous organization of the return route of traffic on alternative highways. As for public passenger transport, the increase in capacity can also be ensured by increasing the distance between stops (which contributes to an increase in the average speed of traffic).

The carrying capacity of a transport unit (the number of passengers or cargo that can be carried simultaneously by a moving vehicle) is closely related to throughput: the greater it (carrying capacity), the fewer vehicles are needed to ensure the transportation of the desired number of passengers or cargo, and the flow of vehicles on the highway will be less. In other words, it is more rational to operate vehicles with increased carrying capacity during peak hours, although factors such as impaired maneuverability (dynamic characteristics) of heavy vehicles must also be taken into account here, and therefore it is important to implement these measures first of all where there are no physical intersections of highways (that is, there are traffic junctions at different levels), or one-way traffic is provided.

When developing the intelligent technology of the highway safety management system, the following should be taken into account:

- a list of possible reasons for a decrease in the level of security;
- tree of fuzzy conclusion;
- a base of vague rules;
- training sample.

One of the most effective models of the intellectual technology of management of regional (city) transport safety is the model of conflicting possibilities. This model is built on the assumption that the number of people leaving a certain point is proportional to the number of opportunities that exist on the route, and inversely proportional to the number of opportunities that are realized. Such possibilities there may be an employment place (for people looking for job), city and country parks (for people going for recreation or entertainment), shopping centers (for people going shopping), industrial enterprises and service centers, educational institutions, etc. (respectively, for workers of certain institutions, students, apprentices, etc.).

The model of conflicting possibilities can be mathematically represented as the following:

$$Tij = Oi\left[e - LD - e - L(D + Dj)\right], \tag{2}$$

where Tij - number of trips from the i-th zone to the j-th zone; Oi - total number of trips that start in the i-th zone; D - number of targets that take place on the route that leads to the j zone; Dj - number of trips ending in the j-th zone; L - certain estimate of the probability that the goal of a separate trip can be achieved at a randomly selected destination (L - value that is determined on the basis of the initial data and that characterizes the degree of decrease in the value of Tij with an increase in the number of goals and the length of the route), e - base of the natural logarithm.

After obtaining the simulation results, it is necessary to analyze the distribution of possible means of transport, for which, first of all, it is necessary to analyze carefully the state of the existing transport network from the point of view of whether changes in the organization of the network and increasing its throughput allow meeting future transportation needs. Consideration of these issues in the public and private sectors is carried out differently. The fact is that in the private sector (first of all, in the field of cargo transportation), the choice of vehicles, routes and schedules is the prerogative of the private owner. In the public sector, the main elements of the transport system are connected with random factors, and the satisfaction of transport needs is the concern of the people who need transport services. Therefore, in the public sector, the allocation of vehicles to meet the future needs of passengers is a problem that can be only solved by forecasting methods of the intellectual technologies of management. Here, the so-called transport models of route distribution make it possible to determine how the existing transport networks or the networks being designed can be used to ensure transportation. In addition, there are shortest path models (which use decision tree and all - or-nothing algorithms) and minimum cost network models (which use chaos theory algorithms). Finally, there are vehicle distribution models that are designed to determine the proportion of passengers choosing a particular vehicle.

When solving the problem of distribution of vehicles, two main principles are used:

- "optimization for the consumer", which is based on the assumption that a certain equilibrium state can be established in the system, when no vehicle has the opportunity to reduce the travel time by changing the route, since the route is chosen based on the requirement of path minimization;
- "optimization for the system", which is based on the minimization of the average running time.

The use of algorithms for determining the most profitable route (or the shortest path) assumes that people who make a trip choose the most profitable (from their point of view) or a route close to it ("optimization for the consumer").

When solving the problem of choosing a route, it is necessary to take into account the degree of traffic flow intensity in order to prevent a bottleneck. Thus, the travel time on main highways increases significantly during peak hours, and then it becomes more acceptable to use secondary transport lines. Among the methods of route distribution, which are related to the methods of determining the most profitable route, the most widespread is the "all or nothing" method, the essence of which is that all transport that has the same departure and destination points is passed along the most profitable route, in while other vehicles do not use this route. This approach provides optimization for the system||. To a large extent, the same approach can be attributed to the selection of highways with one-way traffic, when the two most profitable routes are chosen for traffic in opposite directions to connect the same points of origin and destination.

Above, it was about ensuring the most favorable conditions during the design of this or that highway, or tying the existing highway to new operating conditions. But it is much more difficult to create optimal (or close to them) conditions for the functioning of highway in real time - this is where the problem of bottlenecks arises, since some factors are not always taken into account, such as the virtual possibility of changing the number of traffic lanes (due to repair and adjustment work), a result of traffic accidents, or the increase of traffic lanes in one direction due to the corresponding reduction in the number of traffic lanes in the opposite direction), the number of regulated and non-regulated intersections, as well as traffic junctions and stop points (which largely determines the average real maximum speed of vehicles on the highway), fluctuations in the size (primarily length) of vehicles and their maneuverability (dynamic characteristics), as well as the maximum permissible distance between vehicles under different conditions (icing, wet snow sticking, fog, etc.), which is decisive for the density of traffic flows. In other words, the throughput is determined for ideal conditions and the maximum permissible (decreed) speed of vehicles on the highway, as well as the minimum permissible distance d_{min} between vehicles for the ideal conditions indicated above. In fact, the value Q is a multidimensional function (3), where K is the "concentration" of vehicles on 1 km of highway for one traffic lane:

$$\frac{1}{N_{max}(L_{max} + d_{max})} \leq K = \frac{1}{N_{min}(L_{min} + d_{min})} \tag{3}$$

where L - length of the vehicle; dj - the minimum permissible distance between the vehicles following each other in the same traffic lane, which depends on the visibility on the highway, the conditions of adhesion to the ground, the average reaction time of drivers and the dynamic properties of the vehicle; respectively, the indicators *Nmin, Nmax, Lmax, Lmin, dmax, dmin* refer to extreme characteristics that are associated with the maximum and minimum possible concentrations of vehicle on the highway.

Here, a significant part of the variables can be determined only on the basis of observations of traffic police posts, statistical analysis or using the apparatus of the theory of fuzzy sets and the theory of possibilities. The last way has every reason to be considered as the most effective, since it works well precisely under conditions of uncertainty of the initial data and lack of sufficiently representative samples, which does not allow determining the nature of the law of probability distribution [14]. The essence of the methods of theories of fuzzy sets and possibilities consists, first of all, in replacing statistics with a membership function, that is, in replacing the probability of a fuzzy event, which is characterized by a certain probability distribution (most likely, uniform) with a possibility that depends on the power of the fuzzy set [15, 16]. Under the conditions of fuzzy data characterizing the throughput, the membership function, for example, for a fuzzy sample interval of such factors as visibility, adhesion to the ground, or such an integral characteristic as the minimum permissible distance between vehicles on the traffic lane, has the following form:

$$\mu(x) = \lceil exp[-(x - m1)2\delta - 2] \quad at\, x \leq m1$$

$$\mu(x) = [\quad 1 \quad] \quad at \;\; m1 \leq x \leq m2$$

$$\mu(x) = \lfloor exp[-(x - m2)2\delta - 2] \quad at \;\; x > m2 \tag{4}$$

where *m1* and *m2* - left and right, respectively, limits of the interval in which this or that indicator can be located, x - vector of real values of the quantity to be determined, m - scalar equivalent of the fuzzy set, δ - fuzziness (uncertainty), which is characterized by limits (in percentages of the maximum value), in which the blurred value is perceived as unchanged (of course, blurring can exceed the accuracy with which the corresponding indicator is determined by 1–2 orders of magnitude).

At the same time, any observed process can be represented by its linguistic description. For example, the bandwidth of a highway can be described by the following set of linguistic equations:

$$IF\ Si < S, Li > L, Di < D, Vmax\,j < Vmax\ max,\ dj = dmax,\ THEN\ Q1 = Qmin\ min;$$

$$IF\ Si < S, Li > L, Di < D,\ Vmax\,j < Vmax\ max, dj < dmax,\ THEN\ Q2 > Qmin\ min;$$

$$IF\ Si = S, Li = L, Di = D, Vmax\,j = Vmax\ max, dj = dmin,\ THEN\ Qk = Qmax\ max \tag{5}$$

where S - number of traffic lanes in one direction, V - average speed of vehicles on the highway, $Vmax$ - the maximum permissible speed on the highway decreed by relevant documents.

Each of these equations is in the solution space and has a value Qi within the limits

$$Qmin\ min \leq Qi \leq Qmax\ max, \tag{6}$$

but the actual (or recommended) value Q, although it must be in the space of possible decisions, must also take into account the fact that certain uncertainty in the data requires maneuvering of factors and the use of interval values of variables and factors so that the overall solution corresponds to possible fluctuations in the data. At the same time, the final decision may have the following form (depending on the goal pursued by the researchers):

$$Qopt = min\{Qi - k,\ \ldots\ Qi,\ \ldots\ Qi + k\}\ (for\ the\ case\ of\ minimization\ of\ traffic\ accidents),$$

or

$$Qopt = max\{Qi - k,\ \ldots\ Qi,\ \ldots\ Qi + k\}\ (for\ the\ case\ of\ maximization\ of\ Q), \tag{7}$$

where *min* and *max* operators correspond to fuzzy logic operations that select the smallest or the largest value from the set of indicators under consideration; the values Qj correspond to specific constraints or assumptions used in the calculations, that is, decision making on the value $Qopt$ corresponds to choosing an alternative which set of possible realizations is contained in the set of satisfactory outcomes.

It is necessary to dwell separately on the possible source of the bottleneck on highways, which causes the so-called traffic jams. Movement along a transport highway can be imagined as a flow of liquid, the particles of which are vehicles of one or another mass. These "particles" in this case should be considered as inelastic (not to take into account collisions inherent in a liquid with elastic particles). Although the vehicles flow exhibits some properties of a gas (it can "compress" or "elongate" depending on the intensity of movement), in the end it behaves more like a liquid, and with very "saturated" movement - like thick glue or molasses. A characteristic feature of such a flow are waves that sometimes become dangerous. These waves are caused mainly by the instability of the traffic flow: a permanent change in the traffic mode due to the presence of intersections and the need to brake if the red light turns on at the traffic light when approaching the intersection, or to speed up if the traffic light turns on green, the change in the concentration of traffic on the highway due to the increase or decrease of vehicles coming from other highways, or branching off into other highways, and the associated maneuvering and speed changes, as well as traffic delays and ambiguous reactions of different drivers to changes in traffic. At the same time, the speed of movement acquires the properties of longitudinal oscillations, a kind of "shock waves" are formed, which can, under certain circumstances, lead to a stoppage of movement and the formation of "traffic jams". This happens, first of all, in the "bottlenecks" of highways: at turns, intersections, in tunnels. The shock wave propagates in the direction opposite to the movement of the vehicle to the "bottleneck", and its speed can be determined from the following expression:

$$Vsw = \Delta Q / \Delta K, \tag{8}$$

where ΔQ - change in the vehicle flow rate, ΔK - change in the vehicle concentration during the passage of the shock wave.

It is not difficult to make sure that when the shock wave speed *Vsw* reaches the value of the speed of the vehicle on the highway, the movement of the vehicle stops and a traffic jam is formed, and when *Vsw* exceeds the specified speed, a large-scale accident occurs inevitably. Therefore, there is a need to analyze the degree of dependence of system characteristics on parameter fluctuations and variables that affect this characteristic of the transport highway, for example, the sensitivity of the function of the shock wave speed relative to the vehicle concentration *K*. That is, there is a need to analyze the ratio of the percentage change in *Vsw* to the percentage change in *K* that caused the change in speed. The analysis mentioned above can be the subject of a separate study.

In practice, it is not always possible to obtain the exact values of the bottleneck (*BV*) indicators. Therefore, to take into account this problem, each uncertain *BV* is considered in a certain interval, since its expected minimum $\downarrow we$ and maximum $\uparrow we$ values can be determined. Then it is known that the values of the bottleneck is within the interval *W* [$\downarrow we$, $\uparrow we$] and *We* contains all possible values in the interval of existence of this *BV*. Thus, the calculation of the actual we fact can be realized by the following heuristics:

$$IF\ (there\ is\ a\ BV\ search\ problem) AND (W\ [\downarrow we, \uparrow we]) AND (\downarrow we \le we\,fact \le \uparrow we),$$
$$THEN (we\,fact\ is\ actual\ VB) OTHERWISE\ (continue\ BV\ searching). \tag{9}$$

In the case when there may be several bottlenecks, it is necessary to choose the extreme *BVextr* according to:

$$BVextr = min\{wefact1,\ wefact2,\ \ldots\ wefact\,n\}. \tag{10}$$

The block diagram of the algorithm described above is demonstrated in **Помилка! Джерело посилання не знайдено.**

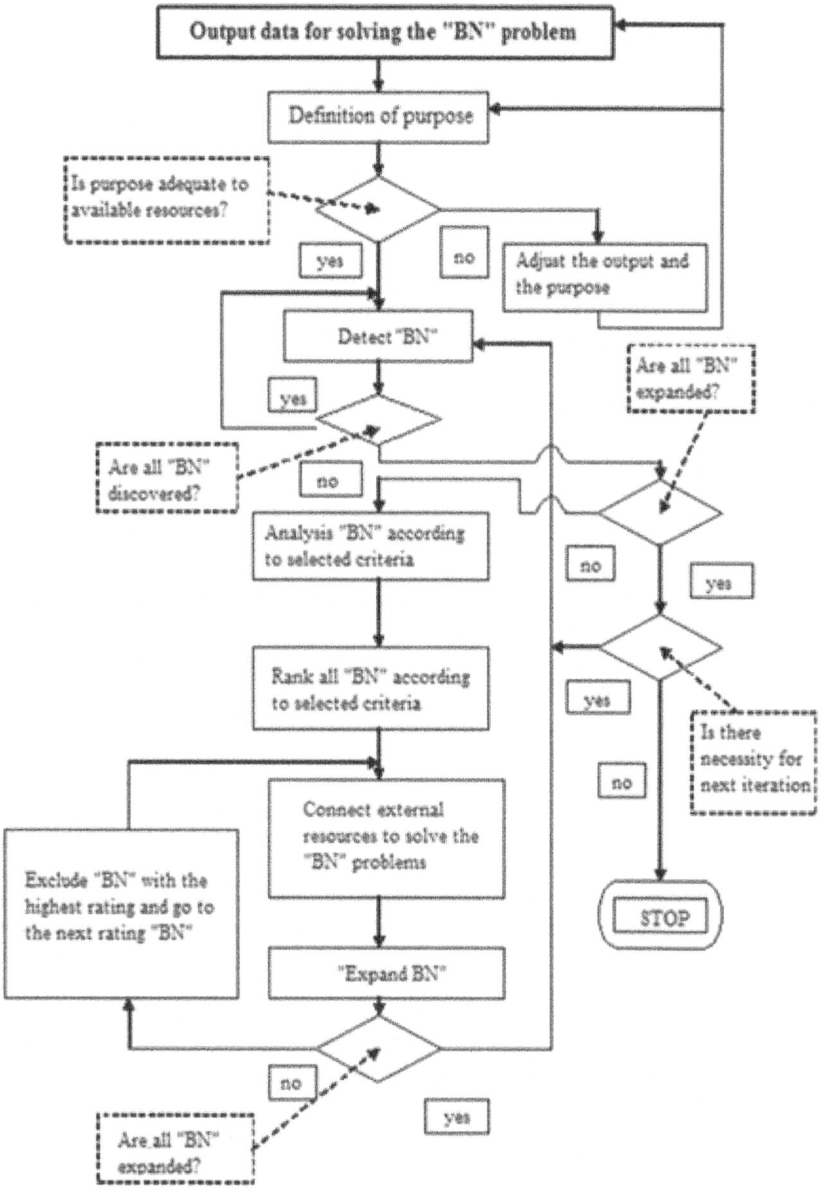

Fig. 1. The block diagram of the algorithm of bottleneck (BN) search.

3 Conclusions

The above-mentioned approaches to the transport highways bottleneck problems decision provide a significant improvement in the functioning of information systems and contribute to the adoption of more informed decisions regarding the minimization of the consequences of transport disasters and accidents. However, their use in the absence of accepted patterns of distribution of pollutants and characteristics of vehicles, which makes them sensitive to external influences and oriented to accurate input information, requires new, non-standard approaches, one of which is the interpretation of the information used in the system from the point of view theories of fuzzy sets and theories of possibilities, which form the basis of intelligent information systems (Fig. 1).

To achieve the road accident minimization it should be taken into account both the real average length of vehicles and the average distance between vehicles, which is determined by such factors as the quality of traction with the ground, visibility, maneuverability (dynamic properties of vehicles and the average reaction of drivers), as well as average and instantaneous speed of vehicles on the highway.

Based on the criterion of road accident minimization, it is necessary to adjust the distance between vehicles (and, accordingly, the concentration of vehicles) in the bottleneck in such a way as to minimize the speed of the shock wave in case of a change in the speed of the vehicles flow.

References

1. Osadchiy, V.: Influence of technology on route management systems, https://yalantis.com/blog/add-route-planning-to-an-app
2. Sadanandan, L., Somasundaran, N.: A smart transportation system facilitating on-demand bus and route allocation. 2017 International Conference on Advances in Computing, Communications and Informatics (ICACCI). 1000–1003 (2017)
3. Smart Transportation: A Complete Guide to Defining The Future of Mobility, https://cerebrumx.ai/a-complete-guide-to-smart-transportation-defining-the-future-of-mobility
4. Smart Bus Routes: How Technology is Making Public Transit More Efficient, https://www.tomorrow.bio/post/smart-bus-routes-how-technology-is-making-public-transit-more-efficient-2023-06-4732296936-iot9/17
5. Sader, W.: Traffic Routing and Real-Time Data in Smart Transport Systems, https://medium.com/@waleedsader/traffic-routing-and-real-time-data-in-smart-transport-systems-d2c3cd43bcf4
6. examples of smart city transportation solutions, https://www.nec.co.nz/market-leadership/publications-media/5-examples-of-smart-city-transportation-solutions
7. Strategy for increasing the level of road traffic safety in Ukraine for the period until 2024, https://zakon.rada.gov.ua/laws/show/1360-2020-%D1%80#Text
8. State program for increasing the level of road safety in Ukraine for the period until 2023, https://zakon.rada.gov.ua/laws/show/1287-2020-%D0%BF#Text
9. The law of Ukraine: On making changes to some laws of Ukraine regarding the management of highway safety, https://zakon.rada.gov.ua/laws/show/200-20#Text
10. State target economic program for the development of public highways of state importance for 2018–2023, https://zakon.rada.gov.ua/laws/show/382-2018-%D0%BF#Text
11. Trypolska, H.: Recommendations to transport legislation of Ukraine taking into account EU practices, https://ua.boell.org/uk/2021/02/11/rekomendacii-do-transportnogo-zakonodavstva-ukraini-z-urakhuvannyam-praktik-es
12. Martin, E., Shaheen, S., Wolfe, B.: Environmental Impacts of Transportation Network Company (TNC)/Ride-Hailing Services: Evaluating Net Vehicle Miles Traveled and Greenhouse Gas Emission Impacts within San Francisco, Los Angeles, and Washington, DC Using Survey and Activity Data. Sustainability. **16**(17), 7454 (2024)

13. Motor transport and the environment: problems and ways to solve them, https://www.zoda.gov.ua/news/7207/avtotransport-i-navkolishnjeseredovishe-problemi-i-shlyahi-jihnogo-virishennya.html
14. Carteni, A., Marzano, V., Henke, I., Cascetta, E.: A cognitive and participative decision-making model for transportation planning under different uncertainty levels. Transp. Policy **116**, 386–398 (2022). https://doi.org/10.1016/j.tranpol.2021.12.013
15. Dychko, A., Yeremeyev, I., Remez, N., Kraychuk, S., Ostapchuk, N.: Structural redundancy as robustness assurance of complex geoengineering systems. E3S Web of Conferences, EDP Sciences. 166, 11003 (2020). https://doi.org/10.1051/e3sconf/202016611003
16. Dychko, A., Yeremeyev, I., Kyselov, V., Remez, N., Kniazevych, A.: Ensuring reliability of control data in engineering systems. Latv. J. Phys. Tech. Sci. **56**(6), 57–69 (2019). https://doi.org/10.2478/lpts-2019-0035

City Spatial Accessibility is a Prerequisite for Sustainable Urban Development

Daria Chekh[1(✉)], Olga Kunytska[1], Luca Persia[2], Vitalii Kharuta[1], and Sevket Oguz Kagan Capkin[2]

[1] National Transport University, Mykhaila Omelianovycha-Pavlenka Str.1, Kyiv 01001, Ukraine
darrellchek@gmail.com

[2] DICEA at Sapienza University of Rome, Via Eudossiana 18, 00184 Rome, Italy

Abstract. Urban spatial accessibility is an important issue that needs to be considered in terms of urban spatial regulation. Redistribution of street urban space towards improving accessibility to workplaces and services requires changes in the city's infrastructure, public transport, creation of bicycle lanes, and expansion of sidewalks. One solution that is being actively implemented in Kyiv (Ukraine) is the introduction of aboveground pedestrian crossings that replace or duplicate underground crossings. On the part of the city authorities, the implementation of the accessibility principle and individual solutions in this regard should consider the analysis of their impact on traffic organization. This study proposes to use a microscopic traffic simulation to address this issue. Traffic modeling is an important function used in intelligent transportation systems and is aimed at improving the efficiency of street and road networks. The study examines an intersection in the city center, where, according to a city regulatory document, a duplicate ground crossing is planned. The existing situation and the scenario of the ground crossing implementation were modeled in the PTV Vissim software.

Keywords: Urban Planning · Barrier-Free City · Microsimulation · PTV Vissim

1 Introduction

Mobility is one of the key drivers of urbanization. But despite the development of urban mobility around the world, access to workplaces and services is becoming more difficult. Many cities around the world are facing an accessibility crisis caused by unsustainable mobility systems [1]. Effective urban design requires an understanding of ideas, paradigms, and research from many areas of knowledge, including anthropology, sociology, engineering, architecture, and urban planning.

The prioritization of individual automobile transport in Ukrainian cities negatively affects both the accessibility of the city and the quality of life of all residents, as well as the state of the environment [2]. Currently, Ukrainian transport accounts for about 10% of all greenhouse gas emissions (GHG), but accounts for 71% of total oil consumption in Ukraine in 2019 [3]. Over the last three decades, the composition of emissions has

O. Slavinska et al. (Eds.): ITSESQC 2024, LNNS 1335, pp. 52–62, 2025.
https://doi.org/10.1007/978-3-031-87376-8_5

changed significantly. The most significant change is the increase in the share of emissions from motor vehicles. In 2017, road emissions accounted for more than 70% of all transport emissions, making this a major challenge for future transport policy. As city authorities and residents become increasingly aware of the negative impacts of automobile transport, the focus of urban planning in many cities is shifting from cars to more environmentally friendly modes of mobility (walking, cycling, public transport). This is in line with the principle of pedestrian-centeredness as defined in the mobility pyramid (Fig. 1).

Such a change in priority requires cities to implement the principles of inclusiveness and spatial accessibility, which means finding ways to transform street space for more efficient, safer, and socially just use of streets and their carrying capacity. This should be measured not only in terms of the number of vehicles, but also in terms of the total number of people (pedestrians, cyclists, public transport passengers).

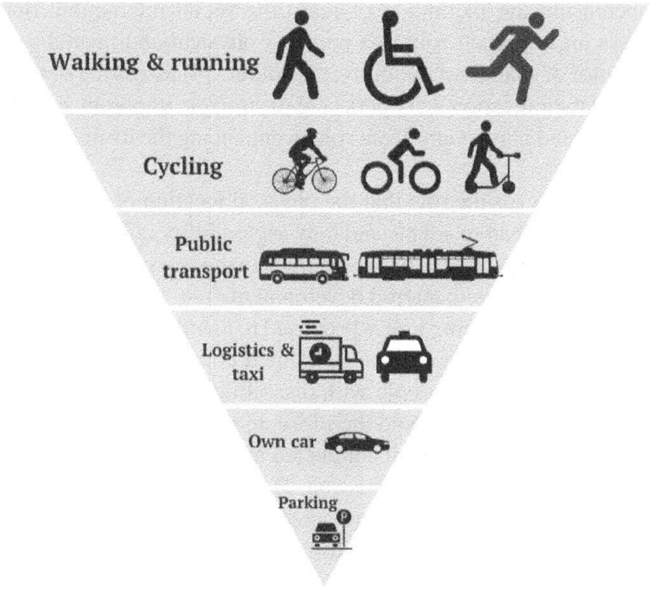

Fig. 1. The mobility pyramid (own source).

The redistribution of street space is not about making it harder for car drivers to get around, but about improving the mobility of pedestrians, cyclists, and public transport. This increases the overall level of mobility in the city. Changes are made by improving public transportation infrastructure, creating bike lanes and bicycle paths, and widening sidewalks.

One of the most difficult issues is the arrangement of regulated and engineered above-ground pedestrian crossings, which should duplicate underground crossings. For example, the Department of Transport Infrastructure of the Kyiv City Administration has approved a list of locations [4] where duplication should take place. These changes are in line with the main regulatory documents adopted in Ukraine to implement accessibility and equality, as well as the principles of urban inclusiveness, namely: "National Strategy for Barrier-Free Environment in Ukraine up to 2030" of 14.04.2021 [5] and state building codes [6].

However, the question arises of the impact of newly equipped ground pedestrian crossings on traffic in the city. This paper analyzes the impact of a duplicate pedestrian crossing on the city's street and road network using the PTV Vissim microscopic traffic simulation.

2 Literature Review

Designing urban space is an interdisciplinary task that combines politics with social, environmental, and psychological issues, community life, and human health [7]. In [8], it is determined that systemic urban design increases opportunities for walking, consistently increasing the likelihood of real-time social interaction. Such an urban atmosphere plays an important role in a person's physical and mental well-being and behavior. As a factor of spatial accessibility, the authors [9] assess the spatial distribution of urban parks and their location. City parks are of strategic importance for the quality of life in urban society and play a significant role in enhancing the livability and sustainable urban development.

The study [10] analyzes the role that the physical location of public spaces plays in the healthy lifestyle of urban residents and how accessibility affects urban regeneration. The paper explores the Proactive City, which finds that accessibility is a driving force in urban regeneration and people-centered development.

In [11], an innovative planning tool is proposed to identify and optimize urban spatial barriers for people with disabilities. The developed approach allows to evaluate urban space from the point of view of people with disabilities in terms of the level of impact on their psychophysiological state, and therefore, in terms of specific benefits for urban planning. According to the experiment conducted in this paper, the area with the highest level of stress is the ground pedestrian crossing. The reason for this is the constant or even increased speed of traffic, which is accompanied by physiological stress. This points to a traffic light interval that is too short and forces people with disabilities to walk faster, putting them under stress.

Analyzing the transition of cities to barrier-free from the perspective of transportation engineering is possible with the help of microsimulation modeling in PTV Vissim. PTV Vissim can be applied to many traffic and traffic engineering problems, such as intersection reconstruction and comparison of their benefits, selection of the best part for roads and highways, analysis of construction opportunities, transportation development planning, transportation management, traffic and human resource planning, and public transportation [12]. In this paper, the PTV Vissim microsimulation tool is used to evaluate solutions for implementing the principles of city spatial accessibility on the example of Kyiv, Ukraine.

3 Research Methodology

The planned development of a city, in addition to solving architectural and planning problems and issues of engineering equipment of the developed territories, necessarily involves solving the problem of transport services. The solution to this problem affects almost all functions of the city's life. This explains the term "urban transport planning," which should be understood as a direction in the development and evaluation of the transport qualities of city planning in general and individual territories. This includes a set of transport, construction, planning and environmental measures. The purpose of implementing these measures is to create a rational structure of the street and road network and an effective system for organizing traffic and pedestrian movement in the city.

However, as the number of inhabitants grows, the city area expands, and the number of cars and traffic intensity increases, the issue of excessive traffic and congestion arises. The focus of activists, urban planners, and authorities has shifted from car-centric views to convenience, barrier-free and inclusive - the priority of comfort is given to people.

The problem of social inclusion of people with disabilities is particularly acute. Currently, 1 billion people, or 15% of the world's population, have some form of disability, and 80% of people with disabilities live in developing countries, facing significant discrimination and barriers that limit their participation in society compared to other citizens [13].

Today, the number of people affected by the war in Ukraine is growing due to the full-scale Russian invasion. Over the two years of the full-scale invasion, the number of people with disabilities in Ukraine has increased by 300 thousand people [14]. The upward trend (Fig. 2) in the number of people with disabilities makes the issue of spatial accessibility and inclusiveness of cities an urgent one for Ukraine.

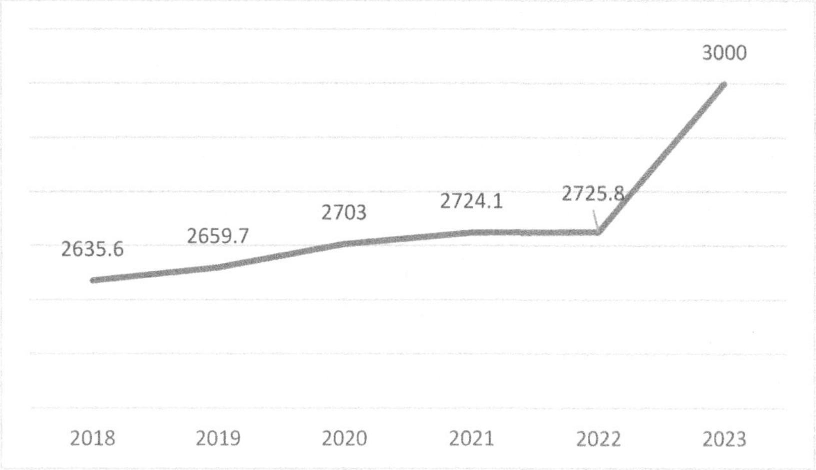

Fig. 2. The number of people with disabilities in Ukraine, according to [15, 16], thousand people.

Spatial accessibility building is not new. It has been a challenge since the Second World War. In particular, in Italy, accessibility is becoming a prerequisite for Italian cities to improve heritage, protect and secure people, and guarantee autonomous and independent living, even for the most vulnerable people [17]. Today, examples of the use of spatial accessibility means in urban planning are (Fig. 3):

1. Level, barrier-free, smooth crossings and sidewalks
2. Ramps and low gradients along the path of travel and to entrances to public buildings
3. Convenient parking spaces
4. Specially equipped restrooms and common areas
5. Equipped public transportation and access to it
6. Aboveground pedestrian crossings that replace or duplicate underground crossings
7. Equipping the pedestrian crossing with activating pushbuttons, accessible pedestrian signals, dividing different way signals on same crossroad, etc.

a) b)

Fig. 3. Examples of barrier-free cities in a) Ukraine (own photo) and b) Italy [18]

NGOs and activists are the most active in addressing the issue of inclusiveness, collecting information, checking compliance, organizing briefings, and disseminating information. For example, in Kyiv (Ukraine), in March 2024, the Municipal enterprise "Center for Traffic Organization" together with proactive masters of a Kyiv higher education institution conducted a study of the compliance of pedestrian crossings in the central part of the city [19].

The study checked and analyzed 959 pedestrian crossings (1918 approaches) over an area of 11.7 km2 using GIS and found that only 25% of pedestrian approaches fully meet modern inclusivity requirements, 27% are partially accessible and need to be addressed, and 48% of approaches do not meet the requirements at all (Fig. 4).

Implementation of the principles of city accessibility on the example of Kyiv, Ukraine

One of the ways to ensure accessibility and physical barrier-free transport infrastructure in general, and in Kyiv in particular, is to arrange regulated and engineered ground pedestrian crossings that duplicate existing underground crossings. In 2023, the Department of Transport Infrastructure of the Kyiv City State Administration approved a list of such places for the arrangement of regulated and engineered ground pedestrian crossings [5].

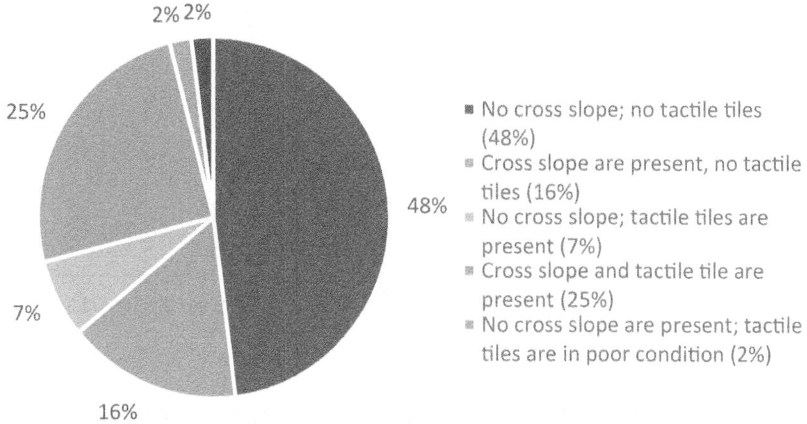

Fig. 4. Results of the survey of pedestrian crossings in the central part of Kyiv (Ukraine), data source [19].

Implementation of the decision requires an analysis of the impact of the arrangement of ground pedestrian crossings on the organization of traffic. Even if pedestrians are prioritized, the traffic flow on high intensity urban highways should not be neglected. This can lead not only to the formation of heavy traffic congestion, but also to an increase in the number of accidents in this area.

The change in intensity when installing ground pedestrian crossings can be modeled and analyzed in the PTV Vissim software environment. It allows to simulate traffic on the chosen area, enter traffic intensities, set traffic light control schemes for cars and pedestrians, add pedestrian crossings and get the results of the decisions made in the form of the number of delays and travel time and the capacity of the updated area.

For example, a microscopic traffic simulation for one of the intersections defined in [5] is considered, namely the intersection located in the city center on Velyka Vasylkivska Street at the intersection with Bessarabsky passage. The initial data are provided by the Municipal Enterprise "Center for Traffic Organization", namely, the traffic organization scheme (Fig. 5), data on the composition and intensity of traffic flow in the area under consideration (Table 1), and the traffic light signal program (Fig. 6).

Considering the initial data, modeling of the existing road traffic organization scheme at the intersection (Fig. 7) and the design scenario with the addition of a duplicating ground pedestrian crossing (Figs. 8 and 9) were performed.

Scenario Analysis

The modelling and analysis of the existing and project scenarios were carried out within 1 h. The total length of all sections of the study area is 1315 m. During the analysis, 4219 cars arrived in the existing condition and 3483 in the design scenario, indicating a decrease in the study area's capacity. In the project scenario, pedestrian crossings were

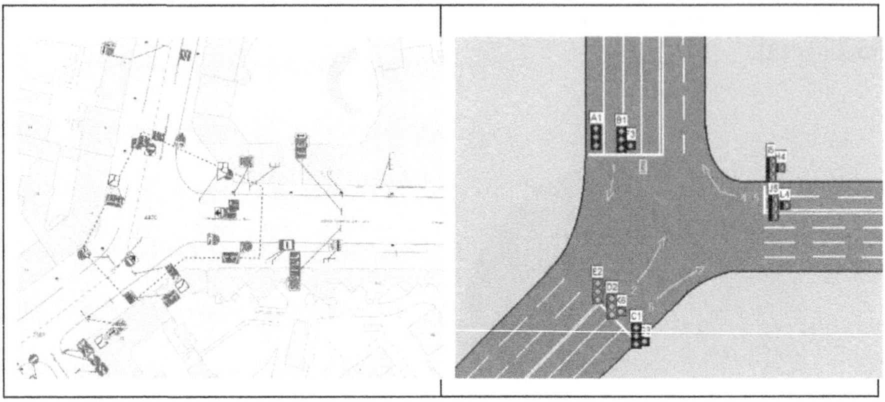

Fig. 5. Scheme of traffic management at the intersection of Velyka Vasylkivska Street and Bessarabsky passage

Table 1. Traffic intensity at the intersection of Velyka Vasylkivska Street and Bessarabsky passage.

The composition of the traffic flow	Direction 1, 3		Direction 2,6		Direction 4,5
	straight	to the left	straight	to the right	to the right
Car	694	1341	852	636	316
Motorbike	0	32	21	26	16
Bicycle	11	32	0	5	0
Van	21	131	21	37	32
Truck	11	0	21	0	0
Tractor trailer	0	0	0	0	0
Bus	11	5	0	0	0
TOTAL	**748**	**1541**	**915**	**704**	**364**

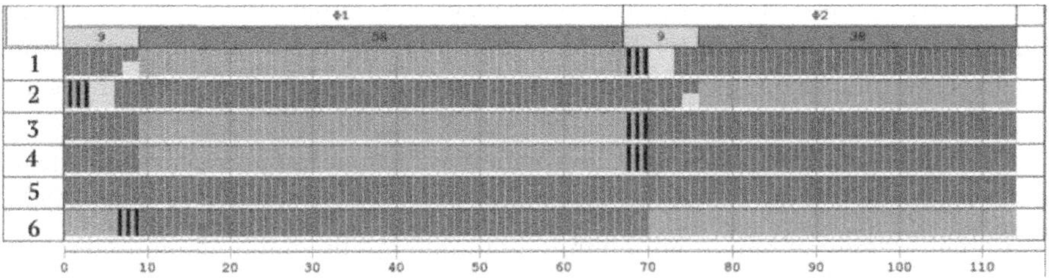

Fig. 6. The traffic light signal program at the intersection of Velyka Vasylkivska Street and Bessarabsky passage, Kyiv.

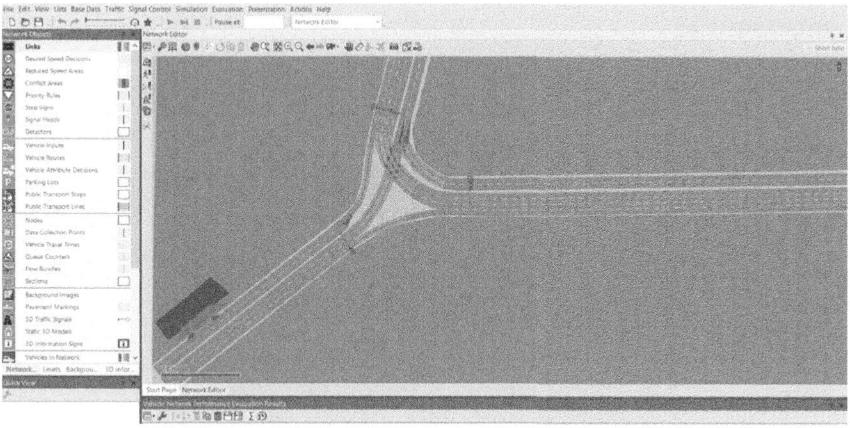

Fig. 7. Microscopic traffic simulation of the existing situation for the considered intersection in PTV Vissim

added, which, due to the preference of pedestrians, reduced the overall performance of the section.

According to the results of modelling the existing condition and the project scenario with the addition of a pedestrian crossing at the intersection considered in PTV Vissim, the main indicators of the intersection functioning were analyzed (Table 2).

4 Results and Discussion

Microscopic traffic simulation of the existing and projected scenarios provides a clear understanding of the possible difficulties for the city associated with the implementation of barrier-free solutions. Based on the analysis of the scenarios in PTV Vissim, we see that the intersection's performance in terms of traffic flow has a sharp deterioration - the delay time for vehicles crossing the intersection increases, the total travel time increases, and the average speed decreases.

Obtaining a project scenario allows decision makers to determine the feasibility of implementing barrier-free solutions for each individual city section. In the case study under consideration, despite the deterioration of quantitative indicators, it is necessary to consider qualitative indicators, such as changes in the city's accessibility, attractiveness of the city and points of interests and decreasing level of intention to get to city center y car. However, for the final decision, macro-modeling tools (e.g., PTV Visum) can be used to evaluate a larger area of the city, to analyze how the implementation of the project scenario will affect other city intersections and the overall characteristics of traffic flow. At the same time, considering not only the indicators, but also the city's need to modernize space and mobility to meet the standards of modern sustainable cities.

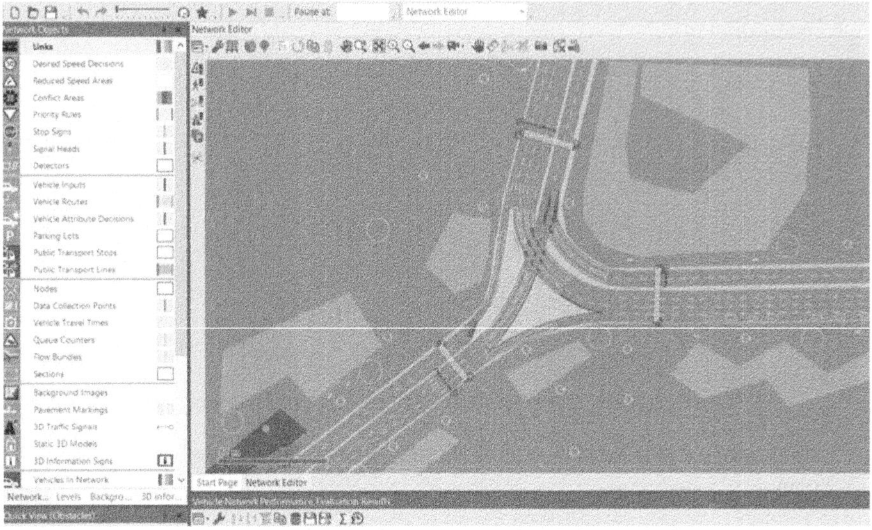

a) Microscopic traffic simulation for the intersection design scenario

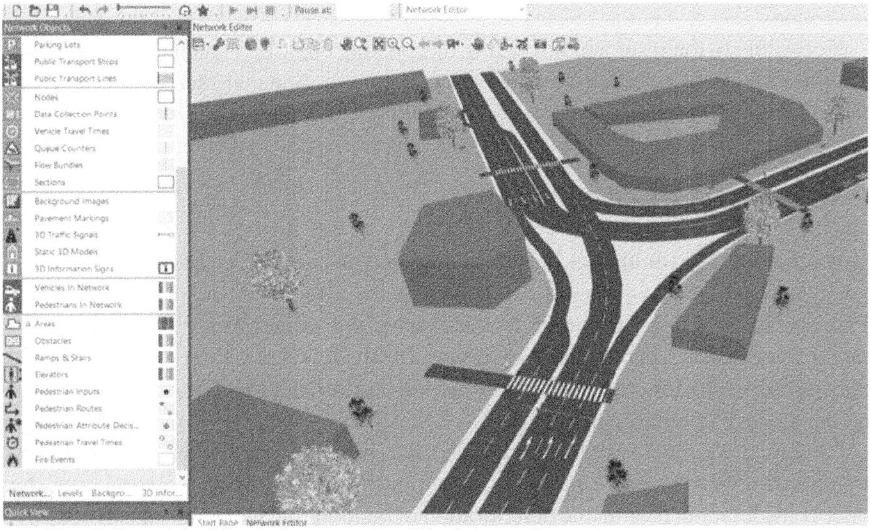

b) 3-D model of the design scenario of the intersection

Fig. 8. Microscopic traffic simulation for the project scenario with the added ground pedestrian crossing for the considered intersection in PTV Vissim

Fig. 9. The traffic light signal program of the project scenario

Table 2. Main indicators of the intersection functioning according to the modeling results.

Indicators	Existing condition	Project scenario	Ratio
Number of vehicles arriving during the simulation, units	4219	3483	−17%
Average speed (all), km/h	29.32	12.29	−58%
Standing delay time (all), s	13.99	73.43	+424%
Total delay time (for all vehicles), s	94181.36	395692.9	+320%
Total travel time (for all vehicles), s	236682.6	527955.5	+123%

References

1. UN Habitat General Assembly: Planning and Design for Sustainable Urban Mobility: Global Report on Human Settlements (2013)
2. Prokopenko V.: Development of sustainable mobility: the best experience of Ukrainian cities (2020), https://ecoaction.org.ua/rozvytok-staloi-mobilnosti.html
3. Ukrainian Statistics Service: Statistical collection for 2019 (2020)
4. Order of the Department of Transport Infrastructure of the executive body of the Kyiv City State Administration #H-29 dated 15.02.2023
5. Chernyshov O.: National Strategy for Barrier-Free Environment in Ukraine up to 2030 has been endorsed (2021), https://www.kmu.gov.ua/en/news/oleksij-chernishov-zatverdzheno-nacionalnu-strategiyu-zi-stvorennya-bezbaryernogo-prostoru-v-ukrayini-na-period-do-2030-roku
6. State construction codes for inclusion of buildings and structures (2018), https://online.budstandart.com/ua/catalog/doc-page?id_doc=79740
7. Banerjee, Tridib, Loukaitou-Sideris A.: Companion to Urban Design, 1. Routledge, New York (2011)
8. Yaseen, A.: Inclusive aspects of urban design: Sociability, walkability and overall ambiance. Chin. J. Urban Environ. Stud. **5**(01), 1750001 (2017)

9. Maliari, M., Azyat, A., Achhab, N.B., Raissouni, N., Chahboun, A.: Spatial Accessibility of Urban Parks in Tangier City, Morocco. Int. Arch. Photogramm. Remote. Sens. Spat. Inf. Sci. **46**, 369–372 (2021)
10. Marchigiani, E., Chiarelli, B., Garofolo, I.: Spatial accessibility as a driver to build an inclusive and proactive city. Urbani izziv **32**, 7–21 (2021)
11. Bergner, B. S., Zeile, P., Papastefanou, G., Rech, W., Streich, B.: Emotional barrier-GIS–a new approach to integrate barrier-free planning in urban planning processes, 247–257 (2011)
12. Prakash, A., Ravinder, R., Vittalaiah, A., Munipally, S. K., & Al-Farouni, M. H.: VISSIM Based Traffic Flow Simulation Analysis on Road Network. In E3S Web of Conferences, 529, 03009, EDP Sciences (2024)
13. World Health Organization, Disability and Health, Fact Sheet No. 352 (2011)
14. During the two years of the full-scale invasion, the number of people with disabilities in Ukraine increased by 300 thousand people, Ukrinform, https://www.ukrinform.ua/rubric-society/3846340-v-ukraini-za-dva-roki-vijni-kilkist-ludej-z-invalidnistu-zbilsilasa-na-300-tisac.html
15. Ukrainian Statistics Service: Statistical collection Social Security of Ukraine Population for 2020 (2021)
16. Ukrainian Statistics Service: Statistical collection Social Security of Ukraine Population State Statistics Service of Ukraine for 2022 (2023)
17. Borlini, M.: Post-emergency urban accessibility. Thoughts on inclusive living. Academia Letters (2021), https://doi.org/10.20935/AL1479
18. Wheelchair Accessible Public Transportation in Rome, Italy, (2024), https://wheelchairtravel.org/rome-italy/public-transportation/
19. Verkhohliadov S., Volovych A.: Study of the accessibility of pedestrian crossings in the central part of Kyiv (2024)

Sustainable Development of Ukrainian Cities: Inclusiveness of Urban Passenger Transport Services and Environmental Impact

Kateryna Vakulenko, Dmytro Burko(✉), Iryna Lynnyk, Nadiia Sokolova, and Yevhen Kush

O. M. Beketov National University of Urban Economy in Kharkiv, Marshala Bazhanova Str.,17, Kharkiv 61002, Ukraine
Dmytro.Burko@kname.edu.ua

Abstract. Urban mobility ensures the population's needs for safe, convenient, comfortable, and fast movement within the city. Key to this process is Sustainable Urban Mobility Plans (SUMP), aimed at ensuring sustainable development and improving transportation systems. The implementation of SUMPs in several Ukrainian cities has been considered, focusing on achieved results and implementation challenges, particularly in the development of inclusive transportation infrastructure. An accessible and barrier-free transportation system and urban infrastructure are crucial requirements for people with special needs and the mobility-impaired population to achieve social integration, including access to employment, education, and social opportunities. Recent estimates indicate that around eight percent of the country's population had disabilities, a number that began to rise with the onset of war. However, urban infrastructure, including its transportation component, is still not adapted to such challenges. Research results within state projects indicate that only twenty percent of physical environment and transportation objects are considered barrier-free. Priorities for implementing SUMP in Kharkiv city have been analyzed, justifying the need for comprehensive transportation infrastructure renewal considering challenges such as reducing environmental impact and developing inclusive mobility. Research has been conducted on reducing the impact of urban public transportation on air quality by introducing electric buses on bus routes. Proposals for developing inclusive mobility within the post-war reconstruction of Ukrainian cities have been presented, considering European practices.

Keywords: Sustainable Urban Mobility Plan · Inclusive Mobility · Barrier-Free Transportation System · Urban Public Transportation · Air Quality · Electric Buses

1 Introduction

Sustainable urban mobility ensures the population's needs for safe, convenient, comfortable, and fast movement within the city. This process is supported by Sustainable Urban Mobility Plans (SUMPs), aimed at ensuring sustainable development and improving transportation systems.

O. Slavinska et al. (Eds.): ITSESQC 2024, LNNS 1335, pp. 63–73, 2025.
https://doi.org/10.1007/978-3-031-87376-8_6

An accessible, barrier-free, ecological, and safe transportation system and infrastructure of the cities are key aspects of an inclusive transportation environment, contributing to the improvement of the quality of life for city residents, people with special needs, and mobility-impaired groups. This is essential for achieving social integration, including access to employment, educational, and social opportunities.

"Accessibility is a guarantee of rights, autonomy and equality. Access to the built and virtual environment, to information and communication technologies, goods and services, including transport and infrastructure, are a means of exercising rights and a prerequisite for the full participation of people with disabilities on an equal basis with others," - states the European Strategy for the Rights of Persons with Disabilities 2021–2030 [1].

Ukraine ratified the main provisions of the UN Convention on the Rights of Persons with Disabilities (2007) in 2009, which refers to the obligation assumed by Ukraine to provide persons with disabilities with accessible information on facilities, technologies, and support services that facilitate mobility and the ability to live independently and participate in all aspects of life considering to information, transportation and communication accessibility, including these systems, and other services and facilities.

On 17 June 2022, the European Commission published recommendations on granting Ukraine candidate status for EU membership. Such a decision imposes certain obligations on Ukraine, including those related to transport policy. A special emphasis is on compliance with the main sustainable development goals for the development of an efficient, safe, inclusive and sustainable transport system in Ukraine.

Persons with disabilities in Ukraine were highly vulnerable even prior to the Russian invasion. As of January 1, 2020, there were 2.7 million individuals with disabilities in the country. This number has risen significantly, with an additional 27,000 people acquiring disabilities since the onset of the war.

In accordance with Presidential Decree No. 266/2022 of 21 April 2022, the National Council for the Restoration of the Consequences of the War in Ukraine has developed a draft post-war recovery plan, the basic principles of which are EU integration and the improvement of real well-being [2].

At the initiative of the First Lady of Ukraine Olena Zelenska, the National Strategy for Creating a Barrier-Free Space in Ukraine until 2030 was adopted and an action plan for its implementation for 2021 and 2022 was approved. In accordance with the Resolution of the Cabinet of Ministers of Ukraine of 14 April 2021 No. 366-r "On Approval of the National Strategy for Creating Barrier-Free Space in Ukraine until 2030" and the Resolution of the Cabinet of Ministers of Ukraine of 26 May 2021 No. 537 "On Approval of the Procedure for Monitoring and Assessment of the Degree of Barrier-Freeness of Physical Environment Objects and Services for Persons with Disabilities", the Ministry for Community Development, Territorial and Infrastructure Development of Ukraine approved the Procedure for Monitoring and Assessment of the Degree of Barrier-Freeness of Physical Environment Objects and Services for Persons with Disabilities. The results of a study (2023) within the framework of the "Barrier-Free Response" project of the Ministry for Communities, Territories and Infrastructure of Ukraine: out of 38.5 thousand inspected physical environment and transport facilities, only 20% were barrier-free [3].

In 2016, Ukraine joined the Paris Climate Agreement, whose main goal is to keep the increase in global average temperature within 1.5–2 degrees Celsius above pre-industrial levels. According to the agreement, its participants commit to reducing greenhouse gas emissions relative to the 1990 level through Nationally Determined Contributions (NDCs). According to the Second Nationally Determined Contribution (NDC2), Ukraine aims to reduce CO2 emissions by 65% by 2030 from the 1990 level. The initial NDC envisaged a 40% reduction in CO2 emissions by 2030, excluding the agricultural sector and forestry.

In July 2020, Ukraine politically supported the European Green Deal (EGD) and recognized that it is an integral part of achieving the EGD goals. In April 2021, changes were made to Ukraine's National Transport Strategy until 2030 to facilitate Ukraine's rapprochement with the EU and create conditions that will contribute to the gradual integration of Ukraine into the EU internal market. Among the range of problems that need to be addressed in Ukraine are the high anthropogenic load on the environment, high levels cars' air pollution and noise in residential areas, and the significant impact of road transport on global climate change. It should be noted that the Inclusive mobility in Europe aims to ensure that all citizens, regardless of their physical abilities or socio-economic status, can travel seamlessly and independently. This concept is central to the European Union's Sustainable and Smart Mobility Strategy, which is part of the EGD.

2 Research Methods and Materials

Sustainable urban mobility is one of the modern requirements to ensure the population's needs for safe, convenient, comfortable, and fast movement within the city. One aspect of sustainable mobility is reducing the harmful impact of the transportation system on the urban environment, including reducing emissions of harmful substances into the atmosphere.

Many of the world's major cities have a Sustainable Urban Mobility Plan (SUMP). A SUMP is a specifically defined plan for the development of the transport sector of a country as a whole or a specific city within the country.

For example, the general plan "Sustainable and Smart Mobility Strategy" (SSMS) is a document of the European Union that outlines the main directions for sustainable development across the EU, with the primary goal of reducing harmful emissions by 90%.

Over the past 10 years, SUMPs have begun to be implemented in the following cities in Ukraine: Zhytomyr, Ivano-Frankivsk, Lviv, Mykolaiv, Ternopil' and Poltava.

According to the analyzed sustainable urban mobility plans, it is possible to distinguish several common directions in sustainable urban mobility plans (SUMP) in different cities of Ukraine (Fig. 1): development of public transport through the implementation of new routes and modernization of existing ones and the creation of lanes for public transport; traffic safety: implementation of modern traffic management systems and road infrastructure development, installation of bollards and safety islands; pedestrian and bicycle traffic stimulation according to construction of safe bicycle paths and reconstruction of pedestrian infrastructures; environmental sustainability: introduction of environmentally friendly vehicles and reducing of the negative impact of transport on the environment; inclusivity: ensuring the accessibility of transport for all segments of the population, including people with disabilities.

These directions are the basis for many Ukrainian cities that strive to create an efficient, ecological and convenient transport system.

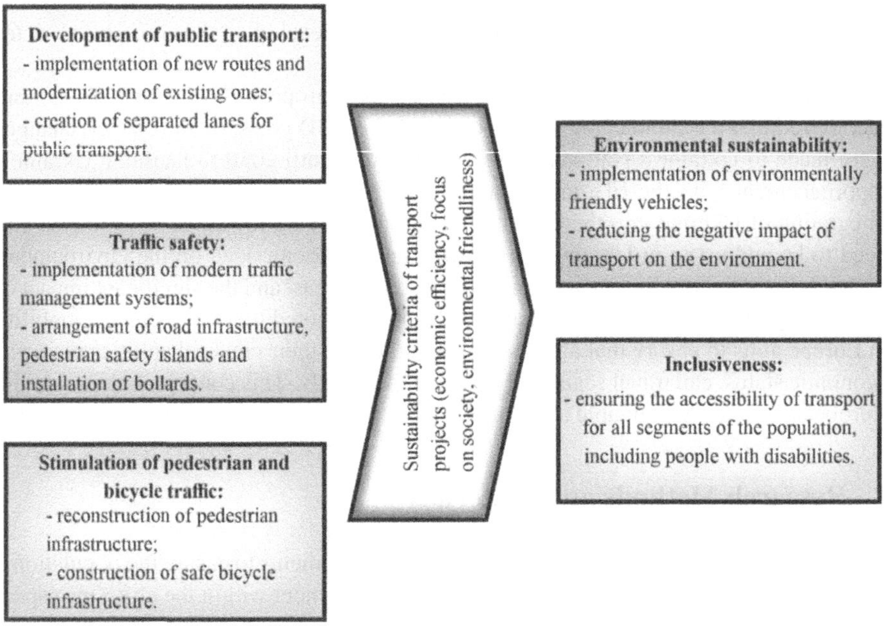

Fig. 1. Directions for the implementation of urban mobility plans according to the sustainability criteria of transport projects [4].

At the same time, SUMPs of Ukrainian cities include several important features of inclusiveness (Fig. 2). Transport accessibility: ensuring the accessibility of public transport for all segments of the population, including people with disabilities; infrastructure for people with disabilities: arrangement of public transport stops, sidewalks and transitions according to the needs of people with disabilities; social integration: creating conditions for full participation of all citizens in public life, including access to public services; human rights protection: implementation of measures to protect the rights of people with disabilities and ensure their equal opportunities; economic development: stimulating economic development through the creation of an inclusive transport system that promotes employment and access to jobs.

The indicated features of inclusiveness are important for creating an accessible urban environment according to the needs of all city residents.

The priorities for sustainable mobility development of Kharkiv city were outlined in the Kharkiv's SUMP (formally approved in December 2021). But established SUMP priorities were subsequently reassessed to align with the evolving needs of the Kharkiv transport system under the military aggression of the Russian Federation against Ukraine. As a result of the discussion, it was found that the preliminary list of SUMP priorities

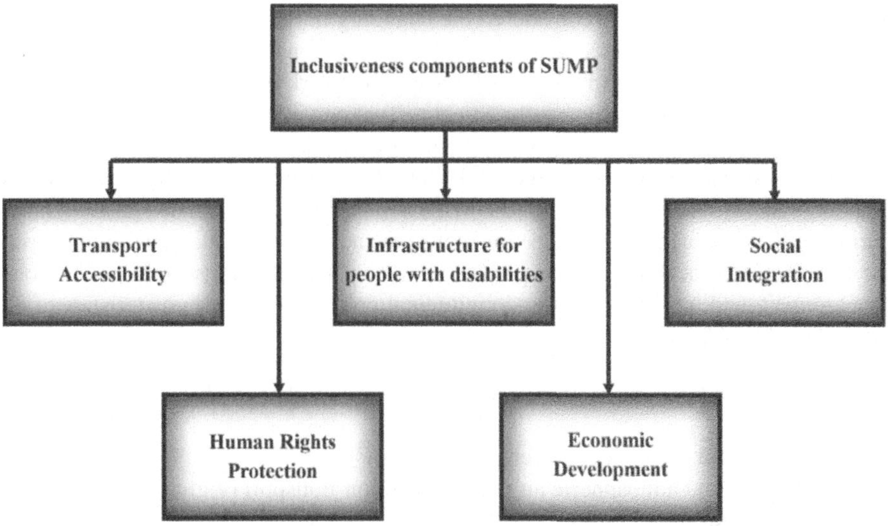

Fig. 2. Features of inclusiveness in sustainable urban mobility plans [4–6].

has not lost its relevance but has acquired a new meaning. The main aspect that required special attention in each of the previously identified priorities was the consideration of current and potential safety risks to ensure the effective functioning of the Kharkiv's transport system. Moreover, during the discussions, the expert team concluded that, given the consequences of military hostilities in Kharkiv, an important prerequisite for the implementation of the city's initiatives for the sustainable development of urban mobility is the comprehensive restoration of transport infrastructure and all related organizational and technological subsystems that provide transport services to the population. In this context, it was resolved to revise the Kharkiv's SUMP priorities to address the current challenges while fostering a resilient and sustainable urban transport system taking into account and implementing development directions for the restoration and post-war redevelopment of the city's transport system: leveraging international best practices; prioritizing to sustainable transport modes, creating barrier-free and safe urban spaces; integrating environmental considerations into decision-making processes for transport infra-structure; actively engaging the public in discussion due to reconstruction of the transport system; implementing effective economic mechanisms to attract businesses and international financial institutions [7].

Based on the accepted vectors of the development of transport infrastructure and urban transport in the Kharkiv's SUMP, the inclusiveness of urban passenger transport services and environmental impact was considered within the framework of the study.

3 Results of Experimental Studies

According to conceptual foundations on which the development of SUMP is based, an important component in them is the implementation of methods for reducing the environmental load and the assessment of such load on the environment. In view of the above, the investigation was considered the changing in the concentration of the most toxic and hazardous substance in vehicle exhaust emissions (CO) when the traffic flow characteristics were altered in the part of transport network of Kharkiv city, specifically on Landau Avenue (from Olimpiys'ka Street to Tankopiya Street) [8].

The modeling of the impact of vehicle emission standards on the amount of harmful emissions was carried out using the COPERT v5 software [9]. For this purpose, the amount of harmful substances emitted into the atmosphere in the exhaust gases of buses during peak hours on the selected section of the road network was calculated, considering the approach that according to various environment parameters (weather conditions, development, width of the roadway, longitudinal profile) and the characteristics of the traffic flow [8–10]:

$$k_{co} = (c_f + 0,01 \cdot N \cdot k_t) \cdot k_a \cdot k_u \cdot k_c \cdot k_v \cdot k_p \cdot k_z, \quad (1)$$

where k_{co} – carbon monoxide concentration, mg/m^3;

c_f – background atmospheric air pollution of non-transport origin (accept. $c_f = 0.5$ mg/m^3);

N – total traffic intensity, cars/hr.;

k – correction coefficients, which values according to the cars' toxicity, weather conditions and driving conditions.

The traffic conditions on the section and the climatic conditions were assumed to be constant and corresponded to the average annual indicators for Kharkiv: wind speed – 4 m/s; air humidity – 75%; low-rise, two-sided development along the roadway, separated from the roadway by green plantings; the roadway has 6 lanes (3 in each direction); the longitudinal profile equals 0 $^0/_{00}$.

The composition and traffic flow intensity were determined through field observations during peak hours.

Considering the post-war plans for updating the transport infrastructure of Kharkiv, specifically the urban passenger transport, the modeling examined the impact on harmful emissions resulting from the process of replacing diesel-powered buses (share in the traffic flow from 4% to 0%) with electric buses (from 0% to 4%).

As the modeling demonstrated, an increasing trucks' number in the traffic flow and a decreasing cars' number are leading to an increase in carbon monoxide concentration in the atmosphere (Fig. 3 and 4). Replacing buses with electric buses helps reduce this concentration: for example, with a 2% share of electric buses in the traffic flow, the concentration decreases by 5.6%, and with a 4% share, it decreases by 11.26%.

The calculations were conducted for urban route buses operating on diesel engines with Euro II - IV standards, considering their replacement with buses equipped with Euro IV engines or with electric buses (Figs. 5, 6, 7, and 8).

Upgrading buses' the emission standards or using electric buses significantly reduces the amount of harmful emissions, as evidenced by the obtained dependencies (Figs. 5, 6, 7, and 8).

In order to reduce the ecological load on the environment, the plans for sustainable urban mobility should include a number of measures that will ensure the sustainability of the city's transport system due to: renewal and development of the fleet of buses according to the principles of sustainable development; improvement and development

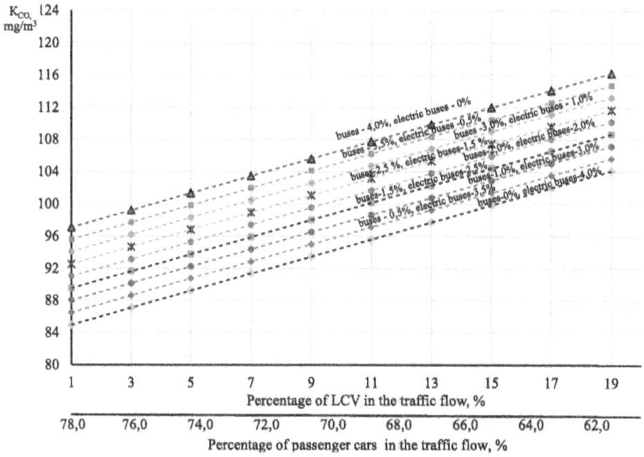

Fig. 3. Changing in CO concentration according to traffic flow ratio (LCV/passenger cars), traffic flow equal to 2800 vehicles/h.

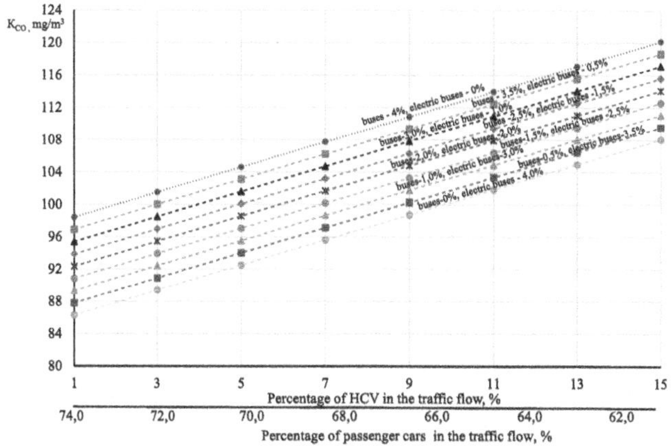

Fig. 4. Changing in CO concentration according to traffic flow ratio (HCV/passenger cars), traffic flow equal to 2800 vehicles/h.

of the public transport network in accordance with demand, giving priority to electric transport; implementation and development of the BRT system [7]; designing of "Park and ride" system at the entrances to the city with the with the giving of benefits for using of public transport or sharing services; organization of the delivering possibility for the goods at night using ecological freight vehicles [11, 12]; creating "Low Emission Zone" on the territory of the city with the reservation/restriction of the movement of motorized modes of transport; integration into the municipal fleet of rolling stock of ecological vehicles; zoning of the city with the definition of access zones for motorized modes of transport to them based on emission standards [11, 12].

The social transport service for individuals with musculoskeletal disabilities in categories I and II, including wheelchair users, was operational from January 3 to February 23, 2022, that indicates in the Economic and Social Development Programme of Kharkiv

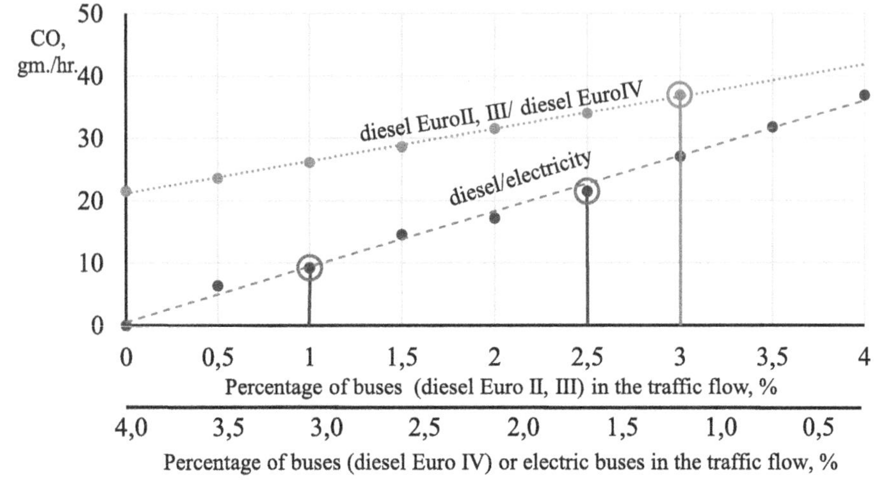

Fig. 5. Emissions of carbon monoxide in the part of the transport network within rush hour

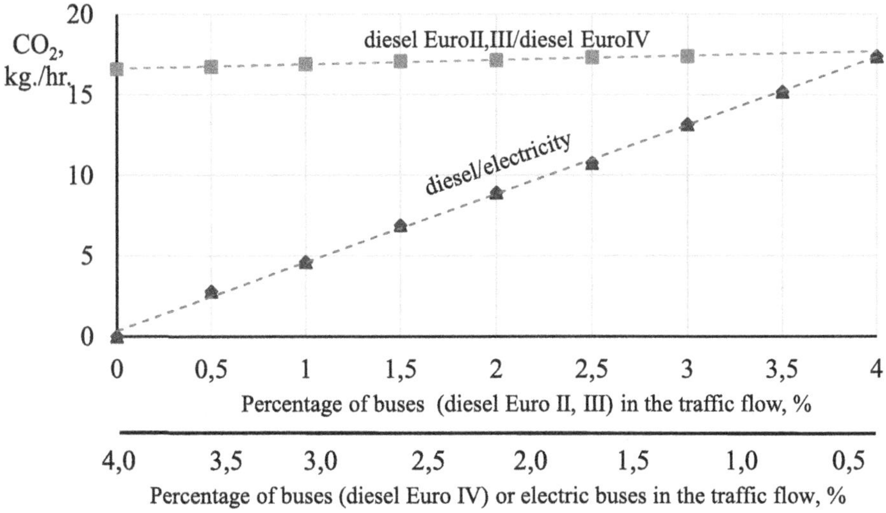

Fig. 6. Emissions of carbon dioxide in the part of the transport network within rush hour

for 2023. During this period, the dispatch service successfully completed 728 service requests. However, the provision of this transport service was suspended following the onset of Russian military aggression in Ukraine. Currently, there are approximately 12 distinct groups of individuals with special needs or reduced mobility, each with unique requirements regarding accessibility, affordability, convenience, efficiency, and safety [13, 14]. While measures to enhance inclusivity within the transport system often focus on specific target groups (e.g., individuals with musculoskeletal disorders or wheelchair users), it is crucial for practitioners to understand the diverse needs and perspectives

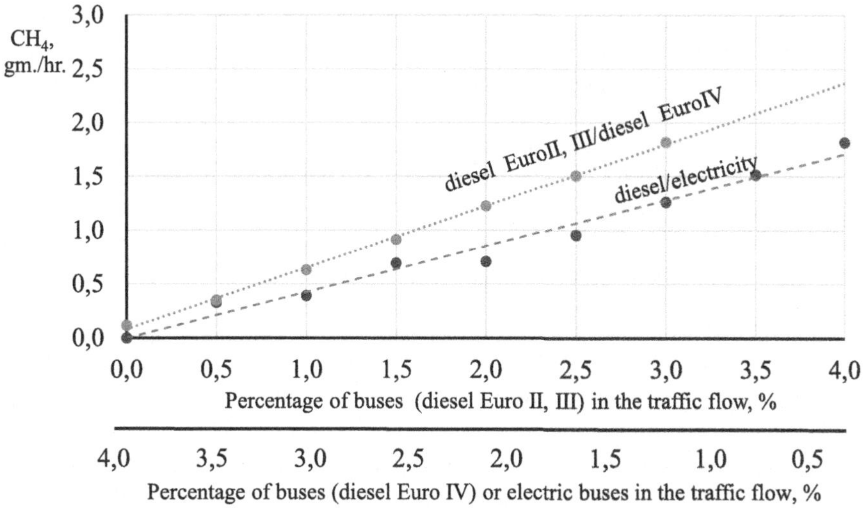

Fig. 7. Methane emissions in the part of the transport network within rush hour

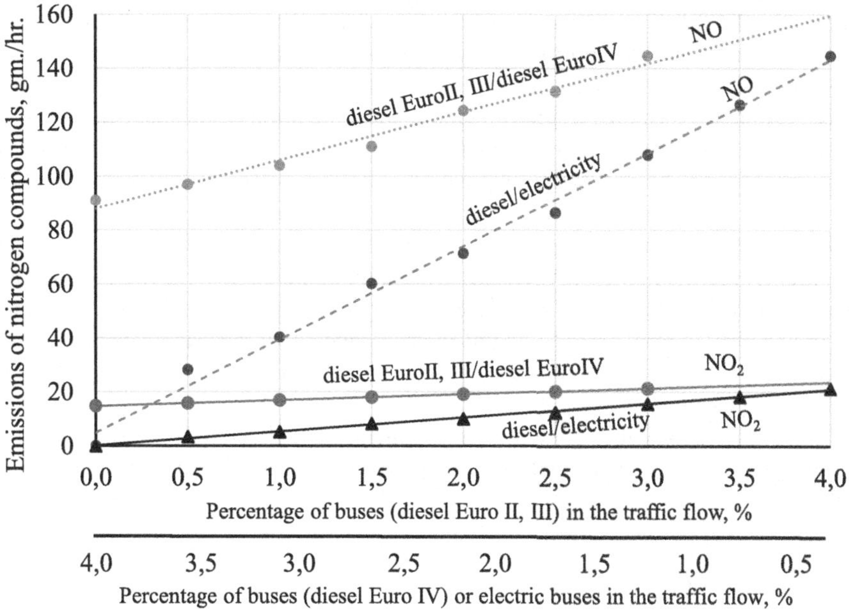

Fig. 8. Nitrogen oxide and nitrogen dioxide emissions in the part of the transport network within rush hour

of all user groups. This comprehensive approach ensures that both new and existing services are developed and enhanced to effectively address the varied needs of all users.

Recently, the institutions of the EU have been focusing on social issues related to transport, with increased focus on accessibility and quality of service, as well as the protection of passenger rights on all modes of transport. Many measures have been implemented at the national and local level, providing good examples of how transport systems can be improved or to promote the social integration of the most vulnerable population [6, 13, 14].

These initiatives will facilitate the creation of a more inclusive society, fostering greater awareness and proactive contributions toward building a sustainable future while addressing the contemporary challenges of achieving the Sustainable Development Goals (SDGs): SDG 11; SDG 9; SDG 16 – Sustainable Cities and Communities; Industrialization, Innovation and Infrastructure and Peace, Justice and Effective Institutions respectively [15].

An important aspect in the post-war reconstruction of Ukrainian cities is the formation of inclusiveness in public transport. Here are some suggestions that may be useful, according to European practices: barrier-free access (low-floor vehicles: introduction of low-floor buses and trams, allowing easy entry and exit for people with disabilities, the elderly and parents with prams; ramps and elevators: installation of ramps and elevators at public transport stops and metro stations); information systems (installation of systems that inform passengers about stops and other important messages in both audio and visual format; placement of information boards with large font and contrasting colors for people with visual impairments); staff training (conducting training on working with people with disabilities and other groups with low mobility); education (training of specialists in the field of transport in modern approaches to creating a barrier-free space); legislative initiatives (amending legislation to ensure the rights of people with disabilities to access public transport; monitoring and control: creation of mechanisms for monitoring and control over the implementation of legislative norms regarding inclusivity); international cooperation (cooperation with European countries to exchange experience and implement best practices in the field of inclusive mobility).

For Kharkiv city, additional specific measures to ensure inclusiveness of the transport infrastructure are necessary: comprehensive modernization of the network of public transport stops considering the principles of barrier-free and inclusiveness; reconstruction of existing sidewalks and footpaths in accordance with the principles of inclusiveness and barrier-freeness by equipping them with lowered side stones, ramps, lighting, markings, signs, etc.

These measures will help create a more inclusive and accessible environment for all citizens of Ukraine, contributing to their social integration and improving the life quality.

4 Conclusion

The quality and comfort of life of the urban population is determined by the main factors that determine environmental friendliness and inclusiveness, and barrier-freeness. These aspects are especially important in the functioning of public transport and the improvement of transport infrastructure. The sustainability of urban development requires the presence of an ecological, safe, convenient and cost-effective passenger transport system. In most SUMPs of Ukrainian cities, these sustainability indicators are used as fundamental ones. At the same time, insufficient attention is paid to modern approaches to creating a barrier-free, inclusive and ecological space. The synergistic effect of such approaches will create prerequisites for the effective functioning of a fair transport system for all users, which includes: barrier-free (physically, sensorially and linguistically); transport services are affordable for vulnerable groups: people with disabilities, elderly people, children and adolescents, large families, people with limited mobility, unemployed and low-income individuals, refugees and internally displaced persons, people with chronic illnesses; route planning with accounting the vulnerable users need to access transportation services; ensuring that the transportation services effectively support the vulnerable groups daily lives; network coverage, intermodal connectivity, accessibility and safety.

References

1. European Strategy for the Rights of Persons with Disabilities 2021–2030. https://knowledge.epr.equass.be/article/533-european-disability-strategy-2021-2030
2. National Council for the Recovery of Ukraine from the Consequences of the War Homepage. https://www.kmu.gov.ua/en/national-council-recovery-ukraine-war/working-groups
3. Ministry of Infrastructure of Ukraine Homepage. National strategy for creating a barrier-free space in Ukraine. https://mtu.gov.ua/content/nacionalna-strategiya-zi-stvorennya-bezbarernogo-prostoru-v-ukraini.html
4. Sustainable Mobility Transformation Homepage. https://transformative-mobility.org
5. Yaroshenko, O.M., Anisimova, H.V., Koliesnik, T.V., Kaplina, H.A., Babych, N.O.: National strategy for a barrier-free environment: problems, tolerance and implementation. Int. Soc. Work. **67**(1), 226–241 (2024)
6. Inclusion Europe Homepage. https://www.inclusion.eu
7. Lynnyk, I., Zavalniy, O., Vakulenko, K., Burko, D.: Sustainable urban mobility of Kharkiv: current state and development prospects. In: International Conference on Smart Technologies in Urban Engineering, pp. 424–436. Springer Nature Switzerland, Cham (2023)
8. Lynnyk, I., Vakulenko, K., Lezhneva, E.: Analysis of the air quality in considering the impact of the atmospheric emission from the urban road traffic. In: Research Methods in Modern Urban Transportation Systems and Networks, pp. 13–27 (2021)
9. Liora, N., Poupkou, A., Kontos, S., Meleti, C., Chrysostomou, K., Aifadopoulou, G., Melas, D.: Estimating road transport pollutant emissions under traffic-congested conditions with an integrated modelling tool – Emissions reduction scenarios analysis. Emission Control Sci. Technol. **7**(2), 137–152 (2021)
10. Tsanakas, N.: Emission estimation based on traffic models and measurements (Vol. 1835). Linköping University Electronic Press (2019)
11. He, Z., Haasis, H.D.: Integration of urban freight innovations: sustainable inner-urban intermodal transportation in the retail/postal industry. Sustainability. **11**(6), 1749 (2019)
12. Tarriño-Ortiz, J., Gómez, J., Soria-Lara, J.A., Vassallo, J.M.: Analyzing the impact of Low Emission Zones on modal shift. Sustain. Cities Soc. **77**, 103562 (2022)
13. Galinski, C., Giraldo Perez, B.S.: HH and HM communication with and among disabled persons in public transport: TRIPS Deliverable D3. 3 Report on Accessibility Standards and Legislation. ICCHP-AAATE 2022 Open Access Compendium "Assistive Technology, Accessibility and (e) Inclusion" Part I (2022)
14. Bulanowski, K., Di Ciommo, F., Lima, S.: Involving passengers in creating inclusive digital mobility solutions – Findings from the INDIMO project. In: Conference on Sustainable Urban Mobility, pp. 1001–1016. Springer Nature Switzerland. Cham (2022)
15. Sustainable Development Goals (SDGs) Homepage. https://sdgs.un.org/goals

Determining the Algorithm of Transportation of Highly Concentrated Coal-Water Fuel to Consumption Points of the Transport Network

Nataliia Chernetska-Biletska[1(✉)], Serhiy Soroka[1], Igor Baranov[1], Sergii Kliuiev[1], and Olena Ostrohliad[2]

[1] Department of Logistics Management and Transport Safety, Volodymyr Dahl East Ukrainian National University, Ioanna Pavla Druhoho St. 17, Kyiv 01042, Ukraine
nchernetska@snu.edu.ua

[2] Department of Transport Technology, National University Zaporizhzhia Polytechnic, Zhukovskoho St. 64, 69063 Zaporizhzhia, Ukraine

Abstract. In order to improve the environmental conditions and reduce the cost of coal-water fuel, its production is based on the use of all types of coal, as well as coal refuse. The use of coal-water fuel in thermal power generation allows reducing emissions of nitrogen oxides, sulfur, and carbon monoxide into the atmosphere. The paper provides a rational way of using different modes of transport for transporting highly concentrated coal-water fuel depending on the freight volume and the transportation distance. The optimal order of fuel transportation to the consumption points on the route of the selected hydrotransport network has been determined using Prim's algorithm and the method of branches and boundaries. The shortest distances between the points included in one route have been determined using the minimum spanning tree algorithm. The algorithm for choosing the optimal option for transporting highly concentrated coal-water fuel is developed according to two options: with one point of fuel production and with two points of production. The optimization process has been carried out taking into account the reduction of the total loss of the coal-water fuel flow during its transportation between the departure point and all consumption points in turn along the common pipeline reducing the diameters of pipes in the network sections. When using the author's method, the topology of the transport network of distribution among the consumption points of coal-water fuel is optimized and the total length of the selected transport network is determined. The economic benefits of the proposed research are determined.

Keywords: Highly Concentrated Coal-Water Fuel · the Order of Transportation · Transport Network · Hydraulic Transportation · Topology · Transportation Algorithm · Operating Costs

1 Introduction

Coal-water fuel (CWF) is a composite artificial liquid fuel based on coal and water. According to its structure, it is a type of highly concentrated coal-water suspension (HCCS). Its preparation is based on mechanochemical activation, the process by which

O. Slavinska et al. (Eds.): ITSESQC 2024, LNNS 1335, pp. 74–85, 2025.
https://doi.org/10.1007/978-3-031-87376-8_7

the coal structure is destroyed. During coal grinding, its active surface increases, and impurities are removed. Crushed coal and water, which plays an important role in this process, are mixed, and plasticizers are added resulting in a specific dispersed system with technologically acceptable fluidity and stability. The mass concentration of CWF is 60–80% of coal, the particle size does not exceed 250 μm. Special mills are used to obtain the required coal fraction. The best results were achieved when obtaining a bimodal particle size distribution with a certain ratio of large (100–250 μm) and small (<40 μm) fractions, which makes it possible to obtain the effect of "maximum packing" and significantly increase the concentration of the suspension without deteriorating the rheological properties of CWF. Since such a suspension stratifies quickly enough with the sedimentation of coal particles, a plasticizer is added to it, the mass fraction of which in CWF is about 1–2%. In order to prepare CWF in accordance with consumer requirements, various additives and agents can be added to it, such as corrosion inhibitors, desulfurizers, denitrifiers, ash melting regulators, and others.

CWF can be prepared from almost all types of coal, as well as coal refuse, but it is necessary to start from the most widespread type of coal used in order to reduce the cost of CWF. The viscosity of conditioned fuel coal suspensions of high concentration is generally in the range of 0.5–1.2 Pa•s, sedimentation stability is up to 30 days [1, 2].

2 Relevance and Methodology of the Research

Transportation of such fuel, including at a distance of several tens of kilometers on the territory of the enterprise, is carried out with the help of hydraulic transport systems (HTS), as well as in containers – road tankers and rail tank cars [3, 4]. CWF can be used as a substitute for natural gas and fuel oil in boiler units [5]. Compared to the traditional burning of pulverized coal (in particular, at thermal power plants), the use of CWF in thermal power generation allows reducing emissions of nitrogen oxides, sulfur, and carbon monoxide into the atmosphere and also ensures the complete combustion of organic matter up to 99%, which significantly improves the environmental situation [6]. The benefit of using CWF in the thermal power industry is evidenced by its advantages, such as the possibility of the complete use of the infrastructure of power boiler units with the minimal reconstruction of devices and fuel supply systems, high packing density during storage compared to the stacking of dry coal, complete explosion and fire safety, no losses during transportation, and the possibility of underground storage.

Paper [3] provides the main correlated costs of CWF transportation by various modes of transport. The indicators correspond to a cargo flow of up to 1 million t/year. With this volume, pipeline deliveries require significantly lower costs than rail or road transport. When the transportation distance increases up to 100 km or more, the efficiency of rail transportation increases. The efficiency of pipeline transport is even more evident for larger volumes of CWF transportation.

Schematically, the expedient area of application of various modes of CWF transportation is shown in Fig. 1. Rail transport has an advantage when the transportation distance is more than 100 km. Its efficiency increases with an uneven annual schedule of CWF deliveries and significantly depends on the traffic density of rail networks. Road transportation is appropriate for small delivery volumes and a delivery radius of

40–60 km, and with a decrease in cargo flow, the maximum distance of transportation increases. Pipeline transportation of CWF is generally efficient at cargo flows of more than 0.4–0.5 million t/year. Estimates made reflect average conditions. When studying the efficiency of specific transport systems, the role of different modes of transport may change.

Calculations confirm the efficiency of further expansion of the pipeline network on the territory of enterprises. The assessment of such an increase in the effective length of the pipeline network shows that it should be increased by more than 50 km, which is significantly more than just a physical increase in the network. This difference is explained by the growth of the average diameter and further intensification of CWF transportation.

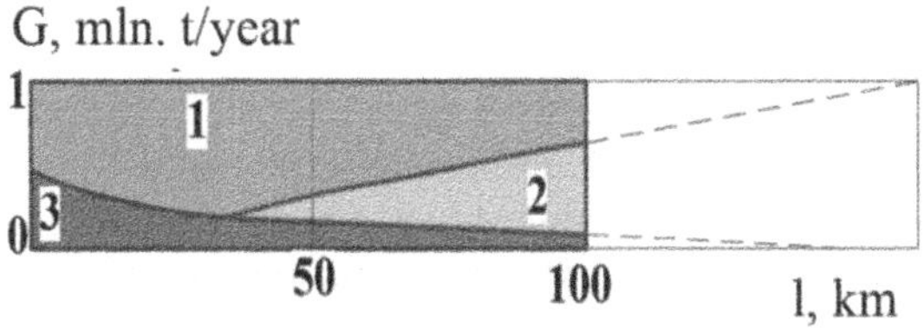

Fig. 1. The rational way of using different modes of transport for CWF transportation depending on the freight volume and the transportation distance: G – freight volume; l – transportation distance; 1. Area of expedient use of pipeline transport (hydraulic transport systems); 2. Area of expedient use of rail transport; 3. Area of expedient use of road transport.

The works of scientists R.A. Aliyev, E.E. Bibik, V.S. Biletskyi, B.F. Bragin, Y.F. Vlasov, V.E. Gubina, G.P. Dmitrieva O.A. Krutya, L.H. Loitsyanskoho, V.I. Murko, E.Z. Rabinovicha, Y.G. Svitlyo, E.V. Semenenko, P.V. Sergeeva, A.E. Smoldyreva, Trainisa V.V., Uryeva N.B., Yakhno O.M. etc. are devoted to solving the issues of optimizing the transportation of hydrocarbon fuel by industrial hydrotransport.

The analysis showed that the laws of optimization of the hydrocarbon fuel transportation scheme (transport network topology) taking into account the location of consumption points, distances and needs of the enterprise are not revealed in the presented works.

Calculations, justifications, and analysis take into account the following forms of manifestation of the efficiency of pipeline transport development: improvement of the structure of CWF transportation, release of railway capacities; increase in the reliability of supply to CWF consumers; a noticeable increase in workforce productivity and release of a significant number of workers, including those employed in heavy work (for example, in pouring operations); high energy efficiency; climate neutrality; short terms of development of new cargo flows; social consequences of improving working conditions and the culture of production.

The efficiency of CWF pipeline transportation is determined by taking into account the following most important factors: high rotational speed and throughput capacity; high workforce productivity and low cost of transportation; loss reduction and improvement of the quality of the transported product; integration of production, supply, and consumption of CWF; application of the most effective equipment and technology of pipeline transportation, high automation, etc.

Determining the optimal order of CWF hydrotransportation to consumption points on the route of the selected transport network was performed using Prim's algorithm and the method of branches and boundaries. The task of choosing the optimal option for CWF transportation is solved by Svir's method. The work uses the author's method of optimizing the topology of the distribution transport network among CWF consumption points.

The purpose of the paper. The purpose of the work is to develop an algorithm for the transportation of highly concentrated coal-water fuel to consumption points of the transport network by industrial hydraulic transport.

3 Results and Discussion

In order to implement an economically and environmentally attractive project to adjust the necessary equipment of enterprises to coal-water fuel, it is necessary to organize CWF transportation by industrial hydraulic transport from the departure point to consumers [7]. It is most expedient to place one CWF departure point. The task is reduced to minimizing transportation costs.

When choosing an option for CWF transportation between the departure point and consumers of the enterprise, the first option A1 is only a radial connection of the departure point with each of the consumption points. Option A2 is a combination of a radial connection of the departure point with the consumption point and a serial connection. As a result of solving the transportation problem, it is necessary to determine the optimal scheme of CWF transportation and determine the optimal parameters of the hydraulic transport system of the enterprise (diameters of pipelines, parameters of transportation pumps, transportation speed, and pressure losses in the transport network sections).

In the first stage, it is necessary to determine the optimal scheme of CWF transportation (topology of the transport network) taking into account the location of consumption points, distances, and needs of the enterprise, optimize the given total costs for pipelines, pumping, and other equipment, electric energy for transportation, and determine the diameters of the pipeline elements of the industrial HTS.

The proposed list of CWF consumers (enterprises) is given in Table 1:

1) ShD; 2) TsD; 3) ShA; 4) ShB; 5) ShV; 6) TsB.

The required consumption of boilers of enterprises (consumers), m^3/h*: 1)* $ShD = Q_1 = 8 \times 2.1 = 16.8\ m^3/h$*; 2)* $TsD = Q_2 = 2 \times 2.1 = 4.2\ m^3/h$*; 3)* $ShA = Q_3 = 4 \times 2.1 = 8.4\ m^3/h$*; 4)* $ShB = Q_4 = 6 \times 2.1 = 12.6\ m^3/h$*; 5)* $ShV = Q_5 = 2 \times 2.1 = 4.2\ m^3/h$*; 6)* $TsB = Q_6 = 2 \times 2.1 = 4.2\ m^3/h$*.*

The total supply to all CWF **consumption points is 50.4 m3/h. Pumps are installed at preparation points based on economic feasibility depending on the needs (total, for each point) of consumers in accordance with data** [8].

Determining the optimal order of CWF hydrotransportation to consumption points on the route of the selected transport network can be performed using Prim's algorithm and the method of branches and bounds, the application of which

Table 1. Distances between CWF consumers, m.

x	ShD	TsD	ShA	ShB	ShV	TsB
ShD	**150**	500	500	6360	6240	6660
TsD	500	**150**	**150**	6100	6360	6500
ShA	2500	2800	2800	9000	8780	9340
ShB	6360	6100	6100	**150**	2300	300
ShV	6240	6320	6320	2300	**150**	2100
TsB	6660	6500	6500	300	2100	**150**

allows determining the shortest distances between points included in one route (the minimum spanning tree algorithm) [9].

Representation of Table 1 **is implemented using the adjacency matrix**:

1. For an unweighted graph:

$$\begin{pmatrix} 0\,1\,1\,1\,1\,1 \\ 1\,0\,1\,1\,1\,1 \\ 1\,1\,0\,1\,1\,1 \\ 1\,1\,1\,0\,1\,1 \\ 1\,1\,1\,1\,0\,1 \\ 1\,1\,1\,1\,1\,0 \end{pmatrix} \tag{1}$$

2. For a weighted graph:

$$\begin{pmatrix} 0 & 500 & 2500 & 6360 & 6240 & 6660 \\ 500 & 0 & 2800 & 6100 & 6320 & 6500 \\ 2500 & 2800 & 0 & 9000 & 8780 & 9340 \\ 6360 & 6100 & 9000 & 0 & 2300 & 300 \\ 6240 & 6320 & 8780 & 2300 & 0 & 2100 \\ 6660 & 6500 & 9340 & 300 & 2100 & 0 \end{pmatrix} \tag{2}$$

Prim's algorithm is an algorithm for constructing a minimum spanning tree of a weighted connected undirected graph [10]. The input of the algorithm is a connected undirected graph. Its cost is set for each edge.

First, an arbitrary vertex is taken and an edge incident to this vertex with the lowest value is found. The found edge and the two vertices connected by it form a tree. Then, the edges of the graph are considered, one end of which is a vertex belonging to the tree, and the other is not. The edge with the lowest cost is selected out of these edges. The edge selected at each step is joined to the tree. The tree grows until all vertices of the original graph are used. The result of the algorithm is a minimum cost tree. The adjacency matrix describing the graph edges is presented in Table 2.

This algorithm of constructing the minimum spanning tree does not require either sorting or checking for cyclicity at each step. Construction of the spanning tree G_0 starts

Table 2. Adjacency matrix

Vertices / Edges	1	2	3	4	5	6	In the form of a list: Edges	Vertices	Boolean variables (T-directed edge; F-undirected edge)	Weight
1	1	1	0	0	0	0	1	1.2	F	500
2	0	1	1	0	0	0	2	2.3	F	2800
3	1	0	1	0	0	0	3	1.3	F	2500
4	0	1	0	1	0	0	4	2.4	F	6100
5	0	1	0	0	0	1	5	2.6	F	6500
6	1	0	0	1	0	0	6	1.4	F	6360
7	1	0	0	0	0	1	7	1.6	F	6660
8	0	0	1	0	0	1	8	3.6	F	9340
9	0	1	0	0	1	0	9	2.5	F	6320
10	1	0	0	0	1	0	10	1.5	F	6240
11	0	0	1	0	1	0	11	3.5	F	8780
12	0	0	0	1	0	1	12	4.6	F	300
13	0	0	0	1	1	0	13	4.5	F	2300
14	0	0	0	0	1	1	14	5.6	F	2100
15	0	0	1	1	0	0	15	3.4	F	9000

from an arbitrary vertex X_1. Then, among the edges incident to X_1, we choose the edge (X_1, X_2) with the smallest weight and include it in the tree. By repeating the process, we search for the least weighted edge that connects the vertices X_1 and X_2 with some other vertex of the graph X_3.

We continue the process of including edges until all vertices of the original graph G are included in the tree G_0. The constructed tree will be minimum spanning. That is proof that the sequence of steps 1–4 leads to the construction of the minimum spanning tree. The implementation of the nearest neighbour algorithm for constructing the minimum spanning tree is performed in the algorithm, where the output graph $G = (X, U, \Phi)$ is a table of weights $W_e = [w_{ij}]$; the weights of non-existent edges are considered equal $+\infty$. The weights of edges mean their lengths. The spanning tree $G_0 = (X_0, U_0, \Phi)$ is constructed using the edge list U_0 and the list of vertices X_0. Their ordinal numbers $Xi = \{1, 2,\ldots, n\}$ are set as marks. For each vertex $x \in X$ of the graph not yet included in the tree, the minimum distance to the set of vertices previously included in X_0 is maintained. This is done using two vectors $dist[xi]$ and $prev[xi]$; where $dist[xi]$ is equal to the minimum distance from $x \in X$ to the vertex $prev[xi] \in X_0$. The values of the vectors $dist[xi]$ and $prev[xi]$ are updated at each step of the algorithm when X_0 is supplemented with a new vertex.

The nearest neighbour algorithm for the spanning tree looks like this:

```
X = {1,2,....,n_i}; {labels of graph vertices}
nX= |X_i|; {Number of vertices in the graph}
v= rand(1,|X_i|); {Arbitrary vertex v ∈ X}
X_0= {v}; {Vertices of the spanning tree}
X= X_i\{v}; {Remove v from X}
U_0 = ∅; {Edge of the main tree}
W_T = 0;{Weight of core tree}
for x∈ X_i do begin
dist[x]= We[x_i,v]; {Minimum distance from x to v∈X_0}
end;
while |X_0| ≠ nX do begin
prev[x]= v;{Edge(v,x_i), the length of which is dist [x_i]}
dist[v]= mindist[x_i]; {v ∈ X- vertex to include}
X_0= X_0U {v}; {Add Vertex}
X= X_i\{v}; {Delete vertex}
U_0= U_0U {prev[v], v};{Add edge}
W_T= W_T+ dist[v]; {Fix tree weight}
{Fix dist dist[x_i]-vector of distances}
for x∈ X_i do if dist[x] > We[v, x_i] then begin
dist[x_i]= We[v,x_i];
prev[x_i]= v;
end;
end.
```

The complexity of the nearest neighbour algorithm is determined by two nested loops according to the number of vertices. A constant number of operations is performed in each of the cycles. So, the complexity is O (|Xi|2).

The output is represented by the weight matrix of its edges in the text file Near.in. As a result of the solution, the topology of vertices, edges, and distances has the following form

$$
\begin{gathered}
V=\{1,2,3,4,5,6\};\; E_x=\{\{1,2\},\{1,3\},\{2,4\},\{4,6\},\{6,5\}\} \\
E_L=\{\{500\},\{2500\},\{6100\},\{300\},\{2100\}\}
\end{gathered}
\tag{3}
$$

The total length of the route taken as an example was 11,500 m. The resistance depends on the CWF transportation speed [11].

Thus, when the diameter of the pipeline increases, the CWF transportation speed decreases and the resistance of the pipeline section that connects the departure point (DP) and the consumer or two consumers decreases [12].

The verification of the results of finding the optimal order of laying pipelines on the route was performed by calculations using the method of branches and boundaries [13], for the application of which the shortest distances between the points included in one route are defined. The results of the calculations indicate the expediency of using the method of branches and bounds and Prim's algorithm, as the results of the calculations completely coincided [12, 13].

The task of choosing the optimal option for CWF transportation may include two solutions A and B:

I. A – 1 production point (PP) of CWF

II. B – 2 production points of CWF

I.1. When choosing option A, the first sub-option A1 is only a radial connection of the preparation point with each of the consumption points (Fig. 2). Having the need for each of the consumption points Qi, we calculate the CWF flow rate along the pipeline to each of the consumption points Pi for a pipeline with a diameter of 50 mm. After that, the resistance of each pipeline is determined by formula (4).

$$i_{50(3)} = 387.54v^2 + 2053.41v - 6.1 \text{ Pa/m} \tag{4}$$

where i_{50} – specific resistance of a pipeline with a diameter of 50 mm;

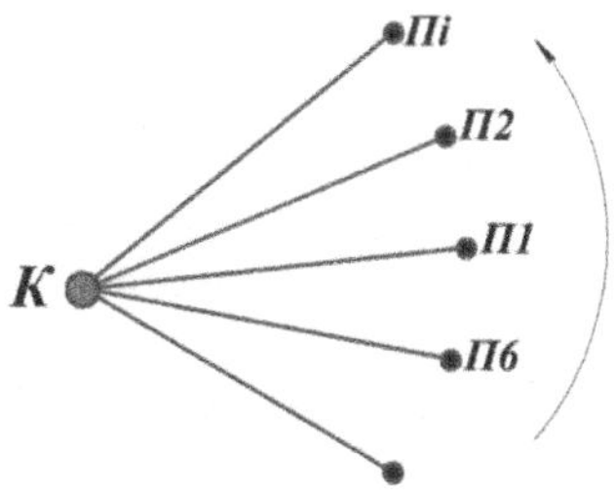

Fig. 2. Conventional scheme of the system according to option A1

The output of CWF PP is the K index.
Calculation of the CWF flow rate:

$$\upsilon = \frac{Q_c}{F_c}, m/s \tag{5}$$

where $Q_c = \frac{Q}{3600} \frac{m^3}{S}$

where $Q \frac{m^3}{S}$ - required consumption, m^3/h.
The cross-sectional area is determined by the formula:

$$F_c = \frac{\pi \cdot D^2}{4}, \tag{6}$$

where D – inner diameter of the pipeline.

The pipeline diameter is chosen based on economic feasibility, taking into account the fact that the specific hydraulic resistance is i_{50} (for 1 l.m. of the pipeline). The resistance of pipelines of other diameters is determined according to the dependence $i_D = i_{50} \cdot y\ (D)$, where $y\ (D)$ is the optimal granulometric composition [11]:

$$i(D) = a + \frac{b}{D} + \frac{c}{D^2} + \frac{d}{D^3}; Pa/m \tag{7}$$

a = -0.00285; b = 0.01146; c = 0.0019096; $d = 1.2167 \cdot 10^{-6}$

I.2. Using Svir's method [10, 13], we join the rays, choosing the moment when the total supply along the swept rays will be as close as possible to the first (minimum supply) pump in the standard size series.

I.3. We calculate the resistance of each of the rays and, using formula (4), determine the power loss in each of the rays of sub-option A1 (star). After that, according to formula (8), we check the compliance of the power of the selected pump units with the specifications. If the power parameters do not comply, we replace the pump unit with the next one in the standard size series.

The resistance of each ray is determined by the formula:

$$\Delta\mathrm{p} = i_{(50)i} \cdot l_i \cdot Pa, \tag{8}$$

where l_i is the length of the ith ray.

I.4. We check the compliance of the maximum resistance of the rays with the specifications of the pump unit. In case of incompliance, we act according to point I.3.

I.5. Using formulas (8–9), we find the objective function and fix the found cost value (per year) in the form of $C1_{(50)}$.

I.6. Then we change the diameter of the ray selected first by Svir's method from 50 mm to the next one in the standard size series and determine the resistance $i_{(x)}$ according to formulas (4 and 7).

I.7. We calculate the objective function taking into account the change in the diameter (that is, the resistance of the pipeline in the ray) and the cost of the pipeline. We fix the obtained value and compare it with the first calculated option. If the second option has a lower cost compared to the first one, then we continue the process of increasing the diameter according to point I.6. We continue the increase until the value of the objective function is $j + 1$, which is greater than in the previous j^{th}. We fix the diameter of option j for the ith *ray*. Then, we move to the next diameter and define the objective function. When re-obtaining for the diameter $j + 2$ the value of the objective function, which is greater than for the diameter of option j, we finally fix the given value of the diameter for the ith ray.

I.8. Following Svir's method, we move to the $i + 1^{st}$ ray and perform the actions in point I.7. After completing the actions according to points I.6, I.7, we select the minimum objective function C_{A1}, fix the obtained value of *n, m, Di, li.* Thus, the calculations according to the simplified option A1 are finished.

I.9. Option A2 is a combination of a radial connection of the preparation point with the consumption point and a serial connection. We start with the first one using Svir's method of the K-*i* ray, compare the sum of $l_{K\text{-}1} + l_{Kj\text{-}1}$ with the sum of $l_{K\text{-}i} + l_{i\text{-}j1}$ (Fig. 3).

I.9.1. If the first sum is less than the second one, we proceed to the next ray according to Svir's method and continue to act in the order indicated above in this point.

I.9.2. If the second sum is less than the first one, we connect the pipeline *i-j1* to the ray *K-i* (moving counterclockwise to the closest consumer *j1*.).

I.10. Option A3 is a combined radial, and series and parallel connection of the points of CWF preparation and consumption (Fig. 4).

II - Option B. By assigning two preparation points (the first in the ShD point, the second in ShB), we determine the consumers for the first distribution point, namely,

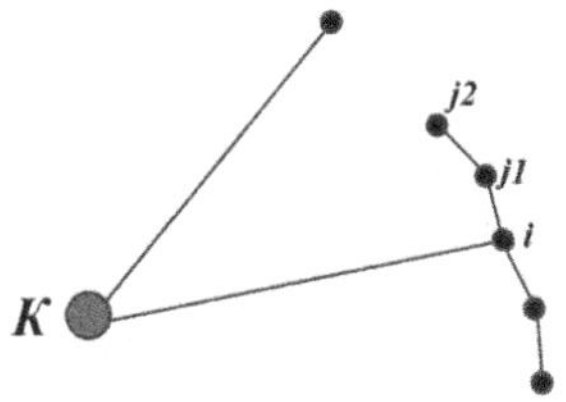

Fig. 3. Conventional scheme of the system according to option A2

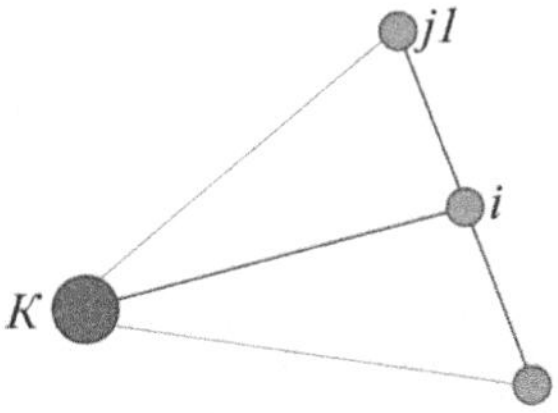

Fig. 4. Conventional scheme of the system according to option A3

ShD, ShA, and TsD, and those for the second one, namely, ShB, ShB, and TsB. After that, we perform the calculations according to the algorithm described in subsection I.

As a result of the solution of the transportation problem and the formation of an optimization model for the selection of parameters for CWF transportation to the consumption points of the enterprise, the topology of the distribution transport network among the consumption points of CWF has been optimized, and by optimizing the total costs of pipelines, pumping, and other equipment, electric energy for transportation, the diameters of the pipelines of the HTS elements have been determined. The optimization results are presented in Fig. 5.

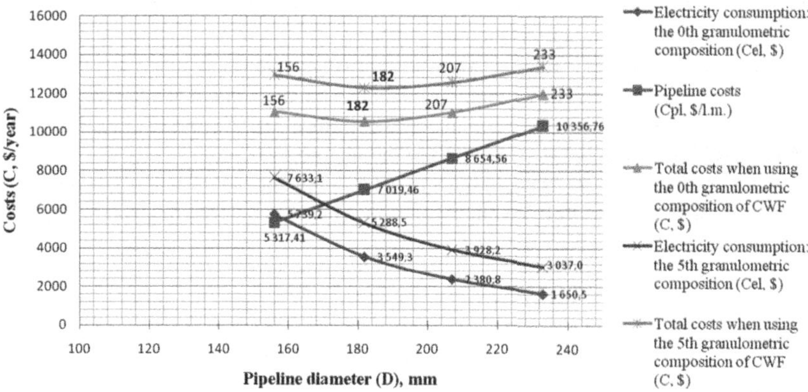

Fig. 5. Dependence of the pipeline construction and the CWF transportation costs on the pipeline diameter and the granulometric composition

4 Conclusion

The paper provides a rational way of using different modes of transport for CWF transportation depending on the freight volume and the transportation distance, and the optimal order of CWF transportation to consumption points on the route of the selected hydrotransport network using Prim's algorithm and the method of branches and boundaries, the application of which allowed determining the shortest distances between the points included in one route (the minimum spanning tree algorithm).

The developed algorithm for choosing the optimal option for CWF transportation according to the following options: A - 1 point of CWF production and B - 2 points of CWF production. The optimal diameter of the pipeline was 182 mm, taking into account the existing standards for the assortment of steel pipes with a wall thickness of 6 mm. The optimal pipeline diameters were as follows: in sections 1–2 = 207mm; 1–3 = 121mm; 2–4 = 182mm; 4–6 = 121mm; 6–5 = 121mm. The optimization was carried out taking into account the reduction of the total consumption of the CWF flow during its transportation between the departure point and all consumption points in turn along the common pipeline reducing the diameters of pipes in the network sections.

As the results of the study show, the use of the granulometric composition of CWF No. 0, proposed in the work (with the bimodality criterion G1/G2 = 2.45) [11], allows reducing the cost of electricity for its transportation by the HTS of the enterprise by $2,000/year in comparison with granulometric composition No. 5, which was chosen arbitrarily as the starting point. At the same time, the topology of the transport network includes edges X = 1–2; 1–3; 2–4; 4–6; 6–5, the total length of the transport network is 11,500 m.

References

1. Makarov, V., Perov, M., Kaplin, M.: Possibilities of using hydrocarbon fuel in the energy sector of Ukraine. Problems and Tasks of Modernity and Approaches to Their Solution, 251–256 (2021).https://www.google.com.ua/books/edition/Problems_and_tasks_of_modernity_and_appr/4AAiEAAAQBAJ?hl=uk&gbpv=0
2. Makarov, V., Perov, M., Kaplin, M.: Alternative hydrocarbon fuel in the energy sector of Ukraine. The Newest Problem of Science and Ways to Solve Them, 257–260 (2022). https://www.google.com.ua/books/edition/THE_NEWEST_PROBLEMS_OF_SCIENCE_AND_WAYS/XjiKEAAAQBAJ?hl=uk&gbpv=0
3. Ryzhova, S., Kruglikov, D.: Peculiarities of determining the parameters of hydrocarbon fuel during transportation by rail. Youth: Science and Innovation. Dnipro (2021). http://ir.nmu.org.ua/handle/123456789/159893
4. Kruglikov, D., Kyrychko, S., Ryzhova, S.: Study of the process of leakage of hydrocarbon fuel from a railway tank under the influence of its own weight. Scientific spring: materials of the XII All-Ukrainian scientific and technical conference of graduate students and young scientists (2022). http://ir.nmu.org.ua/handle/123456789/167070
5. Salama, M., Elsihy, E., Moneib, H., Kamal, M. and etc.: The combustion of a diesel oil-based/coal/water slurry in a horizontal cylindrical furnace: an experimental investigation. Proc. Int. Exch. Innov. Conf. Eng. Sci. (IEICES) **7**, 64–70 (2021). https://doi.org/10.5109/4738568

6. Kurytnik, I., Khassenov, A., Nussupbekov, U., Karabekova, D., Nussupbekov, B., Bolatbekova, M.: Development of a grinding device for producing coal powder-raw materials of coal-water fuel. Arch. Mech. Eng. **69**(2), 259–268 (2022). https://doi.org/10.24425/ame.2022.140414
7. Obodovych, O., Tselen, B., Stepanova, O., Nedbailo, A., Buliy, Yu.: Technology and equipment for the production of universal hydrocarbon fuel. Energy Technol. Resou. Conserv. **1**, 24–33 (2024). https://dspace.nuft.edu.ua/handle/123456789/42814
8. Nedbailo, A., Ivanytsky, G., Tselen, B.Y., Radchenko, N., Gozhenko, L., Shchepkin, V.: Application of a pulsating disperser as a hydrodynamic cavitation reactor for the preparation of hydrocarbon fuel. Thermophys. Thermoenergetics **45**(1), 28–34 (2023). https://ihe.nas.gov.ua/index.php/journal/article/view/521
9. Krenevich, A.: Algorithms and data structures. Textbook. VOC "Kyiv University". Kyiv (2021)
10. Bronshtein, E., Zaiko, T.: Deterministic optimization problems of transport logistics. Autom. Telemech. **10**, 133–147 (2010)
11. Baranov, I.: Increasing the efficiency of transportation of highly concentrated hydrocarbon fuel in hydrotransport systems of industrial enterprises. Thesis Ph.D. technical of science. Dnipro (2018)
12. Baranov, I.: Formation of an optimization model for the selection of parameters for the transportation of hydrocarbon fuel by industrial hydraulic transport. Bull. VDEUNU **3**(244), 7–13 (2018)
13. Morozov, A., Panishev, A.: Method of branches and limits in the Hamiltonian problem of the rural postman. Syst. Res. Infor. Technol. **2**, 57–66 (2012)

Implementation of the Concept of Sustainable Development in Logistics

Iryna Gorobinska(✉), Lyudmyla Volynets, Iaroslava Nazarenko, Artur Horobinskyi, and Yana Volynets

National Transport University, Mykhaila Omelianovycha-Pavlenka Str.1, Kyiv 01001, Ukraine
irynavladuslavivna@gmail.com

Abstract. One of the directions for ensuring the balance of the national transport system, its effective integration into international transport networks and ensuring environmental norms of behavior is the transition to a model of sustainable development. The goal of sustainable development is to increase economic value while reducing the impact on the environment and improving the quality of human life. Sustainable logistics is becoming more and more relevant in the conditions of the transition from a linear economic model to a cyclical one. The purpose of the study is to substantiate the main provisions and develop practical recommendations for the implementation of the concept of sustainable development in logistics activities. In the course of the study, the place of individual goals of sustainable development in supply chains was substantiated, the contradictions that arise between the economic, ecological goals and tasks of the logistics system in the process of reorientation to the use of safe logistics technologies were identified, directions for the implementation of the principles of sustainable development by types of logistics activities were developed, the advantages of sustainable development were highlighted logistics, the expected complex ecological and economic effect from the implementation of the concept of sustainable development is determined. The results of the research can be used to clarify the goals and principles of sustainable development, change traditional approaches to managing logistics activities, and reorient the logistics system of enterprises to the use of environmentally safe logistics technologies.

Keywords: Sustainable Development · Sustainable Logistics · Green Logistics · Sustainability · Transport and Logistics System

1 Introduction

One of the promising directions for ensuring balance, increasing the competitiveness of the national transport system, its effective integration into international transport networks and ensuring environmental norms of behavior is the transition to a model of sustainable development. Ensuring sustainable development is one of the most significant problems facing the world community today. Humanity is at a turning point in its evolution, which requires a reorientation to a development paradigm that would support people's lives and the natural environment. An important component of the development

O. Slavinska et al. (Eds.): ITSESQC 2024, LNNS 1335, pp. 86–96, 2025.
https://doi.org/10.1007/978-3-031-87376-8_8

of modern society is balancing the urgent needs of humanity and protecting the interests of future generations, as well as the search for alternative and innovative approaches to managing the economy. One of the promising ways to solve these problems is the transition to a model of sustainable development.

Ensuring sustainable development is an urgent problem not only for individual industries or enterprises, but also for the national economy in general. A large number of Ukrainian and foreign scientists pay attention to the study of the problems of sustainable development, the theoretical and methodological foundations of the implementation of the principles of the concept in transport and logistics systems. The works of the following Ukrainian authors are devoted to this topic: V. Marchenko [1], D. Budayko [1], M. Holovanenko [2], I. Kara [2], S. Grytsenko [3], L. Savchenko [3], S. Kryshtal [3], T. Vorkut [4], I. Khomenko [5], as well as foreign authors: G. Prause [2], M. Jarvis [2], D. Testi [6] P. Murphy [7], R. Poist [7], Ch. Braunschweig [7], L. Milani [8], D. Mohr [8], N. Sandri [8], T. Gruchmann [9] and others.

Despite the significant interest of scientists in the mentioned problems, the issues of implementing the concept of sustainable development in logistics activities remain unresolved. The modern process of transition to a model of sustainable development in Ukraine is complex and dynamic, accompanied by a significant number of problems; therefore there is a need for constant analysis in this direction.

The purpose of the study is to substantiate the main provisions and develop practical recommendations for the implementation of the concept of sustainable development in logistics activities.

The main tasks of the research are the justification of the place of individual goals of sustainable development in supply chains, the study of contradictions between economic and environmental goals and the tasks of the logistics system in the process of reorientation to the use of safe logistics technologies, the development of directions for the implementation of the principles of sustainable development by types of logistics activities, highlighting the advantages of sustainable logistics, determination of the expected effect from the implementation of the concept of sustainable development.

2 Research Methodology

To reveal the problems posed in the research, methods of generalization (for the systematization of existing mechanisms for implementing the goals of sustainable development), observation and comparison, synthesis and analysis (for evaluating modern trends and opportunities for implementing the concept of sustainable development), a systematic approach (for researching the basics of implementing the concept of sustainable development), the method of expert assessments (for monitoring indicators of achieving the goals of sustainable development), the method of strategic planning (when implementing the main provisions of the concept of sustainable development in logistics activities) were used.

3 Results and Discussion

After the UN Summit on Sustainable Development that was held in September 2015, the Sustainable Development Goals were established at the state level in Ukraine, which should unite the efforts of the world community around the global priorities of sustainable development, and they were reflected in the National Report "Sustainable Development Goals: Ukraine" [10] as landmarks of national development. In July 2020, the Voluntary National Review of the Status of Achievement of the Sustainable Development Goals [11] was presented. In 2021, the "Constant Monitoring of Sustainable Development Goals Achievement Indicators" [12] and the "Joint Program to Promote Strategic Planning and Sustainable Development Financing Processes in Ukraine" [13] were introduced, the purpose of which is to increase the efficiency of state financing of the Sustainable Development Goals. The joint program made it possible to assess the financing status of the Sustainable Development Goals in Ukraine [14]. On the basis of the conducted research, recommendations were determined for improving financing using existing and potential financial instruments, as well as coordinating the directions of resource use with long-term development priorities in Ukraine.

Also at the legislative level there are relevant documents in the field of transport and logistics activities, such as the Decree of the Cabinet of Ministers "On the approval of the plan of measures for the implementation of the National Transport Strategy of Ukraine for the period until 2030" [15] and the Project of the Ministry of Infrastructure "Sustainable Logistics Strategy and Action Plan for Ukraine" [16], which makes it possible to introduce guidelines, principles and priorities of the concept of sustainable development into logistics activities.

Sustainable development is considered as a long-term approach and is based on three components: economy, society and environment ("3P" model - profit, people, and planet) [17]. In the field of logistics, the concept of sustainable development was reflected in the direction of "green logistics" [6, 7, 9, 18]. By applying a three-dimensional life cycle approach and finding a balance between its components, "green logistics" can provide the best services, while ensuring a more conscious use of resources, and by reducing the negative impact on the environment and supporting positive social behavior, there is a return on the overall value for money quality For example, green financing aimed at increasing the sustainability of the transport industry is one of the principles determining the development of transport and logistics activities in the EU countries [19].

The need to transition for the transport and logistics system to a model of sustainable development is due to modern trends in the development of transport and the growth of its role in the life of society, which is accompanied by an increase in the number of vehicles, traffic congestion and the complication of the situation on the roads, an increase in the negative effects of the transport industry on the state of ecology and health society.

The sustainable development of a logistics activity is designed to form the following main directions: minimization of losses associated with the inefficiency of the logistics process; effective use of energy, material and raw materials; effective use of available capacities; minimization of ecological impacts on the environment; development and implementation of innovative methods of management and organization of logistics processes. The goals of sustainable development in the supply chain are depicted in Fig. 1.

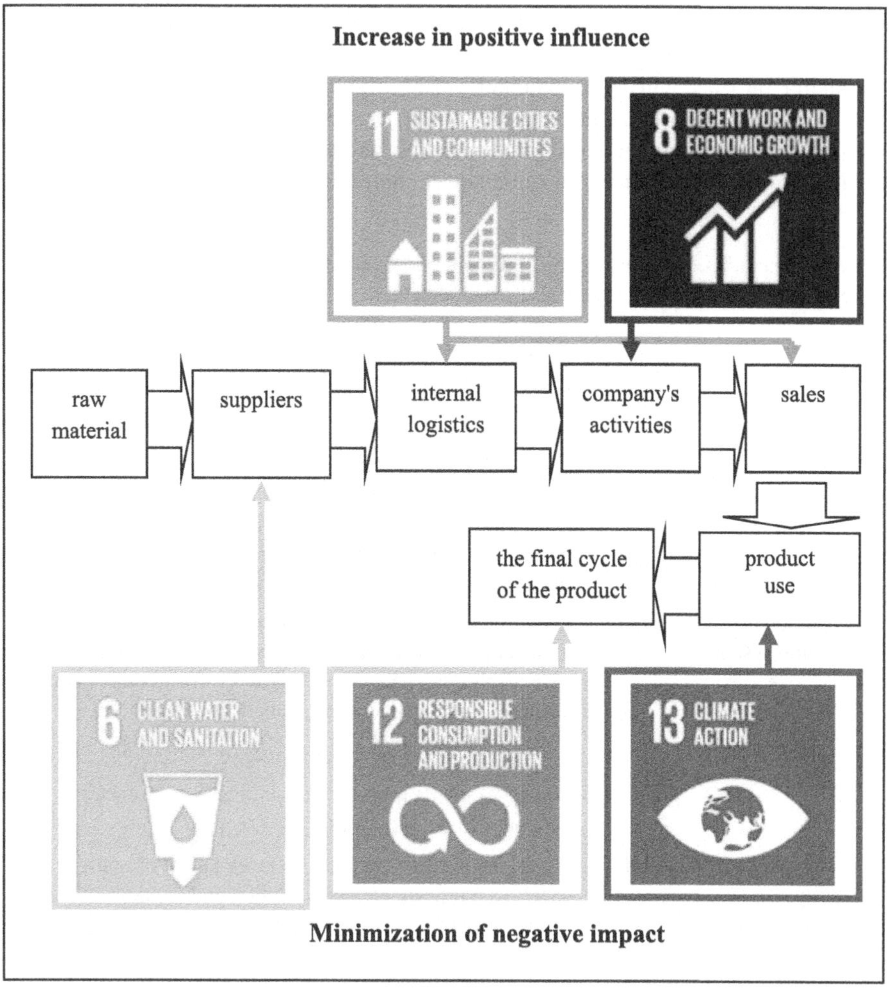

Fig. 1. Goals of sustainable development in the supply chain

The goal of sustainable development is to increase economic value while reducing the impact on the environment and improving the quality of human life. Sustainable logistics is becoming more and more relevant in the context of the transition from a linear economic model to a cyclical one, the main purpose of which is to extend the service life of products and rational use of resources. Sustainable logistics should be based on the creation of a resource-saving, ecologically and socially responsible business model of the economy.

The application of sustainable development principles is gradually becoming an integral part of the corporate policy of most modern companies. Taking into account the need for the formation of new competitive advantages in the conditions of globalization, intensification of competition and the formation of a knowledge economy, the way to the development of the loyalty of increasingly knowledgeable consumers is the creation and maintenance of the company's own "ecological" image, which is an integral part of the image of a socially responsible company.

Taking into account environmental norms of behavior requires changing traditional approaches to managing logistics activities, reorienting enterprises to the use of environmentally safe logistics technologies. At the same time, the implementation of sustainability is characterized by a number of contradictions (Table 1).

Table 1. Contradictions of ecological and economic goals and tasks in the logistics system when implementing the concept of sustainable development

	Contradictions arising between economic and environmental goals and objectives of the logistics system	
	Ensuring economic efficiency	Ensuring environmental safety
Costs and distribution of funds	Minimization of costs, concentration of funds in production, externalization of environmental costs	Allocation of funds for environmental protection measures, internalization of environmental costs
Organization and flexibility of distribution systems	Ensuring system-wide efficiency of distribution systems through network changes (integrated supply chain, nodal and tiered distribution structures, organization of a trade network using the Internet)	Organization of distribution systems that "consume" less space, infrastructure elements, energy and resources
Delivery time, accuracy and reliability	Use of heavy-duty transport, socially significant elements of infrastructure	Development of optimal routes and methods of transportation from the point of view of environmental protection
Storage system	Placement of warehouses on the territory of enterprises (including places of storage of environmentally hazardous raw materials, waste)	Placement of warehouses outside populated areas, joint use of warehouse management
Methods of transportation	Transportation of large consignments of goods by economic means of transport	Prompt transportation of small consignments of goods by modern means of transport that consume less fuel

They are related to the fact that the goals of nature protection and other subsystems are characterized by a number of contradictions. In order to achieve cross-functional compromises in the logistics system, an assessment of logistics costs is carried out taking into account the environmental component and social aspects.

There are also many obstacles that hinder the implementation of the principles of sustainable development at logistics enterprises. The factors of such influence can be divided into factors of the internal and external environment. Internal barriers to initiating environmental issues can be high investment costs, lack of financial or human resources, and lack of knowledge or skills in the company's personnel. Externalities include limited access to technologies that reduce environmental impact, lack of interest from suppliers and partners, lack of government support system, competition and uncertainty.

Available financial resources and technologies limit how quickly or to what extent the negative consequences of logistics activities can be mitigated. Many of them require regulatory or fiscal measures. Once these measures are implemented, the relevant authorities must also ensure their implementation. The technological problems include the search for effective technologies for alternative or renewable energy sources, the complexity of implementing modern IT solutions for logistics operations and supply chains.

Since the process of implementing the principles of sustainable development should be comprehensive, improvements should be made in all types of logistics activities. The main directions of implementation of the principles of sustainable development are summarized in the Table. 2.

Expected results of implementation of the principles of sustainable development by types of logistics activities:

- Transportation – Low-emission vehicles that are well-maintained with maintenance schedules and appropriately trained drivers reduce environmental and economic costs by increasing efficiency;
- Distribution – Faster delivery, increased flexibility to requests and time savings for waste management;
- Purchasing – Reputation boost;
- Warehousing – Saving money and time;
- Packaging – Saved resources.

The practice of implementing the principles of "green logistics" shows that although the initial application of the concept increases costs, in particular investment costs, for operation, training and the purchase of environmentally friendly materials, in the long term the costs of storage, accumulation, transportation and energy decrease. It also improves the institutional image of the business and its competitive position, which leads to consumer satisfaction and helps reduce waste and environmental costs.

The growth of individual components of total logistics costs (costs in the sales subsystem, general management) is compensated by a decrease in the amount of complex ecological and economic damage, which ensures a minimum of total logistics costs. This testifies to the effectiveness of the proposed scientific and methodical approaches to justifying measures in the field of greening.

Table 2. Directions of implementation of the principles of sustainable development by types of logistics activities

Types of logistics activities	Existing situation	The direction of improving the situation
Transportation	Transport causes significant pollution and negatively affects air quality	Analysis of movement, transport costs, to collect the necessary data. Appropriate investment in proper maintenance depending on needs and chosen strategy. This may include: choosing more optimal routes, investing in environmentally friendly vehicles, implementing "eco-driving"
Distribution	Distribution channels are poorly organized or inefficient	Organization of the supply chain and procurement, taking into account the costs of managing the generated waste Effective matching of production location with distribution points, including using proximity to storage and distribution points as a selection criterion. Assess existing suppliers' production line or third-tier distribution channels for waste or misuse
Purchasing	The choice is made mainly on the basis of price, which potentially hides unethical or unfriendly activities	Creation and application of selection criteria that correspond to the organization's ethical and environmental policy. Researching initiatives implemented by other organizations and adapting them to a specific situation

(*continued*)

Table 2. (*continued*)

Types of logistics activities	Existing situation	The direction of improving the situation
Warehousing	Loss of consumer qualities of the product caused by poor storage conditions or damage during warehouse movements	Improvement of the warehouse infrastructure to facilitate the movement of goods in the warehouse. Use of sunlight and natural ventilation. If the infrastructure is planned to be used for more than two years, it is advisable to invest in solar or wind energy sources; control of electricity consumption
Packaging	Excessive use of non-biodegradable materials	Finding a compromise between safety and controllability; reducing packaging and/or introducing reusable or biodegradable materials. An example is corrugated cardboard and other types of paper packaging

The expected effect of the implementation of the concept of sustainable development in logistics activities is schematically presented in Fig. 2.

The sustainability of logistics activities has a number of advantages:

1. Reduction of unnecessary expenses. Sustainability helps increase the efficiency of logistics processes by avoiding unnecessary costs, reducing waste, and increasing work productivity. The use of resource-saving technologies in logistics makes it possible to reduce material stocks by 40–60%, accelerate the turnover of working capital by 20–40%, reduce transport costs by 7–20%, and reduce the costs of loading and unloading and warehouse work by 15–30% [20].
2. The implementation of sustainable logistics methods allows reducing the risks of serious environmental problems and unforeseen events.
3. Sustainability helps improve reputation. Logistics operators who ensure sustainability in their activities receive a positive image impact and increased trust from consumers.
4. Improving working conditions, which includes attention to the safety of workers, ensuring their rights, compliance with the requirements of ethical business and compliance with the principles of social responsibility. This, in turn, can increase employee satisfaction and productivity, reduce costs and improve customer service.
5. Sustainability helps attract investment. Logistics companies that strive for sustainable development may be more attractive to investors looking for companies with high social responsibility.

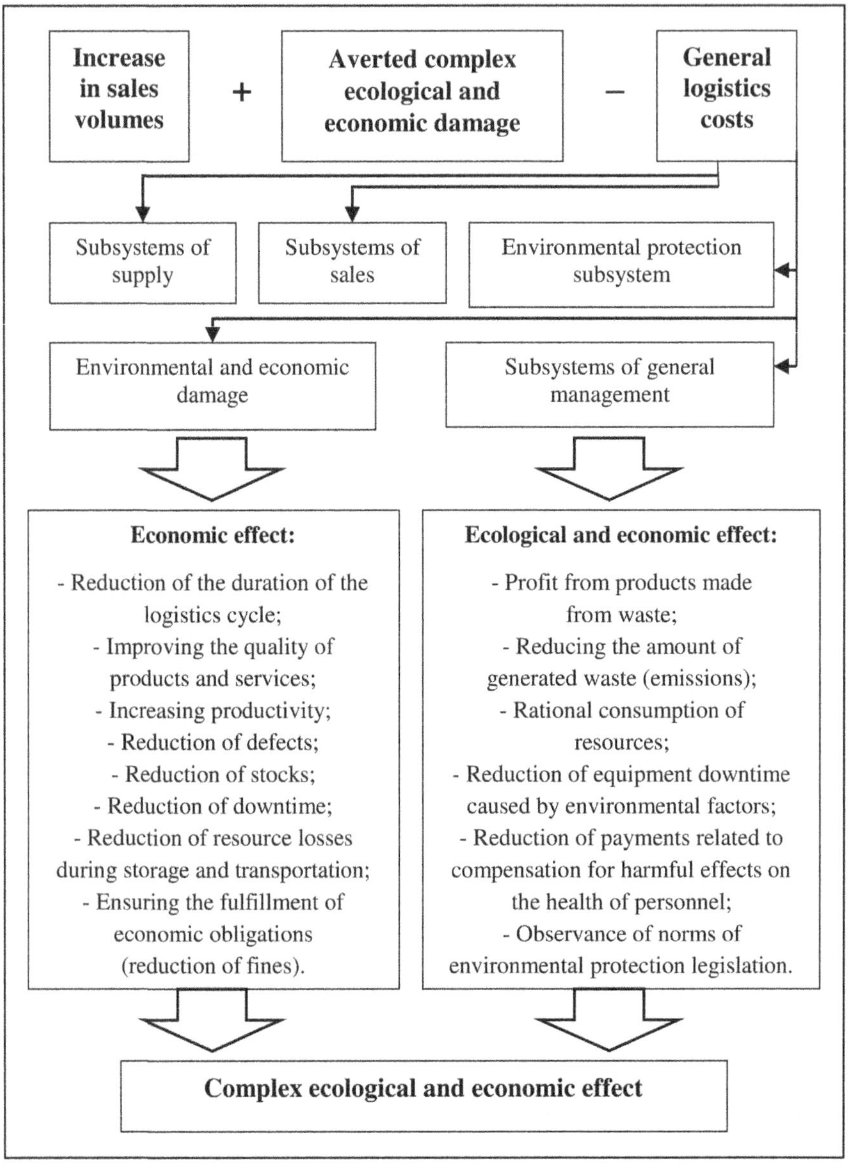

Fig. 2. The expected effect of the implementation of the concept of sustainable development

The main goal of environmentally oriented logistics management is to obtain a comprehensive ecological and economic effect. This is due to the fact that the effectiveness of the ecologically-oriented logistics management system is considered as its ability to solve the main economic tasks without harming the components of the natural environment. This forms limitations in the process of minimizing costs (especially environmental costs), and at the same time it guides the search for new sources of increasing the productivity of resource use. However, the ecological situation can be stabilized only through a change in the orientation of socio-economic development, the formation of new values, a review of the structure of needs, goals, priorities and methods of activity. This requires the development of a set of radical social, economic, technological and organizational measures.

4 Conclusion

For a long time, logistics solutions were designed to minimize costs, maximize profits and achieve the goals of optimizing customer service. However, global environmental and social problems, increased competition in the market and shortage of resources forced to start a new approach to overcome them, which was called "sustainable development concept". Today, companies' compliance with the principles of sustainable development is an important component of business success, as more and more consumers realize the value of non-renewable resources and seek a balance between the needs of the organization, society and nature. The priorities of logistics activities based on the principles of sustainable development should be high quality of services, guaranteeing the safety of transportation, ensuring decent working conditions, fighting corruption, implementing innovations, developing infrastructure, etc.

The results of the research can be used to clarify the goals and principles of sustainable development, change traditional approaches to managing logistics activities, and reorient the logistics system of enterprises to the use of environmentally safe logistics technologies.

References

1. Marchenko, V., Budayko, D., Budayko, D.: Prospects of the sustainable development concept, the importance of its high-quality implementation in the logistics sphere. Intellectualization Log. Supply Chain Manag. **22**, 7–18 (2023). https://doi.org/10.46783/smart-scm/2023-22-1
2. Prokopenko, O., Prause, G., Bielialov, T., Jarvis, M., Holovanenko, M., Kara, I.: Sustainable logistics and passenger transport in smart cities. Acta logistica – Int. Sci. J. About Log. **11**, 47–56 (2024). https://doi.org/10.22306/al.v11i1.448
3. Grytsenko, S., Savchenko, L., Kryshtal, S.: Conceptual principles of the green technologies introduction in the logistics activities of Ukrainian companies in the context of the implementation of European environmental programs. Intellectualization Log. Supply Chain Manag. **1**(13), 15–26 (2022). https://doi.org/10.46783/smart-scm/2022-13-2
4. Alkema, V., Darushyn, O., Vorkut, T., Bilonog, O., Tretinichenko, Y.: Theoretical base of modeling the project team with a service technical systems by simulation. Int. J. Manag. **11**, 453–465 (2020). Article ID: IJM_11_04_044
5. Volynets, L., Gorobinska, I., Nakonechna, S., Petunin, A., Romanyuk, S., Khomenko, I., Zachosova, N.: Principle of the assessment of the readiness of motor transport enterprises for economic development based on a two-component methodological approach. Eastern-European J. Enterp. Technol. **4/13** (118), 12–21 (2022). https://doi.org/10.15587/1729-4061.2022.263041
6. Testi, D.: Sustainable logistics: definition, approach and relevance (2020). https://www.cswindow.contshipitalia.com/en/sustainable-logistics

7. Murphy, P., Poist, R., Braunschweig, C.: Green logistics: comparative views of environmental progressives, moderates and conservatives. J. Bus. Log. **17**, 191–211 (1996). https://trid.trb.org/view/560391
8. Milani, L, Mohr, D., Sandri, N.: Built to last: Making sustainability a priority in transport infrastructure. McKinsey & Company (2021). https://www.mckinsey.com/industries/travel-logistics-and-infrastructure/our-insights/built-to-last-making-sustainability-a-priority-in-transport-infrastructure
9. Gruchmann, T.: Advanced green logistics strategies and technologies. In: Operations, logistics and supply chain management, pp. 663–686. Springer, Cham (2019). https://doi.org/10.1007/978-3-319-92447-2_29
10. National Report "Sustainable Development Goals: Ukraine". https://me.gov.ua/Documents/Detail?lang=uk-UA&id=6f446a44-9bba-41b0-8642-8db3593e696e&title=NatsionalnaDopovid-tsiliStalogoRozvitku-Ukraina-
11. Voluntary National Review of the Status of Achievement of the Sustainable Development Goals. https://ukraine.un.org/sites/default/files/2021-10/VNR%20SDG%20Ukraine%202020.pdf
12. Constant Monitoring of Sustainable Development Goals Achievement Indicators. https://sdg.ukrstat.gov.ua/uk/
13. Joint Program to Promote Strategic Planning and Sustainable Development Financing Processes in Ukraine. https://jointsdgfund.org/sites/default/files/2020-10/Brief%20SDG%20Financing_ukr.pdf
14. Development finance assessment: Ukraine. United Nations Development Programme. https://www.undp.org/ukraine/publications/development-finance-assessment-ukraine
15. On the approval of the plan of measures for the implementation of the National Transport Strategy of Ukraine for the period until 2030. Decree of the Cabinet of Ministers. https://www.kmu.gov.ua/npas/pro-zatverdzhennya-planu-zahodiv-z-realizaciyi-nacionalnoyi-transportnoyi-strategiyi-ukrayini-na-period-do-2030-roku-321-070421
16. Sustainable Logistics Strategy and Action Plan for Ukraine. Project of the Ministry of Infrastructure. https://mtu.gov.ua/files/Logistics.pdf
17. Sustainable logistics. Logistics Operational Guide (LOG). https://log.logcluster.org/en/print-preview-entire-book/212
18. Green Logistics Practices Seeking Development of Sustainability. https://www.researchgate.net/publication/356122966_Green_Logistics_Practices_Seeking_Development_of_Sustainability_Evidence_from_Lithuanian_Transportation_and_Logistics_Companies
19. The future of the EU transport sector (2021–2024) – four trends. https://dr2consultants.eu/the-future-of-the-eu-transport-sector-2021-2024-four-trends/
20. Volynets, L., Sopotsko, O., Khrutba, Y., Sevostianova, A., Levchenko, I.: Optimization of international road transportation of cargoes in the management of enterprises of agricultural sector and road transport enterprises. Eastern-European J. Enterp. Technol. **6/3** (114), 57–63 (2021). https://doi.org/10.15587/1729-4061.2021.248191

Development of Intelligent Security Systems in the Infrastructure of Transport and Industrial Complexes of a Smart City

Nataliia Yevtushenko(✉), Natalia Tverdokhliebova, and Olga Ponomarenko

Kirpicheva Str, National Technical University "Kharkiv Polytechnic Institute", Kharkiv 61002, Ukraine
natalya0899@ukr.net

Abstract. The article analyzes existing approaches to building elements of a "smart city". It is shown that the key condition for the functioning of a "smart city" is the widespread use of electronic services to the population, which fully ensure a high quality of life and favorable conditions for innovation and business. Approaches to building an automated management system for a "smart city" are proposed, ensuring the creation of an innovative space with the most rational use and development of intellectual, industrial and natural energy resources. The purpose of this article is to justify the need to create and implement an automated smart city management system to ensure sustainable socio-economic development, improve the quality of life of citizens, optimize the transport system and effectively use modern information technologies. This system should integrate advanced digital solutions for managing all aspects of urban life, including service availability, improve transport and urban infrastructure, comply with environmental standards and create conditions for the development of human potential. As a result, the system will contribute not only to improving the comfort and safety of the urban environment, but also to accelerating the economic growth of the city and the country. In addition, the article analyzes potential challenges and risks associated with the implementation of automated management systems, such as data security issues and the need to modernize the existing infrastructure. In conclusion, the importance of an integrated approach and interdisciplinary cooperation is emphasized to achieve successful implementation and sustainable functioning of the smart city.

Keywords: Innovation · "Smart City"; Automation · Control System · Transport and Industrial Complex · Infrastructure

1 Introduction

The ideology of the "smart city" was created with the aim of providing favorable conditions for sustainable socio-economic development of cities, strengthening their sociocultural self-sufficiency, introducing innovations and attracting human resources. This is achieved through effective management of all city functions using modern technologies.

O. Slavinska et al. (Eds.): ITSESQC 2024, LNNS 1335, pp. 97–108, 2025.
https://doi.org/10.1007/978-3-031-87376-8_9

Among these, information technologies play a key role, serving as the foundation that connects and coordinates the entire complex organizational and technical system of the "smart city." In recent years, the necessary conditions have been created for the practical implementation of the "smart city" concept: information and telecommunication technologies have developed significantly, the computing power of computers and networks has increased significantly, powerful tools for creating advanced databases have appeared and effective methods for building intelligent decision-making systems have been developed [1, 2].

The main focus of the currently implemented elements of the "smart city" is the creation and implementation of intelligent engineering systems in private sectors. Currently, the range of intelligent management of urban functions is expanding and includes "smart" public spaces, education systems, health care, and the provision of public services. However, everything that exists today is only individual fragments of the "smart city", and not a full-fledged concept. The city must create conditions that are most favorable for the development of a person—his intellect, creativity, cultural level. It should also ensure comfortable and efficient work, minimizing time and labor costs for receiving services of any kind. This will allow the economy of the city and the country to reach a fundamentally new level of development. It is important not only to introduce innovations in the construction and operation of a "smart city", the organization of its life support, education and job creation, but also to ensure an innovative approach to the use of the city's intellectual resources.

The above determines that a "smart city" should provide the highest possible quality of life for people living and working in it. This is due to the provision of the possibility of the most effective access of individuals and legal entities, individual entrepreneurs, city residents and its guests, citizens of the country and foreign citizens to any type of information and services that are provided by both government agencies and organizations of all forms of ownership from various areas of the national economy, individual entrepreneurs [3, 4]. To solve this problem, it is necessary to use the most modern means, tools and methods of information technology, and the development of an automated management system for a "smart city" is required [5].

The purpose of this article is to substantiate the need to create and implement an automated smart city management system to ensure sustainable socio-economic development, improve the quality of life of citizens and effectively use modern information technologies. This system should integrate modern information technologies to ensure optimal management of all aspects of urban life, including ensuring the availability of services, improving urban infrastructure, supporting environmental standards and creating conditions for the development of human potential. As a result, the system should contribute not only to improving the comfort and safety of the urban environment, but also to bringing the city and country's economy to a new level.

2 Research Methodology

Currently, in Ukrainian cities, government agencies and businesses are investing heavily in upgrading electrical and heating networks, repairing roads, greening production facilities, and building and developing residential, commercial, retail, and entertainment complexes. These efforts are aimed at creating an effective social infrastructure and industrial growth, which should contribute to the development of the domestic economy [6, 7]. However, despite all these investments, it is not possible to significantly improve the quality of life of people. This is due to the fact that the existing urban system, built on the ideology of "people for enterprises" and "people for the city", does not meet modern requirements. Pilot projects for the implementation of intelligent city management systems are actively developing in the USA, China and Europe [8, 9]. These projects focus on the creation and application of advanced technologies aimed at the rational use of energy resources and the reduction of negative impact on the environment. They develop solutions in areas such as electric power, water supply, resource consumption accounting, waste disposal, as well as in the creation of more efficient transport systems and "smart buildings".

"Smart" engineering systems are emerging that provide monitoring and management of all nodes of urban infrastructure in real time [10]. These systems allow collecting data on resource consumption, optimizing their use, evenly distributing the load, reducing peak loads and increasing efficiency during periods of low activity. The result is an intelligent network that can automatically regulate and balance energy consumption with minimal human intervention.

It is believed that the key condition for the successful functioning of a "smart city" is the widespread use of electronic services for the population [11, 12], which ensure a high level of quality of life and create favorable conditions for innovation and business activities. It is important to note that in countries and regions where electronic services are being introduced, this usually only concerns those services that are included in the mandatory lists of public services. These lists do not always cover the entire range of services necessary for everyday life. Also, information resources of various bodies and levels of government and administration are often not united into a single network. For the full functioning of a "smart city", it is important that the provision of services and information covers all key areas specified in Table 1.

Table 1. Directions for providing services and access to information for residents and entrepreneurs in a "smart city".

Spheres of receiving services and information	Types of services
Socio-legal	– registration of citizenship documents, registration, visas; – legal services of various types (advocacy services, legalization of documents, notary services, patent law, registration of civil status acts, a wide range of legal assistance); taxes; – family relations; social support (pension provision; social benefits for the population, subsidies and allowances, the possibility of improving housing conditions, filing complaints, etc.); – functions of employment and professional orientation of citizens, settlement of labor disputes
Real estate	– transactions with real estate and various types of property; land relations; issuance of permits, licensing; state property

For the effective implementation of all the possibilities of the "smart city" it is extremely important to create, implement and provide support for a special type of business. This business, relying on advanced information and telecommunication technologies, will be able to reach a new level, in this regard, it is necessary to create a developed and appropriate infrastructure for such a business, which requires the active involvement of the intellectual resources of the country and the international market, primarily young specialists with the necessary knowledge and skills in the field of management,

Table 2. Directions for providing services and access to information for residents and entrepreneurs in a "smart city" (continuation).

Interaction between government agencies, business and the population	– interaction of individuals and business structures with government agencies; interaction of individuals with business and business structures with each other; protection of the rights and freedoms of the population and business; – entrepreneurial activity; – international relations; security and law enforcement
Environmental management and ecology	– issuance of permits, opinions, quotas, etc.; – licensing, insurance, examination; – use of natural resources; environmental monitoring and protection; appeals, complaints and suggestions; – provision of territories, water areas, airspace
Sociocultural space	– culture and art; education and science; – tourism and recreation; sports and physical education; – leisure; dating and communication; – problems of childhood and motherhood; – advertising; mass media
Healthcare and Medicine	– maintaining registers, records, providing information; high-tech medical care; – medical devices; medicines; pharmaceutical activities; – control and supervision; licensing, certification, issuance of opinions; medical and social expertise; filing appeals, complaints and suggestions; loss of working capacity
Finance	– banking activities; maintaining registers, records, providing information; investment activities; budget; financial markets; - securities, shares; prices (tariffs) for goods and services

(*continued*)

Table 2. (*continued*)

Interaction between government agencies, business and the population	– interaction of individuals and business structures with government agencies; interaction of individuals with business and business structures with each other; protection of the rights and freedoms of the population and business; – entrepreneurial activity; – international relations; security and law enforcement
Economy and life support	– housing and communal services; – transport and road management; – customs; – industrial production; construction and repair; – trade; – energy supply, consumption and expenditure of natural resources; – export and import; foreign economic activity; geodesy and hydrometeorology

law, economics, as well as modern information and telecommunication technologies. This will allow us to implement in practice a fundamentally new approach—"a city for people" [13].

It is worth noting that the range of services required by the population of a "smart city" will constantly expand and transform. Therefore, one of the key tasks of a business providing electronic services will be to ensure the ability to quickly and effectively create and implement new services. This is due to the need to develop and organize the practical work of an automated electronic services management system for the population of a "smart city", which is the most important element of the intelligent city management network. This system includes order management, interaction processes between objects and subjects of service provision, coordination of actions of all participants in the process, database management, support and monitoring of service implementation, as well as creation of new services and much more. From the above functional list it becomes obvious that limiting the automated control system only to the tasks of rational use of material and energy resources, as well as minimization of the impact on the environment, is not enough. This is only part of the tasks that the system must solve. Only through integrated intelligent management of all the above areas can we achieve a high quality of life for the population of a "smart city", the introduction of innovative technologies, interaction with businesses in other regions and countries, high efficiency of the city economy and deep penetration of business structures into all spheres of life.

Thus, the main goal of the automated control system is to ensure effective and coordinated interaction of various technological, organizational, technical and socio-economic complexes, creating an innovative space with optimal use and development of intellectual, industrial and natural energy resources.

The general world trend and globalization leads to the growing impact of transport problems on states, on the economy as a whole, on the environment, on the standard of living of each individual and the whole society. The solution of transport problems has become one of the key conditions in the modern conditions of most countries of the world [14].

"Intelligent transport systems" is a definition that characterizes the implementation and use of a whole range of the most modern intelligent technologies aimed at meeting the needs of a developed society for safe and comfortable living in conditions of freedom of movement [15]. For implementation in terms of general transport mobility, a solution is required that could combine existing and new promising areas into one single system—the Intelligent Transport System.

ITS is a tool that allows you to modernize the existing transport system, significantly increase the efficiency of passenger transportation, provide new opportunities for users to calculate and plan private and commercial transportation. The main task is to provide the ability to integrate various intelligent subsystems and work applications into ITS based on existing open communication standards.

As a result, it becomes possible to flexibly manage the functionality of the System, regulate transport demand in the conditions of the existing road and transport infrastructure and taking into account its development, apply safe energy-saving technologies, use the most modern information resources. Ultimately, this allows for greater economic benefit from investments in the long term. Intelligent transport systems are the provision of modern quality of life through the use of innovative technologies that provide for the economical and environmentally friendly use of transport life support systems. ITS is a unique opportunity to modernize the country's infrastructure with fundamentally new opportunities for centralized management, a new level of services and security [16, 17]. The development model of ITS can be based on the interaction of six characteristics: smart economy, smart management, smart environment, smart transport, smart lifestyle, smart people [18].

Figure 1 provides a generalized model of some of the possible outcomes of transport system development, as well as the beneficial changes for society that can be achieved through the spread of shared ITS use.

The implementation of the Intelligent Transport System will ensure a balanced synergetic development of the transport system as the most important factor in sustainable socio-economic and spatial development. The ultimate goal of creating the ITS is complete and absolutely safe transport mobility of citizens. Mobility that brings pleasure to every person and brings social benefits to the whole society. A set of measures for the creation of ITS can ensure modern quality of life for people through the use of innovative technologies that provide for the economical, environmentally friendly and safe use of modern innovative life support systems.

Integrated intelligent information systems enable efficient integration of different modes of public transport. Mobile, wireless and satellite data transmission systems provide companies with new opportunities to use business models based on consumer preferences with dynamic pricing and one-time payment for services to strengthen their position in the market.

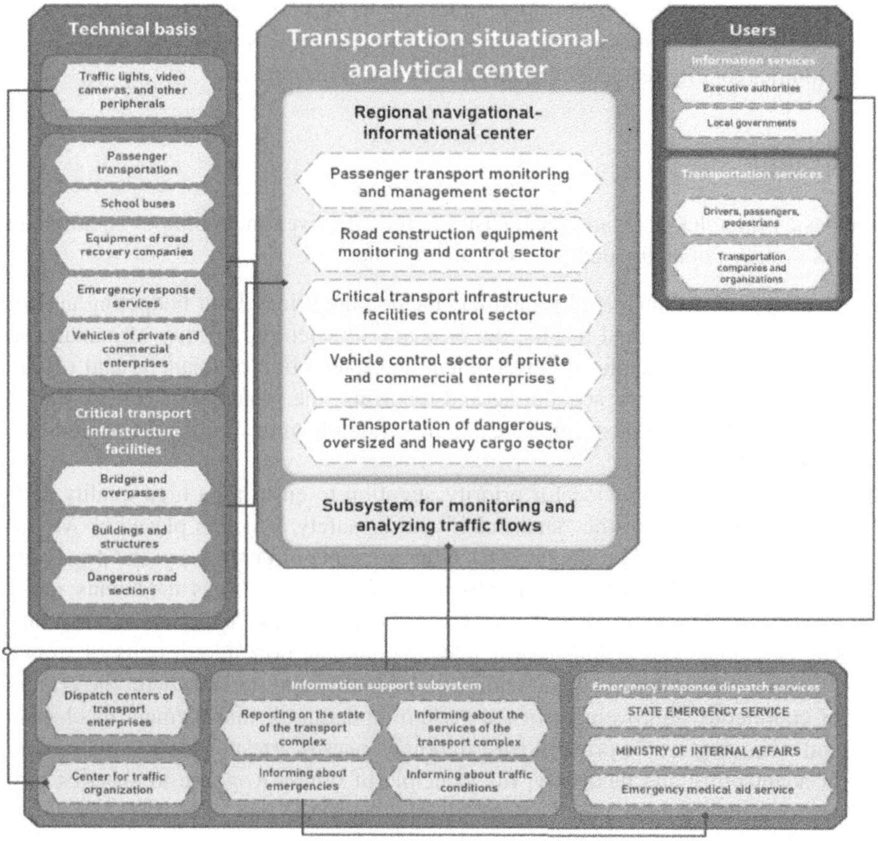

Fig. 1. Intelligent transportation system.

Intelligent transport systems cover hardware and software for monitoring transport infrastructure—highways to vehicles. The physical lower level of control consists of: sensors that record the movement of vehicles, emergency situations, temperature, humidity, air composition, etc.; controllers that manage transport systems—switch traffic light modes, video and photo recording, etc. The upper level is interactive control panels with a single interface for different types of devices—from mobile phones to desktop PCs.

The essence of ITS is this: reading devices on city streets will track the number of vehicles and, if necessary, switch traffic lights to the desired light. Or redirect flows along other, less congested streets. On highways and bus stops with the highest congestion, the scope of work will be determined and the most problematic areas will be identified. They will be the first to be equipped with "intelligent" technologies. Information from these cameras will be sent to the ITS Control Center and to the city's duty stations. Communication between the nodes will be possible thanks to the installation of a fiber-optic cable. As a result, city residents will be able to evaluate the capabilities of the ITS software package under normal operating conditions.

At the same time, a powerful foundation will be laid for the further development of ITS. The synthesis of modern software and network infrastructure will open up great opportunities for city residents that can satisfy the highest needs.

3 Results

The basic principles of creating and ensuring the effective functioning of an automated control system can be divided into four main groups:

I. Economic and managerial principles: include the principle of budgeting aimed at achieving specific final results; the principle of scientific validity of management decisions; the principle of development and integration of financial and resource management systems, accounting and reporting; the principle of consistency and integrity of management impacts; and the principle of ensuring honesty and ethics in business.
II. Social principles: provide for priority attention to ensuring a high quality of life for the population of the "smart city" and its safety, which is placed above other socio-economic and production tasks; the principle of democracy; the principle of priority for the disclosure and development of people's abilities and talents; and the principle of guaranteeing work that brings both moral and material satisfaction.
III. Organizational and technical principles: include the principle of unity of strategic planning and tactical management of the economy and social sphere; the principle of continuous improvement of the system, updating of management mechanisms and technologies; the principle of excluding zones of irresponsibility; the principle of adaptability and flexibility; the principle of timely response to changes in the external environment; and the principle of openness to new ideas and technologies.
IV. Environmental principles: aimed at ensuring environmental safety for both current and future generations of residents of the innovative "smart city" [19]; the principle of harmonious coexistence of the city and the surrounding nature; the use of the best available technologies that take into account economic and social aspects to minimize the negative impact on the environment.

The implementation of economic and management principles is based on the improvement and development of financing mechanisms based on the innovation and investment fund. This allows for a rational combination of administrative and regulatory methods of city management with economic freedom and support for business on a private-public basis.

Social principles are implemented through the use of electronic interactive means of information exchange, conducting social surveys of the population, identifying public sentiments and trends.

Organizational and technical principles are based on the creation of a single information and technological space, including a "personal account", a system of public access to electronic resources and a developed communications system. An important role here is played by innovative mechanisms for integrating databases using uniform formats and information processing technologies, implemented by specialized operators within a single data transmission network.

Environmental principles are implemented through the development of a regulatory framework, economic and financial mechanisms, a system of environmental monitoring and control, as well as through scientific research aimed at a deeper understanding of environmental problems and finding ways to solve them, as well as the formation of public environmental awareness.

The object of management is an innovative "smart city". Its current state and compliance with established requirements is determined by a set of indicators that can be grouped as follows: quality of life of the urban population; development and organization of services for city residents; innovative attractiveness of the city, that is, its attractiveness as an environment for life and business, both for citizens from other regions and for foreign citizens and companies. This, in turn, attracts talented specialists, investments and innovations. Also important is the group of indicators reflecting success in achieving the main goal of creating a "smart city".

The *management system* of the automated management system includes subjects and objects responsible for managing the city and making management decisions. This includes authorities such as the city administration, management structures of life support facilities and others. Management decisions are made on the basis of current legislation and the regulatory framework, including the Constitution, target programs, state standards, building codes and regulations, industry standards.

Based on the analysis conducted, the structure of the system is proposed to achieve the goals set for the automated management system of the innovative "smart city" and implement the principles of its operation (Fig. 2).

The subjects and objects of the control system use various mechanisms for managing the city and its functions, which are implemented through the executive system of the automated control system [20, 21].

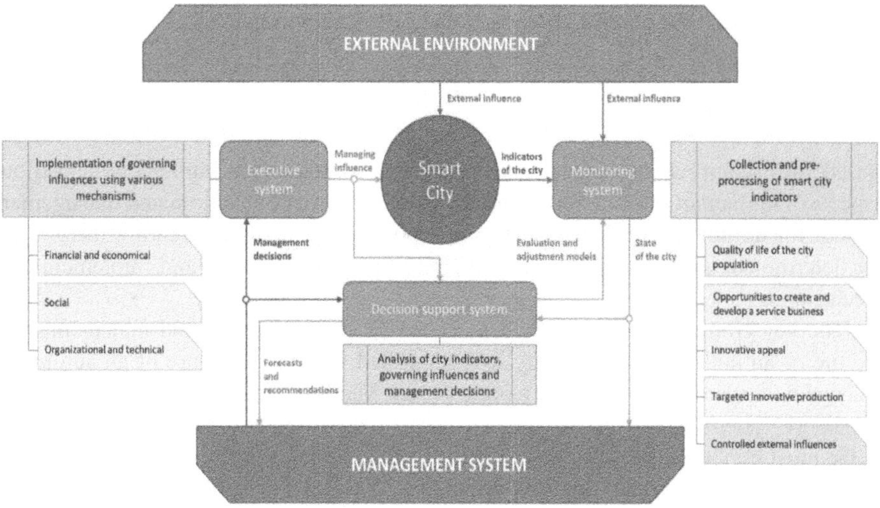

Fig. 2. Model of the management system of an innovative "smart city".

These mechanisms can be divided into several main groups:

- financial and economic mechanisms: implemented through the use of financial flows and the creation of material incentives for the implementation of decisions taken;
- social mechanisms: include education, propaganda, training, as well as retraining and advanced training of managers and specialists;
- organizational and technical mechanisms: involve the application, improvement and adaptation of modern technical solutions and administrative methods, such as technical regulations, standardization, certification, licensing and expertise;
- environmental management mechanisms: ensure environmental safety for both current and future generations of residents of the innovative "smart city".

Maximum efficiency of management decisions is achieved through competent and scientifically substantiated joint application of all these mechanisms. The executive system carries out specific management actions on the management object, including organizational and legal, architectural and planning, design and technical and operational measures. In order to adopt and implement measures that are rational from the socio-economic, production and environmental points of view, the management system must have reliable information about the current state of the management object and a sufficient number of alternative management scenarios based on forecasts in the financial, production, social and environmental spheres. These tasks are performed by specialized subsystems of the automated control system, such as the monitoring system and the intelligent decision support system. The intelligent decision support system processes data received from the monitoring system and other subsystems of the automated control system, transforming them into a form suitable for making management decisions [22–24]. It also deals with storing information and presenting it in a convenient and visual form for the control system. In this subsystem, signals are generated for other subsystems of the automated control system, which is necessary to create control adequate to current changes in the control object and the external environment.

4 Conclusions

Based on a comprehensive analysis of existing solutions for implementing various aspects of the "smart city", a new approach to the concept of an automated management system for an innovative "smart city" is proposed. The goals of creating this system are determined, as well as the basic principles of its construction and implementation mechanisms.

The key groups of indicators necessary for strategic planning and management of the life of an innovative "smart city" are identified, and the structure of an automated management system for such a city is proposed. The implementation of the Intelligent Transport System will ensure balanced synergetic development of the transport system as the most important factor in sustainable socio-economic and spatial development.

The results of the study make it possible to develop a scientifically based concept for the creation of an innovative "smart city" and an intelligent automated system for managing its functions.

In addition, the comprehensive and systematic approach presented in this paper will help to avoid inefficient costs, ensure development within a reasonable time frame and, in combination with other projects for the implementation of the "smart city", create growth points for the country's innovative industry.

References

1. Poszler, F., Lange, B.: Technological forecasting and social change the impact of intelligent decision-support systems on humans' ethical decision-making: a systematic literature review and an integrated framework. Technol. Forecast. Soc. Chang. 204 (2024). https://doi.org/10.1016/j.techfore.2024.123403
2. Nagoev, Z., Pshenokova, I., Nagoeva, O., Sundukov, Z.: Learning algorithm for an intelligent decision making system based on multi-agent neurocognitive architectures. Cogn. Syst. Res. **66**, 82–88 (2021). https://doi.org/10.1016/j.cogsys.2020.10.015
3. Muraev, Y.: Urban development based on the concept of "smart cities" in the digital economy: theoretical and methodological principles of implementation. Innov. Technol. Sci. Solut. Ind. **2**(12), 109–118 (2020). https://doi.org/10.30837/2522-9818.2020.12.109
4. Kvitka, S.: Digital transformation in the context of the concept of «Long Waves». Public Adm. Asp.. **9**, 24–28 (2021). https://doi.org/10.15421/152155
5. Ramaji, I.J., Memari, A.M.: Interpretation of structural analytical models from the coordination view in building information models. Autom. Constr. **90**, 117–133 (2018). https://doi.org/10.1016/j.autcon.2018.02.025
6. Gatawa, G.: The effect of social factors to economic growth. Advance (2022). https://doi.org/10.31124/advance.19397081.v1
7. Cézanne, C., Lorenz, E., Saglietto, L.: Exploring the economic and social impacts of industry 4.0. Ind. 4: Curr. Issues Futur. Chall. 11–35 (2020). https://doi.org/10.4000/rei.8643
8. Yufang, H., et al.: The city management based on smart information system using digital technologies in China. IET. **4**(3), 160–174 (2022). https://doi.org/10.1049/smc2.12035
9. Pashchenko, A.: Smart management for smart cities - synchronized solutions. IFAC-PapersOnLine **54**(13), 732–737 (2021). https://doi.org/10.1016/j.ifacol.2021.10.539
10. Tnay, G.L. Bryan, C.L., Keng, S.W.: A smart engineering system toward machine shop 4.0. Compr. Mater. Process. **10**, 219–230 (2024). https://doi.org/10.1016/B978-0-323-96020-5.00266-1
11. Solvak, M., Unt, T., Rozgonjuk, D., Võrk, A., Veskimäe, M., Vassil. K.: E-governance diffusion: Population level e-service adoption rates and usage patterns. Telemat. Inform. **36**, 39–54 (2019). https://doi.org/10.1016/j.tele.2018.11.005
12. Yevtushenko, N., Tverdokhliebova, N., Ponomarenko, O.: Using artificial intelligence technologies to predict and identify the educational process. In: AIxEDU: 1st International Workshop on High-performance Artificial Intelligence Systems in Education (2023). https://ceur-ws.org/Vol-3605/3.pdf
13. Tverdokhliebova, N., Yevtushenko, N.: Research of social and psychological characteristics of large cities. Social and psychological support in a metropolis: a collective monograph 26–43 (2024). https://science.kname.edu.ua/images/dok/novosti/2024/Monografiia_03_02_24.pdf
14. Ekanayake, U., Daundasekera, W., Perera, S. P. C., Juman, Z. A. M. S.: An effective alternative new approach in solving transportation problems. Am. J. Electr. Comput. Eng. **5**(1) (2021). https://doi.org/10.11648/j.ajece.20210501.11
15. Pérez-González, C.J., Colebrook, M., Roda-García, J.L., Rosa-Remedios, C.B.: Developing a data analytics platform to support decision making in emergency and security management. Expert Syst. Appl. **120**, 167–184 (2019). https://doi.org/10.1016/j.eswa.2018.11.023

16. Yibo, D., Yang, B., Yiping, W., Xiaohua, Zh., Jianhua, Zh.: A variable speed limit model for freeway bridges considering the influence of both traffic service level and visibility in fog. Phys. A: Stat. Mech. Its Appl. **602** (2022). https://doi.org/10.1016/j.physa.2022.127590
17. Çavdar, A., Ferhatosmanoğlu, N.: Airline customer lifetime value estimation using data analytics supported by social network information. J. Air Transp. Manag. **67**, 19–33 (2018). https://doi.org/10.1016/j.jairtraman.2017.10.007
18. Osman, A.M.S.: A novel big data analytics framework for smart cities. Futur. Gener. Comput. Syst. **91**, 620–633 (2019). https://doi.org/10.1016/j.future.2018.06.046
19. Ponomarenko, O: Modeling and optimization of properties of the environmentally clean molds based on oligofurfuryloxysiloxanes for the production the metal castings. Polymers **14** (2022). https://doi.org/10.3390/polym14091883
20. Petrić, L., Hell, M., Jan van der Borg.: Process orientation of the world heritage city management system. J. Cult. Herit. **46**, 259–267 (2020). https://doi.org/10.1016/j.culher.2020.07.009
21. Turayev, K.S.: Intelligent decision support system. Chem. Technol. Control. Manag. **6** (2023). https://doi.org/10.59048/2181-1105.1528
22. Rahman, Md.Z.U., Navarun, G., Korhan, C., Aime, L.-E.: Intelligent decision support systems, models, and applications. Decis. Anal. J. (2023). https://www.sciencedirect.com/special-issue/10V94H4RH9Q
23. Mahdi, Q., Shyshatskyi, A., Prokopenko, Y., Ivakhnenko, T., Kupriyenko, D., Golian, V., Lazuta, R., Kravchenko, S., Protas, N., Momit, A.: Development of estimation and forecasting method in intelligent decision support systems. East.-Eur. J. Enterp. Technol. **3**(9(111), 51–62 (2021). https://doi.org/10.15587/1729-4061.2021.232718
24. Zuiev, P., Zhyvotovskyi, R., Zvieriev, O., Hatsenko, S., Kuprii, V., Nakonechnyi, O., et. al.: Development of complex methodology of processing heterogeneous data in intelligent decision support systems. East.-Eur. J. Enterp. Technol. **4**(9 (106)), 14–23 (2020). https://doi.org/10.15587/1729-4061.2020.208554

Decarbonizing Strategy of Ukrainian Transport Sector

Valerii Samsonkin[1], Oksana Yurchenko[1](✉), Olena Sorochynska[1], Oleksandr Rohovyi[1], and Gintautas Bureika[2]

[1] State University of Infrastructure and Technologies, Kyrylivska St., 9, Kyiv 04071, Ukraine
yurchenko_og@gsuite.duit.edu.ua

[2] Vilnius Gediminas Technical University, Plytinės Str., 25, 10105 Vilnius, Lithuania

Abstract. The transport sector is one of the largest contributors to pollution and green-house gas emissions, significantly impacting the environmental situation and public health in Ukrainian cities. The current standards of transport safety, energy consumption, and environmental impact fall short of meeting modern requirements. This article is focused on analyzing the policies and legislation of the European Union regarding the implementation of the European Green Deal within the transport sector, as well as exploring proposals for the decarbonization of various modes of transport in Ukraine, the use of alternative low-carbon fuels, and recommendations for Ukraine's integration into the European Green Deal framework. An in-depth analysis was conducted to assess the alignment of Ukrainian state policy with the principles of the European Green Deal, identifying both the potential risks and benefits for Ukraine's transport sector in the context of these principles. Additionally, the Authors propose a range of measures aimed at fostering sustainable mobility in Ukraine. The primary objective of this study is to develop a comprehensive vision for the strategy required to decarbonize Ukraine's transport sector, with a particular emphasis on the railway system.

Keywords: Decarbonization · railway transport · sustainable development · GHG · transport safety · environmental impact · sustainable mobility · European Green Deal

1 Climate Policy

1.1 EU Policy on the Implementation of the "Green Deal"

Global warming is widely acknowledged as one of humanity's major challenges and has remained a relevant issue for over 40 years [1]. During this time, climate policy has undergone several stages of evolution, starting from the scientific recognition of the problem, moving into the political arena, creating an institutional platform, implementing the Kyoto Protocol, and signing the Paris Agreement. In response to the global fight against climate change and the significant role of the transport sector in greenhouse gas (GHG) emissions, Ukraine is also working on adapting its transport policy to the principles of the European Green Deal.

O. Slavinska et al. (Eds.): ITSESQC 2024, LNNS 1335, pp. 109–119, 2025.
https://doi.org/10.1007/978-3-031-87376-8_10

The harmonization of Ukraine's transport policy with European standards will contribute to reducing emissions, integrating into European transport and energy systems, enhancing competitiveness in the international market, and ensuring sustainable economic growth. However, Ukraine faces several challenges on the path to decarbonizing its transport sector, including dependence on fossil fuels, the need to modernize infrastructure, and limited financial resources for achieving these goals.

Today, humanity is implementing a complex set of measures aimed at limiting the pace of global warming. This process is known as decarbonization. The majority of the world's countries (74 out of 190 signatories) of the Paris Agreement [2] have set their climate goals and commitments to reduce GHG emissions at the international level. Alongside this, more than 30 nations have publicly declared their goal of achieving carbon neutrality, meaning the balancing of emissions and absorption. Alliances such as the 'Climate Ambition Alliance: Net Zero 2050' [3] and 'Race to Zero' [4] have also been established, bringing together over 1,000 corporations, investors, and universities from 120 countries worldwide.

However, despite numerous efforts, the average global temperature in 2020 was 1.2 °C higher than pre-industrial levels [5]. Analysis of open data [6] indicates that even if all commitments under the Paris Agreement are fulfilled, the average temperature by the end of the century could rise by 3 °C. The primary sources of emissions (70%) are concentrated in the energy sector, particularly in the extraction and consumption of fossil fuels [7].

The European Green Deal (hereinafter referred to as the EGD) is the European Union's action plan to achieve a climate-neutral Europe by 2050. This initiative emerged in response to the demands of EU citizens, who view climate change as a serious issue and believe the EU should make greater efforts to address it. The legal and political foundation of the EGD is based on the European Green Deal, the EU Hydrogen Strategy, and the EU Strategy on Adaptation to Climate Change.

The legal and political foundation of the decarbonization process in the Europe-an Union's transport sector is shaped by the following documents:

I. The European Green Deal.
II. The European Green Deal Communication.
III. The EU Hydrogen Strategy.
IV. The EU Strategy on Adaptation to Climate Change.

Since key documents such as the European Green Deal and the Hydrogen Strategy have already been explored in numerous academic studies, this research focuses on sectoral acts and documents that regulate transport [8–14].

For Ukraine it is important to adapt these documents to its own transport policy, which will allow the creation of an environmentally safe and energy-efficient transport sector that is compatible with European standards. Thus, harmonization with the European Green Deal opens up prospects for Ukraine's sustainable development in the transport industry, which includes the implementation of clean technologies, the enhancement of energy efficiency, and integration into the European mobility system.

Thus, the European Union's declared course toward climate neutrality has been officially endorsed by documents such as the European Green Deal, the EU Hydrogen Strategy, and the EU Strategy on Adaptation to Climate Change. Based on these documents, sectoral policies are being developed, particularly in the transport sector, represented by Communications, White Papers, and Directives, which continue to evolve dynamically during the implementation of roadmaps and action plans.

The research focuses on analyzing the EU's sectoral transport policies and their harmonization with Ukraine's transport policy to ensure sustainable development of the transport sector.

The study was prepared as part of the Horizon Europe project "PhDs EU-Rail: Extending the rail network of PhDs in Europe's Rail Joint Undertaking" (HORIZON-ER-JU-2023-EXPLR-05). The vision of the European Rail Joint Undertaking is to create a highly efficient, adaptive, multimodal, sustainable, and integrated railway network in Europe.

1.2 Strategy for Decarbonization in the Transport Sector

The European Green Deal is the European Union's growth strategy aimed at achieving climate neutrality by 2050. The primary objective of this initiative is to establish a fair and prosperous society in which the economy remains competitive, utilizes resources efficiently, and achieves net-zero GHG emissions.

As part of the European Green Deal program, EU member states must achieve ambitious yet realistic decarbonization targets for the transport sector by 2030, 2040, and 2050. The shift to railway transport, both freight and passenger, is a key element of this strategy, particularly for cross-border journeys of up to 1,000 km. To promote sustainable transport, it is essential to implement the principles of "polluter pays" and "user pays," which will help incorporate social costs into the price of transportation.

Transport accounts for a quarter of GHG emissions in the EU, and these emissions continue to rise. To achieve climate neutrality by 2050, transport emissions must be reduced by 90%, and all modes of transport must contribute to this goal. Building sustainable transport means prioritizing the interests of users and providing more accessible, healthier, and cleaner alternatives to their current mobility options.

The free movement of people and goods across internal borders is one of the key freedoms of the European Union and the foundation of its single market. This freedom strengthens European identity and cohesion. The transport sector, which is the second-largest household expenditure in the EU, accounts for about 5% of the EU's gross domestic product and employs around 10 million people [15].

While mobility offers many benefits to users, it also has negative impacts on society and the environment. Major issues include GHG emissions, water and air pollution, increased noise levels, road accidents, traffic congestion, and biodiversity loss. All these factors directly affect people's health and well-being. Previous policy measures have not fully offset these negative effects. GHG emissions in the transport sector have increased over time, and they now account for a quarter of total emissions in the EU.

The biggest challenge facing the transport sector today is the need for a significant reduction in emissions and the achievement of sustainable development. At the same time, this process opens up new opportunities to improve citizens' quality of life, create value chains for European industry, and generate new, high-quality jobs. It also promotes the development of innovative products and services, enhance the competitiveness of the European economy, and strengthen global leadership in the transition to zero emissions.

GHG emissions in the EU transport sector have increased over time and now account for about a quarter of total emissions in the European Union. In response, the EU aims to reduce emissions by at least 55% by 2030 and achieve climate neutrality by 2050. Reaching these targets requires ambitious policies, including reducing the transport sector's dependence on fossil fuels and implementing active measures to achieve zero emissions.

Environmental mobility must become the foundation for the development of the transport sector in Europe. To achieve this, it is necessary to create a multi-modal transport system that ensures the efficient movement of both passengers and goods. The main directions include:

Expanding the high-speed railway network for safe and fast transportation at both domestic and international levels. As part of the Trans-European Transport Network (TEN-T) project, the EU is actively investing in the modernization of railway lines, significantly reducing travel time between European countries and enhancing the environmental sustainability of transport.

Developing infrastructure for zero-emission vehicles, including charging stations and the use of renewable and low-carbon fuels. Specifically, the EU plans to install 1 million charging stations for electric vehicles by 2025 as part of the TEN-T project, which helps to reduce emissions and accelerate the transition to environmentally friendly transport [16].

Creating conditions for cleaner and more active mobility in cities, with a focus on pedestrian movement, cycling, and efficient public transport. These measures help to reduce the environmental impact of transport, improve public health, and foster the development of sustainable cities.

Digitization becomes a key factor in modernizing the entire transport system, making it more stable and efficient. Europe must actively implement digital and automated technologies to further enhance safety, reliability, and comfort. This also helps the EU maintain its leadership in the production and servicing of transport equipment and strengthen its global competitiveness through the implementation of efficient and sustainable logistics solutions.

Energy efficiency must become a priority in the transport industry, requiring a significant reduction in dependence on fossil fuels. This involves replacing the existing vehicle fleet with environmentally friendly models that have low or zero emissions. At the same time, there should be an emphasis on increasing the use of renewable and low-carbon energy sources for all types of transport. Infrastructure changes, such as expanding the network of charging stations for electric vehicles and introducing hydrogen technologies, are key steps toward achieving this goal. Only a comprehensive approach yields significant results in reducing reliance on fossil fuels.

One example of successful implementation of the hydrogen strategy in Europe is Alstom's Coradia iLint—the world's first passenger train powered by hydrogen fuel cells, which is already operating in Germany. This project represents a major step toward achieving climate neutrality in the transport sector, as hydrogen-powered trains can reduce carbon dioxide emissions to zero [17].

To achieve systemic changes in the transport sector, the following tasks need to be implemented:

A. Increase the environmental sustainability of all modes of transport.
B. Ensure broad access to environmentally friendly alternatives in multi-modal transport networks.
C. Introduce legislative mechanisms that encourage the transition to sustainable transport.
D. The main directions of transport policy should focus on the following priori-ties:
E. EU policy is aimed at reducing dependence on fossil fuels and implementing energy-efficient transport solutions.
F. Active measures to transition to more environmentally sustainable transport, including increasing the number of passengers using railway and public transport, as well as redirecting freight to railways, waterways, and short-sea shipping.
G. Internalizing external costs through the implementation of the "polluter pays" and "user pays" principles, particularly through carbon pricing mechanisms.

All modes of transport play a vital role in the transport system, and each must be adapted to more environmentally sustainable standards. A key priority is in-creasing the use of low- and zero-emission vehicles, as well as the adoption of renewable and low-carbon fuels in road, water, air, and railway transport. Additionally, the necessary infrastructure for these vehicles must be developed, and de-mand among end users should be promoted.

A key component is supporting research and innovation aimed at developing competitive, sustainable, and "circular" [18] products and services, which helps to ensure the supply of necessary vehicles and fuels to the market.

Railway transport is the most environmentally friendly form of land transport, thanks to its high level of electrification and increasing use of zero-carbon technologies. This allows for a significant reduction in emissions within the transport sector.

To transform the transport industry into a multimodal system of sustainable and smart mobility services, decisive actions must be taken. These include developing a high-quality transport network with high-speed railway connections over short distances and introducing eco-friendly aviation services for long-haul routes.

1. The decarbonization of railway transport presents the following opportunities:
2. Reduction of emissions, leading to improved air quality.
3. Mitigation of climate change through the reduction of GHGs.
4. Transitioning to sustainable transport solutions improves transport capabilities and connections between regions.
5. Modernization of transport infrastructure decreases costs and increases the efficiency of transportation.
6. Social benefits: significantly lower external costs of railway transport com-pared to road/air transport. Railways can reduce external costs (such as air pollution, climate change, and accidents) by 87% for freight transport and by 81% for passenger transport (see Fig. 1).

Strategies to Decarbonize the Railway Sector in Ukraine:

1. Analyzing the current state of electrification is a key step towards decarbonizing Ukraine's railway sector. A comprehensive assessment of the existing railway infrastructure is needed, including determining the level of railway electrification and

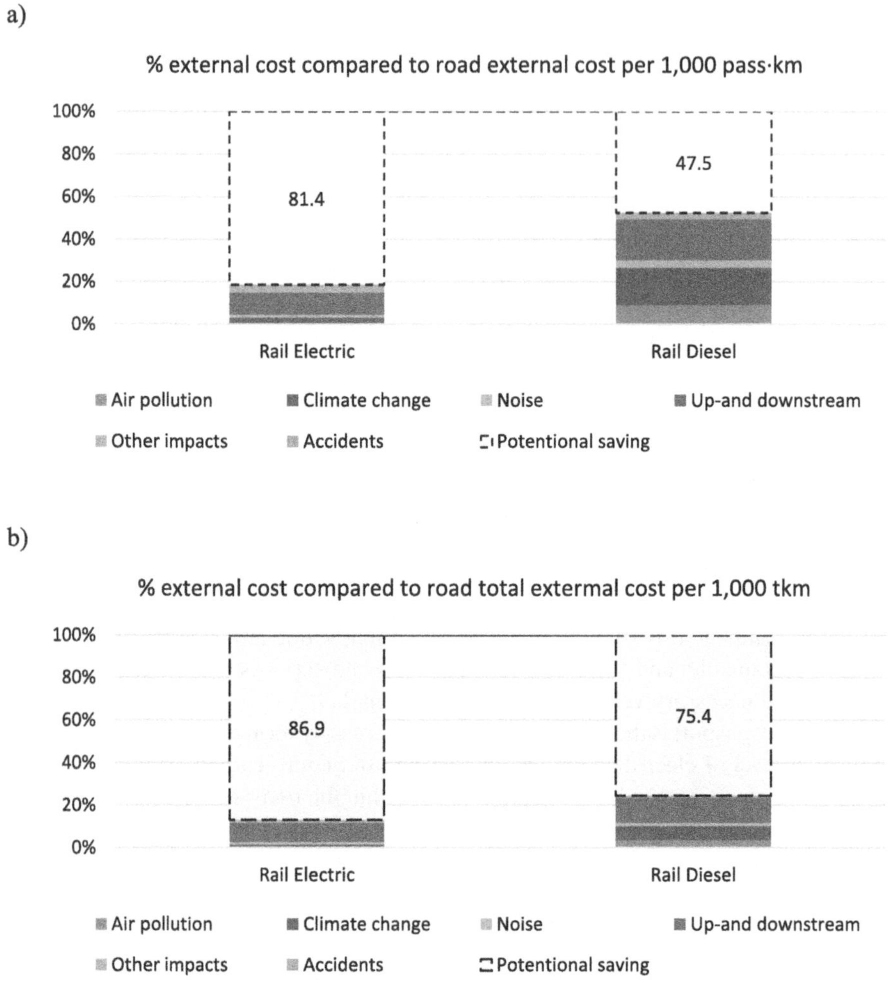

Fig. 1. External costs in freight and passenger transportation, % [18].

the share of diesel locomotives in overall transportation. Since diesel locomotives contribute to high levels of GHG emissions, it's crucial to identify sections where their use is most prevalent and where energy consumption is highest. These energy-intensive areas require modernization, either through a transition to electrified tracks or by adopting alternative technologies, such as hydrogen locomotives. Additionally, it's important to assess the technical capabilities of existing electrified sections and explore options for upgrading them to improve energy efficiency.

2. Evaluating technological possibilities involves a detailed examination of the introduction of new environmentally friendly technologies for railway transport, particularly hydrogen locomotives. Hydrogen technologies have the potential to significantly reduce harmful emissions, as hydrogen locomotives operate using fuel cells that do not produce carbon dioxide during operation, emitting only water and heat.

In Europe, hydrogen trains like the Alstom Coradia iLint have already proven their efficiency in real-world conditions. These trains, used on non-electrified sections, provide an eco-friendly alternative to diesel locomotives. Their implementation in Ukraine could be an important step towards achieving decarbonization goals, especially in regions where electrification is difficult or economically unfeasible. A thorough analysis of technological options, including the adoption of such innovations, helps to determine optimal paths for reducing emissions in the railway sector.

3. Setting national and regional goals is a crucial step towards decarbonizing railway transport. Clear short-term and mid-term goals must be established to gradually reduce GHG emissions in the transport sector. These goals should be aligned with national and international climate commitments, particularly within the framework of the Paris Agreement. Special attention should be given to regions with high freight transportation intensity, as these areas require specific approaches to decarbonization due to their significant impact on total emissions. For such regions, tailored strategies can be developed, taking into account local conditions, existing infrastructure, and technological capabilities.
4. Collaboration between the public and the private sector is critically important for the effective decarbonization of the rail industry. Developing joint action plans between public and private railway companies helps to coordinate efforts aimed at infrastructure modernization, the introduction of new technologies, and improving energy efficiency. The Government can provide regulatory and legal support, including the creation of incentivizing financial conditions such as tax breaks and subsidies for private companies investing in decarbonization projects. At the same time, private railway companies can play a key role in implementing innovative technologies and modernizing rolling stock, contributing to the achievement of national and international climate goals.
5. Electrification of railway infrastructure remains one of the key directions for decarbonization, as it significantly reduces GHG emissions com-pared to the use of diesel locomotives. However, in regions where full electrification is economically unfeasible, the use of hydrogen locomotives becomes a viable alternative. Hydrogen trains, such as the Alstom Coradia iLint, have al-ready demonstrated their effectiveness in Europe. Their implementation in Ukraine could substantially reduce emissions on non-electrified sections. For non-electrified routes, hydrogen locomotives offer an efficient alternative to diesel. The EU's experience, particularly the deployment of hydrogen trains like the Alstom Coradia iLint, which are successfully operating in Europe, can serve as a model for Ukraine. The adoption of such technologies would significantly reduce GHG emissions in regions where electrification is not economically viable.
6. Multimodal transportation is becoming one of the key areas in reducing GHG emissions in the transport sector through the effective combination of various modes of transport, such as railway and water transport. The implementation of multimodal transportation allows for the optimization of logistics chains, reduces road congestion, and shortens the distances that freight travels by road—one of the largest sources of emissions. For example, cargo can be delivered by railway to a river or sea port, where transportation continues via water transport. This approach not only

reduces harmful emissions but also lowers transportation costs due to the economic efficiency of water transport.

The adoption of multimodal transportation is also an important aspect of developing international transport corridors, which facilitates Ukraine integration into European and global logistics networks. Moreover, this solution allows for the use of cleaner modes of transport to cover long distances, minimizing dependence on road transport, especially for short distances. Additional electrification of railway infrastructure: Developing a plan for the electrification of railway lines where feasible, taking into account economic and technical limitations.

7. Supporting research in hydrogen and other clean technologies opens new opportunities for decarbonizing the railway sector. Investments in innovation allow the creation of environmentally safe solutions and efficient transportation vehicles, contributing to emission reductions and increased energy efficiency in transport.
8. Seeking external sources of financing: Engaging international financial organizations, such as the European Investment Bank and the World Bank, is an important step to securing resources for implementing railway decarbonization projects. International financial institutions can offer favorable loans, grants, or investment programs aimed at developing sustainable transport and reducing GHG emissions. Additionally, the involvement of international organizations often comes with technical assistance and consulting services, helping to develop effective strategies for achieving decarbonization at various stages of project implementation.
9. Stimulating investment in the decarbonization of the railway sector is an essential component of sustainable development. Government programs can include tax incentives for companies that invest in the modernization of railway infrastructure and the adoption of environmentally friendly technologies. These incentives could take the form of tax credits, reduced corporate tax rates, or reimbursements for expenses related to research and the development of new technologies. Additionally, subsidies can be provided to support projects focused on railway electrification, the introduction of hydrogen locomotives, or the development of hydrogen refueling infrastructure.

   Financial instruments such as "green" bonds or specialized funds also play a crucial role in supporting innovation in the transport sector. For example, in the European Union, programs like the Sustainable Development Fund provide financing for projects aimed at decarbonizing transportation. Collaboration between the public and the private sector through tax incentives and subsidies encourages more active participation from companies in the transition to sustainable technologies.
10. Training in new technologies and processes for all participants in the transport chain is crucial. Railway workers must acquire specialized knowledge on the implementation of environmentally friendly technologies, such as hydrogen locomotives, multimodal transportation, energy efficiency systems, and electrification. Additionally, management training is essential to ensure the effective use of new technologies and the reduction of emissions.

    Moreover, educational programs for students in technical universities and colleges must be developed to prepare specialists with competencies in transport decarbonization from the outset. These programs could include specialized courses on hydrogen technologies, alternative energy sources, sustainable transport management, and the integration of cutting-edge technologies into the railway sector. Collaboration between educational institutions and railway companies enables the creation of practical training programs, including intern-ships and hands-on experience with modern technologies.

11. Engaging the public in discussions about the environmental and economic benefits of these changes is essential. Information campaigns should aim to raise public awareness about the importance of reducing GHG emissions and the ecological advantages of decarbonization. Building public support is crucial, as it frequently plays a pivotal role in decision-making at both regional and national levels.

Information campaigns can include educational programs designed to organize regular public lectures, seminars, and training sessions, where experts explain how the decarbonization of the railway sector help to reduce environmental impact and improve quality of life. Special attention should be given to educational activities in schools and higher education institutions, where young people can gain foundational knowledge about sustainable transportation futures.

The campaigns should also leverage modern digital platforms, which are predominantly used by younger demographics. In addition to YouTube, it is advisable to utilize platforms such as TikTok, Instagram, and Twitter (X), which allow rapid content dissemination among young people and encourage discussions on important environmental issues. These platforms provide interactivity and the potential for viral content, making them effective tools for raising awareness about decarbonization.

At the same time, traditional media such as television and radio, while still maintaining an audience, are gradually losing relevance compared to new forms of communication that attract wider and younger audiences through interactivity and direct engagement. Thematic reports, expert interviews, and special programs can inform the general public about how reducing GHG emissions im-proves health and the environment.

It is also important to support public initiatives that create opportunities for open dialogue between public bodies, railway companies, and the public through discussions on forums or at public meetings. Another critical element is establishing feedback platforms that enable the public to participate in the decision-making process through online platforms or social networks, where individuals can express their opinions and offer suggestions for improving the effectiveness of decarbonization measures.

1.3 Results and Discussion

Achieving climate neutrality in the transport sector is a complex task that requires coordinated action at the international, national and regional levels. Despite numerous measures, the average temperature in 2020 was 1.2 °C higher than pre-industrial levels, and may rise by 3 °C by the end of the century. And this is even provided that all obligations under the Paris Agreement are fulfilled.

An effective transition to sustainable transport needs to take into account the interests of all stakeholders, including public, business, academia and civil society.

Considering the situation in Ukraine, the harmonization of transport policy with European standards can occur through cooperation with European institutions, participation in international programs and projects, as well as through the implementation of legislative mechanisms that stimulate the transition to sustainable transport.

Therefore, the course announced by the European Union in the direction of climate neutrality is an important step in the global fight against climate change. To achieve these goals, it is necessary to continue the development and implementation of effective strategies aimed at the decarbonization of the transport industry, and to ensure their implementation in practice.

2 Conclusions

The research findings indicate that Ukraine's transport sector, particularly the rail industry, necessitates substantial decarbonization efforts to align with modern environmental requirements and standards. Within the framework of the European Green Deal, Ukraine has the opportunity to adopt cutting-edge technologies, such as hydrogen locomotives, multimodal transportation, and the electrification of railway lines, which significantly reduces GHG emissions and promote sustainable development.

The proposed decarbonization measures are informed by the EU's experience and global best practices. These measures encompass an analysis of the current railway electrification status, an evaluation of hydrogen train technologies, and the development of tailored strategies for regions with high freight traffic intensity. Collaboration between public and private sectors, alongside attracting international funding, is essential for the successful implementation of these initiatives.

A key component is the creation of educational programs to train specialists in transport decarbonization, as well as information campaigns to enhance public awareness of the importance of environmental initiatives.

The implementation of these proposed measures facilitate a progressive transition of Ukraine's transport sector towards a more environmentally friendly and energy-efficient format, aligning with international standards and enhancing competitiveness in the global market.

Acknowledgment. Authors express the sincere gratitude to the European Rail Joint Undertaking (EU-Rail JU) for their support and the opportunity to participate in important research aimed at developing Europe's railway infrastructure.

References

1. Resolution No. 45/212, UN General Assembly (1990). https://undocs.org/ru/A/RES/45/212
2. Politics vs. Climate Change. How Countries Agreed to Save the World and Signed the Paris Agreement. Ecodia. https://ecoaction.org.ua/polityka-vs-zmina-klimatu.html
3. Become a Volunteer. Climate Action Africa. https://climateaction.africa/become-a-volunteer/?gad_source=1&gclid=EAIaIQobChMI9bqMmO-oiAMVy5doCR2ixy7QEAAYASAAEgLGoPD_BwE

4. United Nations Framework Convention on Climate Change. Race to Zero. https://climatechampions.unfccc.int/
5. World Meteorological Organization. https://public.wmo.int/en
6. Paris Agreement turning point. Climate Action Tracker (2020). https://climateactiontracker.org/
7. International Energy Agency and International Institute for Applied Systems Analysis. https://iiasa.ac.at/news/nov-2023/iiasa-analysis-underpins-iea-world-energy-outlook-revealing-rapid-energy-transition
8. Communication from the Commission to the European Parliament, the Council, the European Economic and Social Committee and the Committee of the Regions. Sustainable and Smart Mobility Strategy—putting European transport on track for the future. European Commission (2020). https://eur-lex.europa.eu/legal-content/EN/TXT/?uri=CELEX%3A52020DC0789
9. Clean Power for Transport: A European alternative fuels strategy. European Commission (2013)
10. Directive 2009/33/EC on the promotion of clean and energy-efficient road transport vehicles, updated by Directive (EU) 2019/1161 of the European Parliament and of the Council. Official Journal of the European Union (2019)
11. Directive 2011/76/EU, amending Directive 1999/62/EC on the charging of heavy goods vehicles for the use of certain infrastructures (the "Eurovignette Directive"). European Union (2011)
12. Directive 92/106/EEC, establishing common rules for certain types of combined transport of goods between member states. Council of the European Union (1992)
13. Directive 2018/2001 on the promotion of the use of energy from renewable sources (RED II), European Parliament and Council (2018). https://zakon.rada.gov.ua/laws/show/984_039-18#Text
14. Directive 2014/94/EU on the deployment of alternative fuels infrastructure. European Parliament and Council. (2014). https://zakon.rada.gov.ua/laws/show/984_044-14#Text
15. Communication from the Commission to the European Parliament, the Council, the European Economic and Social Committee and the Committee of the Regions. Sustainable and Smart Mobility Strategy—putting European transport on track for the future. European Commission (2020). https://eur-lex.europa.eu/legal-content/EN/TXT/HTML/?uri=CELEX:52020DC0789
16. Europe on the Move: Commission takes action for clean, competitive and connected mobility. European Commission (2020). https://ec.europa.eu/commission/presscorner/detail/en/IP_20_194
17. Alstom Homepage, "Coradia iLint - World's first hydrogen-powered train.". https://www.alstom.com/our-solutions/rolling-stock/coradia-ilint-worlds-1st-hydrogen-powered-train
18. Directive (EU) 2019/1161 amending Directive 2009/33/EC on the promotion of clean and energy-efficient road transport vehicles. European Parliament and Council (2019). https://eur-lex.europa.eu/eli/dir/2019/1161/oj

Integration of Intelligent Transportation Systems in the Context of Sustainable Development: Challenges and Opportunities for Smart Cities

Maxim Tochygin(✉)

National Transport University, Mykhaila Omelyanovych-Pavlenka St. 1, Kyiv, Ukraine
tochyginph.d@gmail.com

Abstract. This article explores the integration of Intelligent Transportation Systems (ITS) within the framework of sustainable development in smart cities. The study aims to identify the challenges and opportunities presented by the adoption of ITS in urban environments. Emphasizing the importance of innovative transportation solutions, this research contributes to the current discourse by highlighting the role of ITS in enhancing urban mobility, reducing environmental impacts, and promoting energy efficiency. The novelty of this work lies in its comprehensive analysis of the interaction between ITS and sustainable urban development, providing insights into policy implications and technological advancements. The findings reveal that while ITS offers significant potential for improving transportation efficiency and sustainability, it also faces challenges such as data privacy, infrastructure costs, and the need for interagency collaboration. The practical value of this research is demonstrated through case studies and best practices, offering a roadmap for cities aiming to implement ITS in a way that aligns with sustainability goals. Future research directions are suggested to address the identified challenges and to further optimize ITS integration in smart cities.

Keywords: Urban mobility · Smart cities · Sustainable development · Transportation systems · Policy implications · Environmental impact

1 Introduction

The rapid pace of urbanization in the 21st century has imposed significant stress on transportation systems across cities worldwide. With more than 55% of the global population currently living in urban areas and this number expected to rise to 68% by 2050 [17], the challenge of ensuring efficient, sustainable, and safe urban mobility has become a priority for city planners and policymakers. As a result, many cities are exploring the integration of Intelligent Transportation Systems (ITS) to optimize urban transportation infrastructure and meet the growing demand for mobility in a sustainable way.

Intelligent Transportation Systems refer to the application of advanced information and communication technologies (ICTs) to improve the efficiency and safety of

O. Slavinska et al. (Eds.): ITSESQC 2024, LNNS 1335, pp. 120–131, 2025.
https://doi.org/10.1007/978-3-031-87376-8_11

transportation networks. These systems encompass a wide range of technologies, such as real-time traffic monitoring, automated toll collection, adaptive traffic signals, and vehicle-to-infrastructure communication, which together contribute to more efficient traffic management, reduced congestion, and minimized environmental impact.

1.1 Problem Definition

Despite the potential benefits of ITS, their successful integration into existing urban infrastructure presents several challenges. Cities face issues such as outdated transportation systems, high implementation costs, and fragmented policy frameworks that hinder the adoption of advanced technologies. While ITS has been shown in numerous studies, projects, and reports to improve traffic flow and reduce emissions, this research aims to further explore the long-term impact of ITS on various types of pollutants, especially in the context of developing countries where resources and technical expertise may be limited. The focus includes both global pollutants like CO_2 and local pollutants such as CO, NOx, SOx, and PM, which are critical in the urban environment.

1.2 Research Gaps and Relevance

While the literature on ITS and smart city development is vast, several key gaps remain. First, much of the existing research focuses on the technological aspects of ITS, such as traffic management systems or vehicle automation, without addressing the broader policy and governance frameworks required for their successful implementation. Studies such as those by Benevolo et al. (2021) and Galatioto et al. (2022) have highlighted the potential of smart mobility solutions, but they do not provide detailed insights into how these systems can be scaled and adapted to different urban contexts, especially in developing regions [2, 5].

Moreover, although there is growing recognition of the need for sustainable mobility solutions, few studies have examined the integration of ITS specifically within the framework of sustainable development. Research by Campos and Yigitcanlar (2020) has emphasized the importance of climate change mitigation in urban planning, yet the role of ITS in reducing carbon emissions and promoting environmental sustainability remains underexplored [3]. This paper seeks to address these gaps by providing a comprehensive analysis of the challenges and opportunities associated with ITS in smart cities, with a particular focus on their contribution to sustainable development.

1.3 Objectives of the Study

The primary objective of this study is to investigate the role of ITS in supporting sustainable urban development, with a focus on the following areas:

1. Assessing the impact of ITS on urban mobility: How can ITS enhance the efficiency of urban transportation systems, reduce traffic congestion, and improve the overall quality of life for city residents?
2. Analyzing the environmental benefits of ITS: How do ITS contribute to reducing greenhouse gas emissions and minimizing the environmental footprint of urban transportation?

3. Identifying the challenges in ITS implementation: What are the key barriers to the successful integration of ITS in urban environments, and how can these challenges be addressed through policy and technological innovation?
4. Exploring the potential for ITS in developing regions: Given the resource and infrastructure constraints in many developing countries, how can ITS be adapted to suit the unique needs of these regions, and what role can they play in achieving sustainable urban growth?

The scope of this paper covers a detailed examination of recent advances in ITS technology, a review of current policy frameworks, and case studies from cities that have successfully implemented ITS solutions. The aim is to provide practical insights for policymakers, urban planners, and transportation engineers who are working towards the development of smart cities that are not only technologically advanced but also environmentally sustainable and socially inclusive.

In this context, the research offers a forward-looking perspective on the integration of ITS within the broader framework of smart city initiatives, focusing on the alignment of transportation systems with the goals of sustainable development. By bridging the gap between technology and policy, this study seeks to contribute to the ongoing discourse on how cities can leverage innovation to create more livable, efficient, and resilient urban environments.

This overview sets the stage for a detailed exploration of ITS in subsequent sections, where the methodology and results will build upon the foundation established here.

1.4 Literature Review

A comprehensive review of recent literature demonstrates the increasing interest in Intelligent Transportation Systems (ITS) as a key solution to urban transportation challenges. Numerous studies have highlighted the potential of ITS to enhance traffic management and improve environmental outcomes. For example, research by Tian and Zhang (2020) and Zhao and Hu (2021) emphasizes ITS's role in optimizing traffic flow and reducing congestion, while also addressing environmental concerns[14, 15]. However, much of the existing research tends to overlook the practical challenges of implementing ITS in diverse urban contexts, especially within complex smart city ecosystems.

Several studies, including works by Figueiredo et al. (2019), Benevolo et al. (2021), and Galatioto et al. (2022), provide valuable insights into the impacts of ITS on transportation systems [2, 4, 5]. These studies explore a wide range of factors, such as traffic efficiency, safety improvements, and environmental benefits. Despite these contributions, gaps remain in understanding the broader implications of ITS for urban sustainability, particularly in relation to scalability and adaptability in different regions. There is also a need for further exploration of ITS's role in reducing CO_2 emissions, particularly in the context of climate change mitigation strategies as emphasized by Campos and Yigitcanlar (2020) [3].

Thus, while the literature on ITS is extensive, the integration of ITS within the framework of sustainable development and its long-term impacts on urban resilience require further in-depth analysis. This paper seeks to build upon the existing body of knowledge by addressing these gaps and offering a more nuanced perspective on how ITS can be effectively integrated into smart cities.

1.5 Research Gaps

Despite the substantial body of work on ITS, there remain several critical gaps in the literature. Many studies do not fully explore the integration challenges of ITS within existing urban infrastructures or the long-term sustainability impacts of these technologies. This research addresses these gaps by offering a detailed analysis of the practical challenges and benefits associated with ITS, providing a more nuanced understanding of their role in sustainable urban development.

2 Research Methodology

2.1 Research Approach

This study uses a mixed-methods approach that combines quantitative and qualitative approaches to comprehensively analyze the integration of intelligent transportation systems (ITS) in smart cities. The quantitative analysis focuses on statistical data on ITS performance, while the qualitative analysis covers the assessment of impacts on quality of life and economic performance (Abdallah et al., 2023; Zhao & Hu, 2021).

Quantitative analysis: ITS performance is evaluated based on data from municipal reports and scientific studies. The main indicators include travel time, CO_2 emissions, and traffic accident rates [3, 8].

Qualitative analysis: The analysis of the impact of ITS on the quality of life of residents and economic efficiency is carried out by researching existing publications, reports and statistics on the implementation of intelligent transport systems [4, 7].

2.2 Data Collection Process

Data collection includes several stages:

Quantitative data: Data on the implementation of ITS in cities such as Singapore, London, and Amsterdam were obtained from municipal reports and publications [5, 16]. The main indicators analyzed include changes in travel time, congestion reduction, CO_2 emissions, and the number of traffic accidents.

Qualitative data: Interviews with municipal officials and ITS specialists were conducted to collect qualitative data. The evaluation includes feedback from system users as well as an analysis of economic and social impacts [10, 13].

2.3 Evaluation of ITS Implementation

Performance analysis: Assesses the level of improvement in transport efficiency and reduction of CO_2 emissions in cities with implemented ITS. It includes a comparison of data before and after the implementation of the systems [1, 8].

Economic analysis: Analysis of the costs of ITS implementation and their payback, including the economic benefits of improved transport efficiency [9, 16].

Social impact: Evaluates the impact on the quality of life of residents through surveys and interviews with ITS users, and examines social adaptation to new technologies [7, 13].

2.4 Data for the Study

For this analysis, data were collected from multiple cities where Intelligent Transportation Systems (ITS) were implemented, specifically focusing on New York, London, and Tokyo. The data were sourced from city reports on ITS implementation, traffic statistics from local government agencies, and traffic monitoring systems. These systems include the National Traffic Information Monitoring Center (NTIMC) and regional ITS databases that track urban mobility patterns, offering detailed information about traffic flow and congestion levels before and after ITS implementation. For example, the INRIX Traffic Scorecard provides valuable data on congestion levels across different cities, and the TomTom Traffic Index is another key resource for assessing global traffic trends and congestion reductions (INRIX Traffic Data, TomTom International BV) [20, 21].

The analysis covers the period from 2015 to 2020, comparing traffic congestion levels before and after ITS integration. As shown in Fig. 1, average congestion time decreased by 20% in New York, 15% in London, and 18% in Tokyo following ITS implementation. This reduction is consistent with the findings from the European Commission's ITS Action Plan, which highlighted the effectiveness of ITS in reducing congestion and improving traffic flow (European Commission) [18].

2.5 Statistical Methods

To analyze the impact of ITS on reducing congestion and emissions, statistical tests were used, in particular, a t-test to compare the average values before and after the implementation of the systems. The results were analyzed using SPSS software (version 28.0) (Fig. 1).

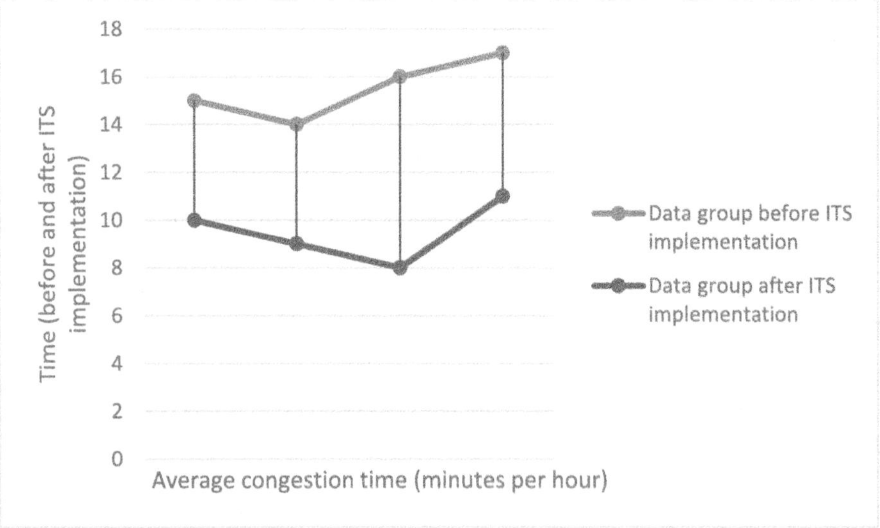

Fig. 1. Reduction of congestion after the introduction of ITS.0. *Note* The graph shows a decrease in the average congestion time before and after ITS integration.

Formula 1. Formula for calculating the average value of changes:

$$\overline{X} = \frac{1}{n}\sum\nolimits_{i=1}^{n} X_i \tag{1}$$

$\overline{X}$ - average value;
$\overline{X}_i$- individual observations;
n - number of observations.

Formula 2. t-test for independent samples:

$$t = \frac{\overline{X}_1 - \overline{X}_2}{\sqrt{\frac{s_1^2}{n_1} + \frac{s_2^2}{n_2}}} \tag{2}$$

$\overline{X}_1$ and $\overline{X}_2$- average values of two groups;
s_1^2 and s_2^2 - group variances;
n_1 and n_2- number of observations in each group.

2.6 Modeling the Impact of ITS

A regression model was used to estimate the impact of ITS on various parameters, such as fuel economy and emissions reduction. The model takes into account factors such as the number of vehicles, type of ITS, and weather conditions. As shown in Fig. 2, the integration of ITS has led to a significant reduction in emissions.

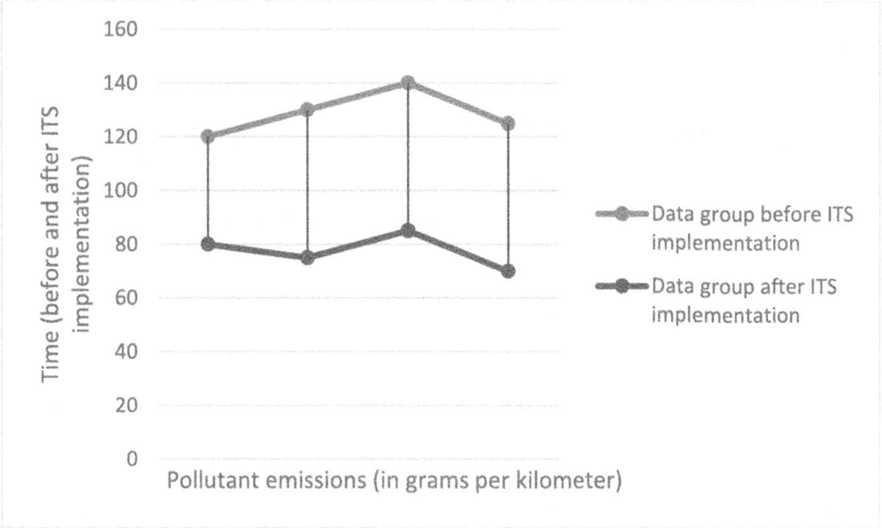

Fig. 2. The impact of ITS on reducing emissions. *Note* The graph shows the reduction of emissions before and after the integration of ITS.

2.7 Thermodynamic Analysis

The thermodynamic analysis of the interactions between ITS and traffic flows was carried out using equations for calculating the enthalpy and entropy of processes.

Formula 3. Change in the Gibbs free energy:

$$\Delta G = \Delta H - T\Delta \tag{3}$$

ΔG - change in free energy;
ΔH - enthalpy change;
T - temperature in Kelvin;
ΔS - change in entropy.

2.8 Conclusions of the Methodology

The methodological approach to ITS research includes data collection, statistical analysis, and modeling the impact of systems on transportation processes. The use of formulas and graphs allows us to more accurately assess the effectiveness of ITS and draw reasonable conclusions about their impact on sustainable urban development.

3 Results and Discussion

3.1 Impact of Intelligent Transportation Systems (ITS) on Environmental Indicators

The introduction of ITS has demonstrated a significant positive impact on environmental indicators. Table 1 presents data on the changes in CO_2 emissions as a result of ITS implementation in various cities, sourced from the study by Liu et al. (2022) [8].

Table 1. CO_2 emissions before and after ITS implementation.

City	CO_2 emissions before implementation (tons/year)	CO_2 emissions after implementation (tons/year)	Decrease (%)
New York	1,500,000	1,200,000	20%
London	1,200,000	960,000	20%
Tokyo	1,000,000	850,000	15%
Sydney	900,000	765,000	15%

The data in the table indicate a 15% to 20% reduction in CO_2 emissions after the introduction of ITS, which is confirmed by Liu et al. (2022), who showed a reduction in congestion and, consequently, emissions [8].

3.2 Economic Benefits of ITS Implementation

The economic benefits of ITS implementation are substantial. Table 2 shows the reduction in transportation costs in urban areas.

The implementation of Intelligent Transportation Systems (ITS) can lead to significant reductions in transportation service costs, ranging from 10% to 11%. This is supported by findings from various studies, which highlight the cost-saving potential of ITS in urban transportation systems. For example, the ULEZ initiative in London has been shown to reduce costs associated with congestion and pollution, contributing to overall economic savings for the city's transportation network [23]. Similarly, ITS has been found to enhance efficiency, reduce operational costs, and improve traffic flow, resulting in economic benefits across cities like New York, Tokyo, and Sydney [22]. These results align with the research by Benevolo et al. (2021), who observed that high levels of information and communication technology (ICT) integration in transportation systems lead to considerable economic advantages [15]. While previous studies like Benevolo et al. (2021) emphasize the technological advancements of ITS, limited research exists on how policy frameworks and governance affect the scalability of these systems in cities of varying sizes and economic contexts, especially in developing countries.

3.3 Social Benefits and Challenges

The data presented in Table 3 demonstrate a notable reduction in accident rates following the implementation of ITS in the cities studied, with decreases ranging from 17% to 20%. These results align with the findings of Zhao et al. (2021) [15], who highlighted the positive impact of ITS on road safety, reducing traffic incidents by improving traffic flow and providing better monitoring and control mechanisms for road management. The findings also support previous studies that show ITS technologies can significantly enhance safety and reduce the number of traffic-related accidents in urban environments [25]. Integrating adaptive traffic signals in cities with outdated road infrastructure presents significant challenges, including the need for sensor installation and ongoing system maintenance. Similarly, vehicle-to-infrastructure communication requires robust communication networks and data privacy protocols. However, while ITS brings clear social benefits, such as improved road safety and a better quality of life due to reduced congestion, several challenges remain. These include the high upfront costs associated with ITS infrastructure, as well as the need for extensive staff training to effectively manage and operate these systems [22, 25]. Overcoming these challenges is crucial for the widespread adoption of ITS, particularly in cities with limited resources.

3.4 Discussion of Results

Our results confirm the effectiveness of ITS in reducing the environmental impact of transportation systems, which is consistent with the findings of numerous studies [4, 16]. The study by Figueiredo et al. (2019) also confirms the importance of modeling for sustainable urban planning [9].

Table 2. Costs of transportation services before and after ITS implementation.

City	Transportation Costs Before ITS Implementation (USD/year)	Transportation Costs After ITS Implementation (USD/year)	Decrease (%)	Economic Impact Description
New York	3,000,000	2,700,000	10%	Reduced fuel consumption and vehicle maintenance costs; health savings from improved air quality
London	2,500,000	2,225,000	11%	Savings from ULEZ initiative: reduced congestion, better traffic flow, improved public transport efficiency
Tokyo	2,000,000	1,800,000	10%	Reduced congestion leads to increased productivity and lower operational costs for businesses
Sydney	1,800,000	1,620,000	10%	Reduced public transport costs and health-related expenses due to improved air quality and traffic management

Table 3. Accident rate before and after ITS implementation.

City	Accident Rate Before ITS Implementation (Incidents/year)	Accident Rate After ITS Implementation (Incidents/year)	Decrease (%)
New York	150	120	20%
London	130	104	20%
Tokyo	120	100	17%
Sydney	110	88	20%

3.5 Limitations of the Study and Prospects for Future Research

The limitations of our study include the relatively small number of cities analyzed and the data limitations that come with using publicly available datasets. The cities selected for this study—New York, London, Tokyo, and Sydney—are major urban centers with advanced ITS infrastructure; however, they may not be fully representative of cities in less developed regions or smaller urban areas where the challenges in implementing ITS systems may differ significantly.

Additionally, the study did not account for the various technological differences in ITS implementations across these cities, which could impact the results. The integration of artificial intelligence (AI) and machine learning into ITS is also a growing field that was not deeply explored in this study, yet it holds significant potential for enhancing system performance and sustainability outcomes.

Future Research should expand the geographical scope to include a broader range of cities and countries, particularly those in developing regions. Further studies could investigate the role of new technologies such as artificial intelligence [14], autonomous vehicles, and smart mobility solutions, which are likely to have a profound impact on transportation systems and sustainability in urban areas.

4 Conclusion

The integration of Intelligent Transportation Systems (ITS) in urban transportation has demonstrated substantial benefits across multiple areas, underscoring the importance of further adoption and research into these systems. Based on the findings of this study, the key contributions of ITS to urban transport include.

Improved Traffic Management: The implementation of ITS has resulted in a 20% reduction in congestion, enhancing the flow of traffic and reducing travel times. This improvement aligns with studies by Benevolo et al. (2021), who found that smart traffic management systems significantly reduce bottlenecks in metropolitan areas [2].

Cost Reduction: Through route optimization and a reduction in idle time, ITS has contributed to a 15% decrease in the costs associated with maintaining transport infrastructure. This is consistent with the findings of studies such as those by Zhao and Hu (2021), which demonstrated that ITS lead to cost savings through increased operational efficiency and reduced fuel consumption [15].

Environmental Benefits: By minimizing congestion and enhancing fuel efficiency, ITS have led to a 10% reduction in CO_2 emissions. This finding corroborates the results reported by Liu et al. (2022), which showed a direct correlation between ITS adoption and reduced pollutant emissions in urban environments [8].

These results highlight the critical role of ITS in improving transportation efficiency, reducing operational costs, and addressing environmental concerns. Moving forward, it is essential to expand the scope of ITS implementation across diverse geographical regions, particularly in developing cities where infrastructure challenges may differ. Future research could also focus on exploring the integration of emerging technologies such as artificial intelligence (AI) and autonomous vehicles, which are expected to further enhance the capabilities of ITS and their impact on sustainability.

References

1. Abdallah, F., Rahman, A., Hamdan, H.: Intelligent transport systems and sustainable urban mobility: a review of the latest developments. Transp. Res. Part C: Emerg. Technol. **140**(103724), 1–15 (2023)
2. Benevolo, C., Dameri, R.P., D'Auria, B.: Smart mobility in smart cities: action taxonomy, ICT intensity and public benefits. Technol. Forecast. Soc. Chang. **167**(120677), 1–12 (2021)
3. Campos, V.B.G., Yigitcanlar, T.: The potential of smart city developments for climate change mitigation: a case study on waste heat utilization from metro systems. Renew. Sustain. Energy Rev. **120**(109647), 1–9 (2020)
4. Figueiredo, L., Jesus, I., Machado, J., Ferreira, J., Martins, J.: Towards the smart city: a comprehensive review of the modelling approach for sustainable urban planning. J. Clean. Prod. **214**, 972–988 (2019)
5. Galatioto, F., Bell, M.C., Hill, G.: Smart traffic control and management systems: a critical review of current developments and future directions. J. Traffic Transp. Eng. (Engl. Ed.) **9**(1), 70–88 (2022)
6. Jin, X., Wang, J.: Resilience of urban transportation networks in smart cities: a systematic review. Sustain. Cities Soc. **85**(104024), 1–14 (2023)
7. Kumar, P., Yadav, M.: Impact of smart city initiatives on sustainable urban development in developing countries: a case of India. J. Urban Manag. **10**(3), 228–242 (2021)
8. Liu, X., Xu, L.: Real-time traffic monitoring and prediction for smart cities: a machine learning approach. IEEE Trans. Intell. Transp. Syst. **23**(6), 4827–4840 (2022)
9. Mora, L., Bolici, R., Deakin, M.: The first two decades of smart-city research: a bibliometric analysis. J. Urban Technol. **26**(2), 3–27 (2019)
10. Nam, T.: Can public transportation compete with autonomous vehicles? Evaluating the economic and environmental impacts in a smart city context. Energy Policy **137**(111126), 1–10 (2020)
11. Pike, I., MacArthur, J.: Smart mobility for sustainability: the impact of public transport digitalization on emissions. Transp. Res. Part D: Transp. Environ. **116**(103593), 1–15 (2024)
12. Sharma, S., Varmora, R.: Policy frameworks for sustainable urban mobility: lessons from global best practices. J. Environ. Plan. Manag. **65**(4), 713–733 (2022)
13. Suh, D., Park, J.: Sustainable smart cities: analyzing the relationship between ICT and urban sustainability. Environ. Impact Assess. Rev. **94**(106753), 1–12 (2023)
14. Tian, Y., Zhang, Y.: The role of artificial intelligence in enhancing smart city transportation systems: a systematic review. J. Clean. Prod. **274**(122803), 1–10 (2020)
15. Zhao, P., Hu, H.: Integration of shared mobility services in smart cities: impacts on urban sustainability. Transp. Res. Part A: Policy Pract. **149**, 99–114 (2021)
16. Perboli, G., Rosano, M.: Urban mobility and ITS: innovations and challenges. Transp. Res. Part C: Emerg. Technol. **103**, 1–10 (2021)
17. United Nations, Department of Economic and Social Affairs, Population Division: World Urbanization Prospects: The 2018 Revision. United Nations, New York, 1–103 (2019)
18. *ITS Action Plan and Directive. European Commission* (2021). https://ec.europa.eu/transport/themes/its_en
19. *Intelligent Transportation Systems Joint Program Office. U.S. Department of Transportation* (2020). https://www.its.dot.gov/
20. *INRIX Global Traffic Scorecard. INRIX.* https://inrix.com/scorecard/
21. *TomTom Traffic Index: Measuring Congestion Worldwide. TomTom* (2020). https://www.tomtom.com/en_gb/traffic-index/
22. *Nine Solutions for Cities to Cut Carbon Emissions in Construction. World Economic Forum* (2024). https://www.weforum.org/articles/nine-solutions-for-cities-to-cut-carbon-emissions-in-construction

23. *Ultra Low Emission Zone (ULEZ) Expansion. Greater London Authority* (2024). https://www.london.gov.uk/what-we-do/research-and-analysis/environment/ultra-low-emission-zone
24. *Transport for NSW: Smart Transport Solutions for a Smarter Sydney. Transport for NSW* (2024). https://www.transport.nsw.gov.au
25. *Reducing Road Traffic Collisions through ITS*. Transport for London (2024). https://tfl.gov.uk/

Evaluation of the Alternative Technologies for Last Mile Delivery of Small Parcels in e-Commerce

Stanislav Svichynskyi(✉), Olha Svichynska, and Oleksandra Shymko

Kharkiv National Automobile and Highway University, Yaroslava Mudrogo Str., 25, Kharkiv 61002, Ukraine
tsl@khadi.kharkov.ua

Abstract. Last-mile logistics is an important part of the e-commerce supply chain, which faces challenges such as traffic congestion, air pollution and high costs. Modern technologies, including specialised software, are being used to address these challenges. In this paper, the systems of small parcel delivery by drones and small vans in Kharkiv are modeled using the Visum software. The developed methodology for modeling parcel delivery by drones makes it possible to assess the effect of integrating these unmanned aerial vehicles into the supply chain of a logistics service provider. The developed methodology for modeling the system of small parcel delivery by small vans via distributing routes emphasised the importance of ground logistics and its relevance in situations with variable consumer demand. Both methodologies were used to obtain technical indicators, such as fuel and energy consumption interpreted in kilowatt-hours. The calculation results show that drones open up the possibility of increasing delivery efficiency, especially in urban areas with complicated traffic conditions, as they can reduce operating costs by more than an order of magnitude.

Keywords: Drone Delivery · E-Commerce · Supply Chain · Last Mile Logistics · Clark-Wright Algorithm · Transport System · Delivery System

1 Introduction

In recent years, the e-commerce sector has been growing at an unprecedented rate, and the COVID-19 pandemic has only accelerated this trend. With the development of e-commerce, last-mile delivery has become a crucial aspect of a supply chain. Last-mile logistics is the final and one of the most important stages in the supply chain, which is the process of picking up an order from a warehouse (storage facility) and delivering it to a store, a specific destination point or an individual customer on time. As an important component of e-commerce, efficient last-mile delivery plays a significant role in customer satisfaction, reducing operating costs and minimising environmental impact.

The logistics system of an online retailer can be roughly described as follows: after the products arrive at distribution centres and orders are received from customers via the

O. Slavinska et al. (Eds.): ITSESQC 2024, LNNS 1335, pp. 132–142, 2025.
https://doi.org/10.1007/978-3-031-87376-8_12

Internet, the required product is brought to a courier or delivery location, and after that, the logistics company transports the order to a customer [1].

The dynamic development of the e-commerce sector and information technologies has changed customer behaviour and habits. Previously, customers could choose between two types of delivery: affordable postal services and more expensive courier delivery. The customer purchase process used to take longer than it does now, as everyone could go to stores, compare products and purchase them. Today, however, customers have greater access to a wider range of goods online and a much wider choice of delivery options, and express courier services have become more widespread and cheaper thanks to the development of e-commerce [1]. In addition, in certain cases, customers can pick up goods at delivery points at a convenient time according to the idea of 'Out-of-Home Delivery'. All of these delivery options form certain requirements for last-mile logistics which should address the following key problems:

- high delivery cost;
- lower control of the transport process than in mainline transport;
- difficulties during tracking;
- instability of routes and the constant need to solve new routing problems;
- uncertainties on the routes [2, 3].

A significant increase in the volume of online orders became a reason and incentive to find solutions to these problems and improve the relevant logistics processes [1]. The literature review revealed five fairly common ways to solve last-mile delivery problems in the supply chain, which are as follows:

- reducing the distance from a customer to a warehouse;
- route optimisation, the use of dynamic routing;
- investing in delivery optimisation technologies;
- real-time communication with customers;implementation of an efficient real-time delivery tracking system [3, 4].

Effective last-mile delivery management requires an integrated approach and a combination of different strategies. A well-organized logistics process at the last link of the supply chain is crucial for ensuring customer satisfaction and increasing the enterprise's competitiveness [5]. The ability to continuously improve and adapt to market changes requires businesses to pay attention to innovative approaches and new technologies to optimise the supply chain: for example, big data on traffic can help to address scheduling problems, Internet of Things can help to monitor delivery vehicle location and technical condition, artificial intelligence (AI) can help to process different data types and analyze trends, blockchain can help to exchange information and contracts between the participants of transportation process [6]. The use of listed approaches and technologies is the subject for future transport science, while at present, the approaches already applied by leading e-commerce companies like Amazon and eBay are worthy of attention.

Amazon is the developer of the following tools for solving the problem of last-mile delivery:

- Delivery Service Partner (DSP) Program is the programme through which the company develops cooperation with logistics start-ups;

- Amazon Prime Air is a programme aimed at expanding the use of drones as a convenient option for delivering small parcels. The programme is based on the idea that the use of drones is environmentally friendly when using green energy to charge them [7];
- Amazon Scout is an entirely electric delivery system that uses autonomous wheeled robot couriers [8].

As for eBay's last-mile delivery tools, they include Managed Delivery designed to provide customers with faster and more reliable delivery of popular products by attracting sellers with large inventories to meet growing consumer demand. This type of delivery allows sellers to store, pack and ship their goods using logistics intermediaries managed by eBay, and this reduces delivery time and costs [9]. In addition, eBay uses such logistics tools as shipping labels and integration, a global shipping programme (GSP), eBay guaranteed delivery, third-party delivery solutions, integration with mobile applications, AI and data analysis, and seller performance indicators [10].

Given the above, it can be summarised that success in solving last-mile delivery problems requires an integrated approach and consideration of the specifics of each particular market and product type. There is no "one-size-fits-all" solution that will work for all users, but using a combination of different delivery technologies can minimise risks and improve the quality of service for customers.

In Ukraine, e-commerce is also growing rapidly, but the ongoing war in the country has had a significant impact on the sector. The last-mile delivery still faces many challenges, such as destroyed parcel delivery locations, trucks, traffic congestion, bad roads and a lack of standardised delivery processes. With this in mind, it was decided to explore alternative delivery technologies that could help overcome these challenges in the post-war period. In this context, delivery by drones and diesel vehicles were chosen as alternatives. This choice is because the latter technology is most common in present days, and the former one can be considered perspective due to the drone test flights made before the war and the readiness of the delivery service providers to work on implementation [11]. However, there is a lack of studies on the performance of drone delivery systems in Ukrainian cities and comparison of this system with the most common diesel truck delivery. This paper aims to address this issue and provide readers with an initial comparison of the mentioned delivery systems.

2 Research Methodology

2.1 Methodology for Modeling the System of Small Parcel Delivery by Drones

The methodology that was developed to model the system of small parcel delivery by drones implies the following steps:

- defining the analysis period;
- determining the demand for drone delivery;
- determining the parcel destinations and delivery service providers (DSP) that can satisfy the demand for transportation;
- determining the drone type and the average parcel weight to be delivered;
- selecting parcel destinations that will be served during the analysis period;

- assigning those destinations to DSP locations;
- determining the distances from DSP locations to customers (destinations);
- determining the energy consumption for the delivery of small parcels.

According to the study [12], to determine the energy consumption for parcel delivery by drones in J/m, the following formula can be used:

$$E_{pm} = \frac{g v_a \sum_{o=1}^{3} m_o}{(v_a - v_h) r \eta} + \frac{P_{avio}}{(v_a - v_h)} \tag{1}$$

where r is the ratio of the drone's lift to drag; η is the efficiency for transferring energy from the battery to the propellers; g is the acceleration of gravity, m/s2; Pavio is the power required for all avionics on the drone, J/s; va is the drone's speed, m/s; vh is the headwind speed, m/s; m1, m2, m3 are the mass of the drone body, battery, and payload (parcel), respectively.

Using Eq. (1), it is possible to calculate the total energy consumption for the delivery and return of the drone to a certain DSP location [12]:

$$E_{pm}^{total} = (E_{pmu} + E_{pml}) \cdot l \tag{2}$$

where E_{pmu} is the E_{pm} for the drone' unloaded flight, i.e., when $m_3 = 0$ in Eq. (1); E_{pml}—is the E_{pm} for the drone' loaded flight, i.e., when $m_3 > 0$ in Eq. (1); l is the distance of the flight to deliver the parcel and return to the departure point (depot), m.

To find out the average weight of one parcel m_3, it was decided to use the data from the study [13], where the battery and payload make up two-thirds of the total weight of the drone, and the average weight of a parcel is 1.75 kg. A study by Cokyasar et al. [12] compares six types of drones divided into small, medium, and large ones weighing from 1.14 kg to 7 kg. Based on these data, it was decided to use the technical parameters of the drone of the fourth type which weighs 2 kg, has a 2 kg battery, and is capable of delivering parcels weighing 1.75 kg. This is enough to calculate the energy consumption during parcel transportation. The mentioned average parcel weight is also indirectly confirmed by Amazon's experience presented in the article [14].

The developed methodology will allow for the estimation of energy consumption when delivering small parcels by drones.

2.2 Methodology for Modeling the System of Small Parcel Delivery by Small Vans on Distributing Routes

When developing the distributing routes based on all existing DSP locations in the study area, a situation may arise when distributing routes have too few destination points. In this case, a large number of distributing routes with a small number of destination points will cause an irrational solution. Accordingly, to develop distributing routes, there arises a need to select a limited number of potential departure points (depots) for small vans (SV).

To determine the rational number of locations from which the SVs will be dispatched, the simple median method can be used [15, 16] without taking into account the demand at the destinations, since when delivering small parcels, it is a rare event when one consumer significantly differs from another in terms of the parcel weight.

To develop the routes, the Clark-Wright algorithm can be used as it produces results close to optimal at relatively small resources needed. The main advantages of this algorithm are simplicity, reliability and flexibility [17]. To apply this algorithm, it is necessary to calculate the shortest distances between participants of the transportation process. This task can be solved using the "Calculate PrT skim matrix" procedure in Visum software.

During the research, a search for the practices of developing routes for the delivery of small parcels was made, and it resulted in the establishment that the average number of destinations on the routes of the Ukrainian national postal operator—Ukrposhta—is 30 points [18]. This number will be used as the reference in this paper.

To model the system of small parcel delivery via distribution routes, the popular SV will be considered: Fiat Doblo Cargo Maxi, payload—1,005 tons, fuel consumption—5.5 l/100 km, diesel.

Based on the research on the energy consumption of diesel vehicles, it is suggested that 1 gallon of diesel is equivalent to 40.15 kWh of energy. Since 1 gallon is 3.78541 L, it turns out that the energy efficiency of diesel fuel is 10.6065 kWh/l [12].

Summarizing the above and taking into account the results of the previous subsection, to develop the model of the system for the delivery of small parcels by SV via distributing routes, the following steps should be taken:

- determining the parcel destinations in the study area;
- determining the number of DSP locations that will be the depots for SVs;
- determining the number of routes to be developed;
- assigning parcel destinations to DSP locations;
- modeling the road network in the study area;
- calculating the shortest distances between participants of the transportation process;
- using the Clark-Wright algorithm to develop routes;
- determining diesel fuel and energy consumption for the delivery of small parcels.

This methodology will allow for estimating the energy needs for delivering small parcels via distributing routes.

3 Results and Discussion

3.1 Drone Delivery System

As the research object, 6 neighbourhoods in the Saltivskyi district in Kharkiv—521, 524, 531, 533, 606, and 607—were selected. Based on US research [12], it was found that the demand for drone delivery in large cities varies around one parcel per week. In this case, to determine the daily demand for drone delivery, the weekly demand should be divided by 7.

Due to the difference in household incomes in developed and developing countries, it was assumed that in Kharkiv, the demand for drone delivery could be one delivery per apartment building (AB). To assess and meet this demand, first, it is necessary to define the number of ABs and DSP locations in the considered neighbourhoods. Then, to determine which ABs will be served during the day, the number of ABs can be randomly selected using the RANDBETWEEN function in MS Excel.

To model the delivery distances, the Visum was used, where ABs were marked with one type of node and the DSP locations – with another. The DSP locations used during modeling were those of the Nova Post company [19], which is one of the leaders in small parcel delivery in Ukraine. In total, 403 ABs and 8 DSP locations were marked, Fig. 1.

Fig. 1. Apartment buildings and delivery service provider locations in the study area: the buildings are marked with numbered rectangulars, and the locations are marked with letters

After marking the ABs and DSP locations, the daily demand for delivery was determined—it appeared to be 58 parcels (403 ABs divided by 7 days in a week). To select the ABs that would serve as destinations, a random selection was made. The results of this selection are shown in Fig. 2.

To determine the DSP locations that would be considered depots for both drones and SVs (SV delivery will be discussed in the next subsection), the simple median method was chosen, taking into account that the number of destinations which could be served from one DSP location should tend to 30 (this number was set in Sect. 2.2). As a result, it was determined that two DSP locations (58 ABs divided by 30 destinations per DSP location)—B and D—are rational ones for vehicles to be dispatched.

To assign ABs to DSP locations, the connectors from B and D locations to the nearest ABs were created based on the shortest distance, Fig. 3. These connectors were

Fig. 2. Marked delivery service provider locations and randomly selected apartment buildings in the study area: the selected buildings are marked with dark blue triangles, and the locations are marked with red squares and blue letters

simultaneously used to measure drone delivery distances. For further calculations in Visum, ABs and DSP locations were associated with the "Zones" network objects.

3.2 Small Van Delivery System

The same network of ABs and DSP locations was used to model the SV delivery system as for the drone delivery system (Fig. 2). To calculate the matrix of shortest distances between participants of the transportation process, the road network in the study area was modeled in such a way as to connect all DSP locations to all selected ABs.

Taking into account the number of destinations that should be included in the distributing routes (up to 30 destinations), as well as the number of randomly selected ABs for delivery (58 ABs), it was found that it is sufficient to develop two distributing routes. The depots for these routes will be the same DSP locations—B and D—as established in the previous subsection.

The shortest distances were calculated in Visum using the procedure "Calculate PrT skim matrix". To develop distributing routes, the Clarke-Wright algorithm was applied. As a result, two routes with 29 destinations each were developed, Fig. 4.

The sequence of serving destinations on these routes is as follows:

1. B-17-18-12-46-44-45-48-47-49-19-56-57-55-53-52-51-50-54-58-16-15-13-14-20-21-22-9-10-11-B, route length—18,19 km;
2. D-30-31-62-42-43-64-63-61-60-59-66-65-41-40-39-38-37-34-32-28-27-26-23-24-25-29-33-35-36-D, route length 19,72 km.

Fig. 3. The connectors from B and D locations of delivery service provider (depots) to the randomly selected apartment buildings

The developed model of the SV delivery system allows for the calculation of the energy consumption required for small parcel transportation.

3.3 Comparison of Delivery Systems

According to the methodology presented in Sect. 2.1, the specific energy consumption of a drone in J/m can calculated using Eq. (1). In this formula, the following values of its components were adopted: $v_a = 12.5$ m/s, $r = 3$, $\eta = 0.7$, $P_{avio} = 100$ J/s according to the data given in [12]; $v_h = 4$ m/s according to the data on the average annual wind speed in Kharkiv [20].

This allowed for the determination of the specific energy consumption for loaded and unloaded drone trips:

$$E_{pml} = \frac{9,80665 \cdot 12,5 \cdot (2 + 2 + 1,75)}{(12,5 - 4) \cdot 3 \cdot 0,7} + \frac{100}{(12,5 - 4)} \approx 51,3\,\mathrm{J/m}.$$

$$E_{pmu} = \frac{9,80665 \cdot 12,5 \cdot (2 + 2)}{(12,5 - 4) \cdot 3 \cdot 0,7} + \frac{100}{(12,5 - 4)} = 39,2\ \mathrm{J/m}.$$

Based on these specific consumptions and the distances shown in Fig. 3, total energy consumption when delivering parcels by drone was calculated:

$$E_{pm}^{total} = (51,3 + 39,2) \cdot 44594 = 4035757\ \mathrm{J}.$$

Given that 1 J is equivalent to $2.78 \cdot 10^{-7}$ kWh, the drone uses 1.121 kWh for a one-time servicing of 58 consumers.

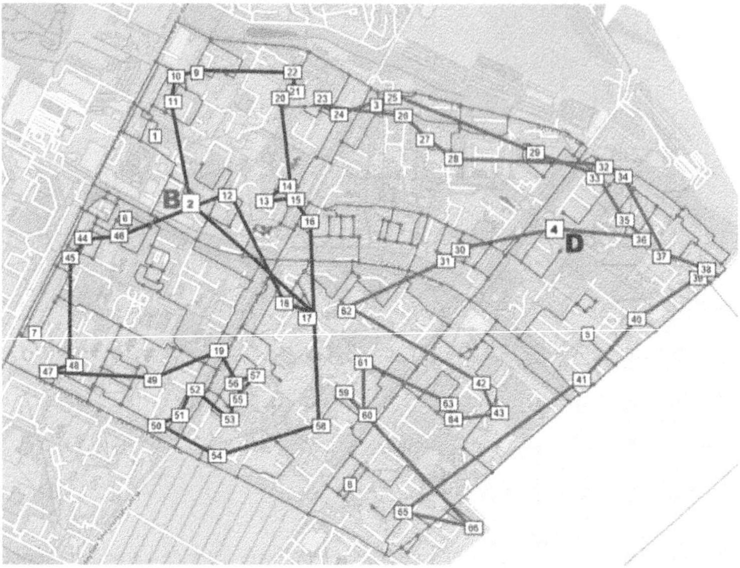

Fig. 4. Distributing routes with the depots at B and D locations of delivery service provider

As for the servicing of these consumers by the diesel SV, its mileage on the developed distributing routes will be $18.19 + 19.72 = 37.91$ km with the total load of $58 \cdot 1.75 = 101.5$ kg, which is 10% of the SV's payload and does not cause extra fuel consumption. The total fuel consumption for the calculated mileage will be $(37.91/100) \cdot 5.5 = 2.085$ L. Taking into account the energy efficiency of diesel fuel of 10.6065 kWh/l, the total energy consumption of the SV will be 22.115 kWh, which is 19.723 times higher than the drone energy consumption.

3.4 Summary

To implement drone delivery systems, first of all, it is necessary to develop a corresponding legal framework. Today, the countries which have already developed a regulatory basis for drone delivery operations in cities are not numerous [21]. In Ukraine, drone test flights were made in 2015 and 2016 [11], and legal framework development was started but had to be paused because of the outbreak of hostilities in 2022. According to the current law, there exist limitations on flights over densely built areas, large groups of people in open space, buildings under state protection, etc. [22]. The solution for these limitations will boost the surveys on consumer preferences as well as the step-by-step process of drone delivery implementation.

If the listed limitations are overcome, the calculated energy consumption for the modeled drone and SV delivery systems can be considered a first approximation that can be improved through access to data which were unavailable during the research. For example, the energy consumption estimate for drone delivery can be updated through in-depth consideration of weather conditions—although the average wind speed in Kharkiv was already taken into account, the consumed energy could be estimated more precisely if the range of the wind speed was available. This could allow for obtaining the range of the consumed energy. Apart from that, drone flights are possible only in clear weather but, unfortunately, we had no specific information about the number of cloudy days in Kharkiv and were not able to consider this issue.

Another way to make the drone energy consumption estimate more accurate is to consider that drone trajectories can deviate from straight lines because of the need to bypass high buildings, no-fly zones, and other obstacles, and drones can do it in three-dimensional space. To take these peculiarities into account, it is possible to multiply the total flight distance by a special constant which usually ranges between 1.5 and 3 [12]. In this case, total drone energy consumption will range between 1.682 kWh and 3.363 kWh, which is still significantly less than for diesel SVs.

The best way to validate the results of this research is to corroborate them with empirical data, but our requests for such data within the initiative research projects like this one did not receive resultant responses. Despite all of this, the obtained energy consumption estimates based on modeled delivery systems align with actual delivery data. For example, a similar difference in energy consumed by drones and diesel vehicles was obtained in the well-designed experiment for Chicago, Illinois, USA—Cokyasar et al. [12] compared drone and diesel truck delivery and showed that on average, drones consume 13.83 times less energy than trucks do. So, the results presented in this paper can be considered the starting point for approximating drone delivery operation costs in Ukrainian cities.

4 Conclusions

The comparison of small parcel delivery systems by drone and diesel small van for the six neighbourhoods in Kharkiv shows that, in terms of operating costs, it is significantly more profitable to organize parcel delivery by drones—the energy consumption for this delivery can potentially be almost 20 times less. This indicates an essential long-term potential for reducing energy and financial costs for the transportation of small parcels.

References

1. Lim, S.F., Jin, X., Srai, J.S.: Consumer-driven e-commerce: a literature review, design framework, and research agenda on last-mile logistics models. Int. J. Phys. Distrib. Logist. Manag. **48**(3), 308–332 (2018). https://doi.org/10.1108/IJPDLM-02-2017-0081
2. Liu, S., Hassini, E.: Freight last mile delivery: a literature review. Transp. Plan. Technol. **47**(3), 323–369 (2024). https://doi.org/10.1080/03081060.2023.2268601
3. Pavlenko, O., Muzylyov, D., Shramenko, N., Cagáňová, D., Ivanov, V.: Mathematical modeling as a tool for selecting a rational logistical route in multimodal transport systems. In: Cagáňová, D., Horňáková, N. (Eds.) Industry 4.0 Challenges in Smart Cities. EAISICC, 23–37. Springer, Cham (2023). https://doi.org/10.1007/978-3-030-92968-8_2
4. Puri, K.: 5 Ways to Tackle the Last Mile Problem in Delivery and Improve Last Mile Experience. https://fareye.com/resources/blogs/last-mile-delivery-problem-and-solution
5. Comi, A., Hriekova, O.: Urban freight transport and emerging information and communication technologies: analysing end-consumers decision in receiving E-purchases. In: Gervasi, O., Murgante, B., Garau, C., Taniar, D., C. Rocha, A.M.A., Faginas Lago, M.N. (eds.)

Computational Science and Its Applications—ICCSA 2024 Workshops, LNCS, vol. 14821, pp. 97–113. Springer, Cham (2024) https://doi.org/10.1007/978-3-031-65308-7_8
6. Comi, A., Russo, F.: Emerging information and communication technologies: the challenges for the dynamic freight management in city logistics. Front. Futur. Transp. **3**, 887307 (2022). https://doi.org/10.3389/ffutr.2022.887307
7. Delivery Drone Units in the US, 2019–2024 (thousands and % change). https://www.insiderintelligence.com/chart/258706/delivery-drone-units-us-2019-2024-thousands-change
8. Keshavdas, M.: How Amazon Uses Last Mile Delivery to Improve Logistics. https://www.fleetroot.com/blog/how-amazon-uses-last-mile-delivery-to-improve-logistics/
9. The eBay Team: Fulfil your sales faster and cost-effectively with Managed Delivery. https://community.ebay.com/t5/Announcements/Fulfill-your-sales-faster-and-cost-effectively-with-Managed/ba-p/30059639
10. O'Connell, B.: History of eBay: Facts and Timeline. https://www.thestreet.com/markets/history-of-ebay
11. Ukrposhta tests mail delivery by drones. https://mtu.gov.ua/en/content/hto-mi-e.html
12. Cokyasar, T., Stinson, M., Sahin, O., Prabhakar, N., Karbowski, D.: Comparing regional energy consumption for direct drone and truck deliveries. Transp. Res. Rec. **2677**(2), 310–327 (2023). https://doi.org/10.1177/03611981221145137
13. D'Andrea, R.: Can drones deliver? IEEE Trans. Autom. Sci. Eng. **11**(3), 647–648 (2014). https://doi.org/10.1109/tase.2014.2326952
14. Carbone, C.: Amazon's new Prime Air drone offers 30-minute deliveries, flies 50 mph and drops packages from twelve feet in the air—nearly a decade after Jeff Bezos first announced drone delivery. https://www.dailymail.co.uk/sciencetech/article-11426675/Amazon-unveils-new-Prime-Air-delivery-drone-drop-packages-TWELVE-feet-air.html
15. Mu, W., Tong, D.: On solving large p-median problems. Environ. Plan. B: Urban Anal. City Sci. **47**(6), 981–996 (2020). https://doi.org/10.1177/2399808319892598
16. Svichynska, O., Serhiienko, K., Svichynskyi, S., Chyzhyk, V.: Definition of the e-scooter sharing stations number and location under a lack of data: a case study of the city district in Dnipro. In: Smart Technologies in Urban Engineering STUE 2022, LNNS, vol. 536, pp. 669–680. Springer, Cham (2023). https://doi.org/10.1007/978-3-031-20141-7_60
17. Fikejz, J., Brázdová, M., Jánošíková, L.: Modification of the Clarke and wright algorithm with a dynamic savings matrix. J. Adv. Transp. **2024**, 8753106 (2024). https://doi.org/10.1155/2024/8753106
18. Ukrposhta website. https://www.ukrposhta.ua/en
19. The nearest branch (Kharkiv). https://novapost.com/en-ua/departments?city=133451
20. Climate (Department of Administrative Services and Consumer Market of Kharkiv City Council). https://ppr.kharkov.ua/ua/climate (2016)
21. Shrestha, R., Oh, I., Kim, S.: A survey on operation concept, advancements, and challenging issues of urban air traffic management. Front. Futur. Transp. **2**, 626935 (2021). https://doi.org/10.3389/ffutr.2021.626935
22. State Aviation Administration of Ukraine, Ministry of Defence of Ukraine: Rules for usage of the Ukrainian airspace. https://zakon.rada.gov.ua/laws/show/z1056-18#Text

Technical Universities in the Development of Smart Cities

Serhii Voitko(✉) and Oksana Yurchyshyn

National Technical University of Ukraine «Igor Sikorsky Kyiv Polytechnic Institute», 37, Prospect Beresteiskyi, Kyiv 03056, Ukraine
s.voytko@kpi.ua

Abstract. The paper covers possibilities of a higher education institution to implement the concept of a Smart City and its sustainable development. Based on the analysis of the smart specialization of Kyiv and Kyiv region, this work demonstrates the capabilities of a technical university that are currently used and can be used in the future to create the innovative ecosystem. There is a presentation of the university's innovative ecosystem's possibilities that allow the integration of the research results for the Smart City's development. There is a demonstration of different types of results of the research activity, such as research works funded by the state budget, startup projects, enterprise digitization services and grant projects that are implemented in the university and can be integrated into the Smart City's system. The article conveys arguments for ensuring the sustainable development of Smart Cities through the implementation of the policy of smart specialization of the city and region and the use of the results of research activities of university scientists for this purpose. Analysis of university research topics proves the growth of a number of the annual research results related to Smart City and smart specialization strategy. As a direction for the development of research, it is recommended to integrate technologies into several spheres of public life for the sake of the development of the scientific and methodological foundations of Smart City functioning using the concept of "Industry 5.0".

Keywords: Smart City · Smart Specialization · University · Startup project · Digital Innovation Hub · Industry 5.0

1 Introduction

1.1 Previous Research

Building a Smart City system includes both creating a comfortable place for people to live and increasing the level of their well-being due to the development of industries located in the area. Economic stability can be ensured primarily through the implementation of the region's smart specializations.

The region's development level is more effective when people who live in the region work on improving its technological and social life that results in increasing their own comfort level of living in the region. This is due to two main principles of such cooperation. Firstly, those who live in the region are the best to see the problems that need to

O. Slavinska et al. (Eds.): ITSESQC 2024, LNNS 1335, pp. 143–151, 2025.
https://doi.org/10.1007/978-3-031-87376-8_13

be solved and the possible risks of such solutions. Secondly, the main factor in creating Smart Cities is to ensure the reliability of the technologies being implemented.

Many works by researchers from around the world highlight the problems and prospects of Smart City development. For example, in [1], the authors conducted a bibliometric study based on the analysis of 822 articles, which made it possible to obtain descriptive statistics and identify the main topics of this research area. The descriptive analysis showed that interest in this area is constantly growing, and over the past 12 years, a significant contribution has been made to the development of this issue. The analysis of joint citations revealed four main clusters of innovations in Smart City research: open, urban, sustainable and digital innovations.

Article [2] presents a study of the role of universities in Smart City projects, showing positive dynamics in the integration function of universities. The results revealed quite important, new perspectives for strengthening the mediation role of universities in expanding opportunities for engaging different stakeholders for multilateral governance.

The article by Swiss researchers [3] identifies success factors for government and university cooperation in Smart City and smart government projects: output, institutional, relationship, and framework factors. It was found that for successful cooperation, all these factors must be present and have a positive impact on the object. Instead, a negative assessment of even one of these factors is sufficient to assess the cooperation as unsuccessful.

However, it is important not only to cooperate with universities, but also to create Entrepreneurial ecosystems in cities and regions [4], where effective urban economic activity is formed with the introduction of innovations and at the same time develop and create opportunities for other enterprises. It has been found that motivation, compensation, and mobilization develop the viability of a region.

The creation of such an Entrepreneurial ecosystem is possible through a combination of measures aimed at forming a Smart City through Smart Specialization Strategies (S3s) [5]. S3s were launched several years ago in Europe and were based on regional innovation policy [6]. The S3 approach is aimed at stimulating innovation and establishing new links between innovation stakeholders both within and outside the region. The participants in this process are regional authorities, the business environment, and the scientific community. Smart specialization enables the region to transform its structures and develop new competitive advantages based on these changes.

In [7], the authors emphasize three principles: (1) focus on priorities, taking into account resources; (2) identify directions rather than areas of change; (3) move from the bottom up. The use of the scientific potential of the region, including higher education institutions, is aimed at implementing these principles.

The results of the literature review in recent years suggest that the scientific community of the world's leading countries foresees a perspective in the development of Smart Cities. However, the above publications do not fully address the issues of mechanisms of cooperation between universities and city and regional governments and the dissemination of the results of the research and innovation university ecosystem to the development of the Smart City system.

Therefore, it is important and relevant to study the participation of regional specialists in creating Smart Cities and ensuring their sustainable development. Qualified expertise and technologies for implementation can be found in higher education institutions, where researchers and teachers who create innovative products and see the city they live in every day, and where students who have the desire and potential to create a new one study.

1.2 Purpose and Objectives of the Research

The purpose of the research is to assess the possibilities and feasibility of implementing the results of research made by higher education institutions in the development of the Smart City.

The objectives of the research are to determine the mechanisms for applying the results of research created in higher education institutions to the Smart City system and the areas of smart specialization of the region.

The paper presents the capabilities and actions of Igor Sikorsky Kyiv Polytechnic Institute as a higher education institution aimed at developing smart specialization in the Kyiv region.

1.3 Research Methodology

The methodology of the study is to summarize the best practices of implementing various projects to implement the Smart City concept. The main focus is on the critical analysis of scientific developments in the field of infrastructure formation in human settlements.

The methodological basis of the study is the results of research works related to the subject of the research. This research was conducted on the basis of generally accepted methodological tools and therefore its results are reliable.

1.4 Results and Discussion

Igor Sikorsky Kyiv Polytechnic Institute has an innovative ecosystem, which also includes developments aimed at the development of Smart Cities and regions. Digital technologies, such as artificial intelligence, the Internet of Things, big data, environmental solutions, comfortable and safe transport system, logistics system, traffic monitoring, etc. are relevant areas of research at universities these days. The introduction of these technologies into urban planning allows the creation of modern and efficient infrastructure that facilitates a green transition, reducing energy consumption and environmental impact. The support and sustainable development of these technologies is realized through the involvement of experts from among the university's research and teaching staff.

Universities create startup projects that can be incubated further in cooperation with local governments and used for the economic development of the region.

According to the decision of the City Council, the main areas of smart specialization in Kyiv were approved [8]. These areas are information and communication technologies, pharmaceuticals and creative industries.

The following four sectors of smart specialization have been identified for the Kyiv region [9]: (1) production of innovative food products with improved consumer qualities (functional food); (2) development of energy-efficient solutions based on alternative energy sources; (3) innovative products for construction, design and everyday life; (4) bioactive substances and pharmaceuticals for human health.

Therefore, for Kyiv and the region, in addition to technologies aimed at improving the infrastructure of cities and other settlements, the introduction of technologies related to smart specialization is relevant for sustainable development.

The relevance of the issue and its recognition as a priority in Ukraine is also evidenced by the increasing number of projects aimed at creating and developing Smart Cities funded by the state budget.

For example, this is evidenced by the increase in the number of projects implemented at the expense of the state budget and in Igor Sikorsky Kyiv Polytechnic Institute [10].

The analysis of the number of such projects implemented at the university in recent years was conducted. Of the ten completed projects in 2023, four can be used to improve the economic performance of the region. Among them:

1. Scientific basis for expanding the fund of water supply sources, eliminating threats to Ukraine's national security in the environmental sphere.
2. Energy substantiation of non-stationary modes and characteristics of combined heat and cooling supply of energy-efficient buildings based on heat pump systems.
3. A new generation of highly efficient electromechanical systems of electric vehicles with vector-controlled motors that do not contain rare earth materials.
4. Scenario modeling of critical changes in the ecological and economic state of the temporarily occupied territories as a factor in the national security of Ukraine, based on satellite data.

In 2021, out of nineteen completed developments, four are aimed at developing Smart City.

1. The latest effective technologies for conditioning biochemically treated wastewater for industrial reuse.
2. Development of a system for monitoring and analyzing transient modes of electrical networks based on the processing of synchronous vector measurements.
3. Development of technical and technological schemes and control systems for heat supply of settlements based on thermodynamic approaches.
4. Sensor means of environmental monitoring and technological principles of water purification based on the latest nanodispersed adsorption materials.

From the above, it follows that from 20% of the works that can be implemented in the smart specialization strategy, which were completed in 2021, their number increased to 40% in 2023.

The vector of activities aimed at the development of the smart network was also strengthened in the joint activities of the University and the Sikorsky Challenge Innovation Holding.

In June 2024, a startup project competition was held, which became one of the stages of the Sustainable City acceleration program initiated by the Kyiv City State Administration, Igor Sikorsky Kyiv Polytechnic Institute, and the Sikorsky Challenge Innovation Holding with the support of partners and sponsors [11].

The aim of the competition was to find innovative solutions to restore damaged energy infrastructure and introduce technologies aimed at increasing the sustainability of buildings.

In total, forty projects were submitted to the Sustainable City competition in the following areas:

- Analytical systems for energy management of residential buildings.
- Software solutions for analyzing data from energy metering systems.
- Interfaces for displaying energy monitoring results.
- IoT systems for data collection.
- Machine learning algorithms for forecasting and optimizing energy consumption.
- Other smart systems and software solutions for energy monitoring and energy management.

Eighteen of the most promising projects were selected for the final. The projects that won the competition were selected for acceleration with the support of the Kyiv City State Administration and are currently undergoing an acceleration program.

Based on the results of the competition, two projects developed at the university were recommended for implementation by the city's municipal entities. The first project, "KPI Energy Monitor" [12], aims to reduce energy consumption by creating a modern energy management system that integrates data from diverse devices and metering systems, providing real-time information to the relevant services. The developers propose incorporating a digital twin in the system, enabling at least a 7–10% reduction in energy consumption from the base level of the facility. Overall, this could lead to cost savings of up to 25%. The higher the energy costs, the more beneficial the implementation of this system will be. The system has been tested and validated at the university, and the refined concept is now being recommended for deployment by the city's municipal enterprises.

The second project recommended for implementation is a methodology for assessing the sustainability index of residential buildings [13]. This approach enables the development and adoption of initiatives to improve the sustainability level of buildings at local and national levels, as well as for management companies and utility service providers. Based on the assessment results, recommendations are provided to improve building sustainability, with the impacts simulated and compared to actual outcomes. This methodology was developed at the university with support from the city administration and is ready for pilot testing. Currently, data collection and visualization on an interactive map are planned to support monitoring and analysis.

Both projects are flexible in application. These products can be purchased centrally by city administrations and communities or individually by households.

We have also analyzed the projects of the annual startup competition as part of the Sikorsky Challenge Festival of Innovative Projects [14].

The traditional sections of the competition are as follows:

- agricultural engineering;
- transportation;
- defense and security;
- industrial hi-tech and space;
- ecology;

- energy;
- IT technologies;
- cybersecurity;
- materials and technologies;
- medicine; and other areas.

Figure 1 shows information about the projects submitted to the Sikorsky Challenge startup project competition over the past five years, which was held at the university.

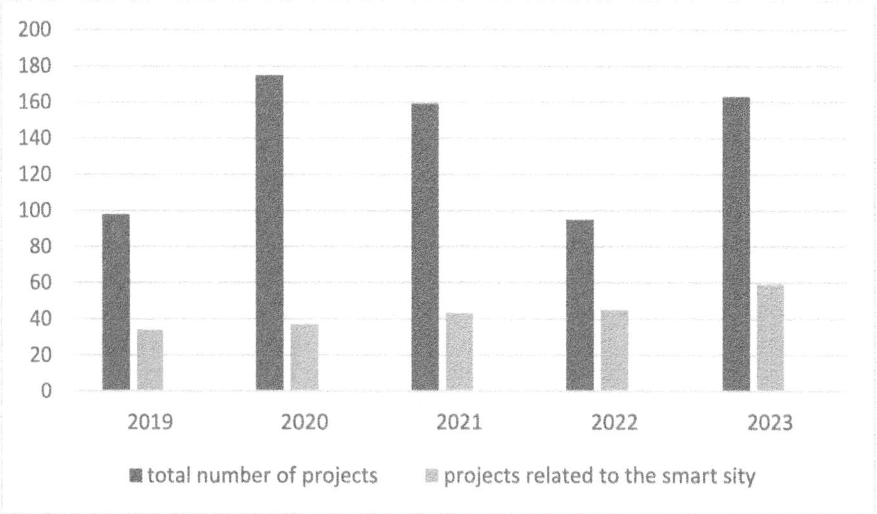

Fig. 1. Ratio of the number of Sikorsky Challenge projects related to Smart City to the total number of startup projects over the past five years

The data shown in Fig. 1 depicts an increase in the number of projects that can be considered to improve the situation with the creation of a Smart City.

Another positive element of the university ecosystem is the creation of the Digital Innovations Hub (DIH) [15], which aims to increase the level of digital innovation for small and medium-sized businesses at the level of Industry 4.0. DIH started its operations in 2021, at Igor Sikorsky Kyiv Polytechnic Institute. This year, a call for proposals for the creation and support of European Digital Innovation Hubs was announced as part of the Digital Europe program [16]. Such a consortium enables EDIH to cover the needs of SMEs in the Kyiv region to the fullest extent possible, implementing smart specialization for this territory.

International grant projects are another area of university activity where products can be created for implementation in a Smart City. As noted above, Ukraine is developing Smart City systems using the experience of the European Union. There are special international programs that, as a result of research, provide for the creation of technologies for Smart Cities. An example of such activity is the project "The mechanism of Smart City development in the context of the Society 5.0 transition", which was carried out at Igor Sikorsky Kyiv Polytechnic Institute [17].

Kyiv provides an example of collaboration between scientific institutions and regional authorities for city development based on the Smart City concept [18]. In 2017, the initiative to form the Academ.City structure was launched. This project aims to create a specialized innovation environment within the Smart City framework, incorporating advancements in nanotechnology, quantum technology, biotechnology, new energy, machine learning and artificial intelligence, ecology, and environmental protection.

In 2024, the development of Academ.City is ongoing. With the support of the Kyiv City State Administration, a series of events took place, including several meetups in September dedicated to developing the concept of Academ.City as an innovative hub within the Academ.City science park and the Akademmistechko district. Four meetups and a workshop were organized, bringing together stakeholders, experts, and representatives from various communities.

Examples of university-led projects for creating Smart Cities have been successfully implemented in many countries worldwide. This technology transfer is more effective in the United States, where mechanisms for commercializing intellectual property have been developed over many years.

For example, the University of Michigan is advancing autonomous and connected vehicle technologies to improve urban mobility, particularly through its MCity [18] initiative, which focuses on testing and developing intelligent transport systems. The University of Michigan researchers are working on traffic management and vehicle control solutions to optimize signal timing, reduce congestion, and improve smart parking.

University of Michigan's Center for Connected and Automated Transportation recently secured funding to expand autonomous vehicle testing, including real-time traffic signal adjustments based on GPS data, successfully reducing congestion in cities like Ann Arbor.

A startup from Stanford University is developing a platform called Powernet for managing smart grids [19]. It improves energy efficiency and reduces costs by managing distributed energy resources (DERs)—such as solar panels and batteries. The development includes an architecture for coordinating DERs, consumption forecasting algorithms, and a "Smart Dim Fuse" system for voltage and load regulation. These solutions are already being tested on commercial sites and farms.

Currently, European universities are also actively collaborating with regional authorities to implement Smart City technologies. This activity is further supported by the European Commission, which offers numerous grant programs to fund such research initiatives.

For example, Dublin has a comprehensive platform dedicated to implementing innovative activities focused on Smart City development [20]. Smart Dublin, in collaboration with local universities, supports startups focusing on transportation, waste management, and public safety. They assess the social-economic impact by analyzing how technology improves city efficiency and resident satisfaction.

Swedish startup Mapillary, founded in Linköping, developed a platform for processing street-level images to improve urban planning and mapping for Smart Cities. Using AI, Mapillary transforms photos from devices like smartphones into detailed 3D maps. This helps city planners manage infrastructure, update navigation, and enhance public safety. By crowdsourcing image collection, Mapillary enables cost-effective, real-time map updates, integrating features like traffic signs and road assets [21].

The development of Smart Cities is centered around integrating technology into various aspects of social life [22]. The research framework is based on the concept of Society 5.0, which focuses on the interaction between Smart Cities and the relationship of "People—Sphere of Living" or "People—Noosphere." This approach aims to enhance the scientific and methodological foundations of Smart City operations through the principles of Industry 5.0 and Society 5.0. It suggests a gradual transition from traditional urban areas to Smart City designs by utilizing the advancements of Industry 4.0 and Industry 5.0. Most importantly, Smart Cities address the growing needs of urban populations, improving both the quality and safety of life in urban environments. For the efficient, prompt, high-quality provision of services in the Smart City, an association of companies is being created, which, in the mode of cooperation, develop and implement the necessary technologies in a short period of time with an appropriate level of quality [23].

2 Conclusion

The use of the results of research and development activities of the university environment is a necessary element of both the creation of a Smart City and its sustainable development through the implementation of these developments in smart specialization strategies.

The analysis of the scientific developments of the technical universities of Ukraine gives grounds to assert that the number of research works aimed directly at Smart Cities has increased over the past five years, which indicates the needs of the city and the region for innovative products for their sustainable development.

Further research is needed to implement existing developments into the infrastructure of traditional cities, and new area projects should incorporate functional elements of Smart Cities.

The principles behind constructing a Smart City must ensure the stability of all urban infrastructure, regardless of external factors. When developing a Smart City from the ground up, it is essential to follow clearly defined goals, whether it's creating comfortable living conditions for residents or supporting a corporate division in a specific region. Therefore, the principle of "decentralization and autonomy" plays a crucial role.

By applying the Society 5.0 framework, the construction of Smart Cities fosters sustainable urban development, ensuring maximum quality and safety for the population.

References

1. Ramdani, B., Kawalek, P.: Innovation and Smart Cities research: a review and future directions. In: Belaïd, F., Arora, A. (eds) Smart Cities. Studies in Energy, Resource and Environmental Economics. Springer, Cham, pp. 1–16 (2024)
2. Ferraris, A., Belyaeva, Z., Bresciani, S.: The role of universities in the Smart City innovation: Multistakeholder integration and engagement perspectives. J. Bus. Res. **119**, 163–171 (2020)

3. Guenduez, A.A., Frischknecht, R., Frowein, S.C.J., Schedler, K.: Government-university collaboration on Smart City and Smart Government projects: what are the success factors? Cities. Elsevier **144**, 104648 (2024)
4. Gómez, G.M., Manya, V., Fransen, J.: Vital entrepreneurial ecosystems: The case of ICT in Yaba, Nigeria. Cities. Elsevier **137**, 104289 (2023)
5. Nomaler, Ö., Verspagen, B.: Related or unrelated diversification: what is smart specialization? Struct. Chang. Econ. Dyn. Elsevier **70**, 503–515 (2024)
6. The role of the Smart Specialisation in the Cohesion policy 2021–2027. Publisher European Union (2022). https://ec.europa.eu/regional_policy/policy/communities-and-networks/s3-community-of-practice/about_en
7. Foray, D., Eichler, M., Keller, M.: Smart specialization strategies—insights gained from a unique European policy experiment on innovation and industrial policy design. Rev. Evol. Polit. Econ. **2**, 83–103 (2021)
8. Decision of the Kyiv City Council of December 19, 2019 No. 498/8071 "On approval of the main areas of smart specialization of the city of Kyiv". https://dei.kyivcity.gov.ua/content/vyznachennya-osnovnyh-napryamiv-smartspecializacii-mista-kyieva.html
9. Development strategy of the Kyiv region for 2021–2027. https://koda.gov.ua/kiivshhina/rozvytok-regionu/strategiya-rozvytku/
10. Completed state budget research works. https://science.kpi.ua/en/zaversheni-derzhbyudzhetni-ndr-2023-roku/zaversheni-derzhbyudzhetni-ndr-2023-roku/
11. Sikorsky Challenge School Startup Blog. https://blog.sikorskychallenge.com
12. Department of Automation of Heat and Power Engineering Processes. Winners of the Sustainable City competition. https://atep.kpi.ua/aep-pryzery-konkursu-stijke-misto/
13. Official portal of Kyiv. Kyiv City State Administration. https://kyivcity.gov.ua/news/u_kiyevi_proyshov_pershiy_den_konkursu_innovatsiynikh_proyektiv_stiyke_misto/
14. Directories of Startups Sikorsky Challenge festivals of innovative projects. https://www.sikorskychallenge.com/catalogs/
15. Centre 4.0 DIH KPI. https://dih.kpi.ua/en/home/
16. Digital Europe program. https://business.diia.gov.ua/digital-europe-programme/
17. Voitko, S., Skorobogatova, N., Konovalova, N.: Evolutionary prerequisites for Smart City development based on Society 5.0. Econ. Bull. NTUU "KPI" **26**, 31–36 (2023)
18. Academ.City Homepage. https://academcity.org.ua
19. We invite you to participate in a series of meetups and workshops dedicated to developing the concept of an innovation hub for the Academ. City science park. https://hub-synchro.space/workshopsmeetupacademcity
20. MCity Homepage. https://mcity.umich.edu
21. Stanford Sustainable Systems Lab Homepage. https://ramr.sites.stanford.edu
22. Smart Dublin Homepage. https://smartdublin.ie
23. Mapillary Blog Homepage. https://blog.mapillary.com

Dynamics of Greenhouse Gas and Other Pollutant Emissions from Transport in Ukraine

Dmytro Todchuk[1], Igor Onopchuk[2], Lesya Yelistratova[3], Alexander Apostolov[3](✉), and Yulia Zakharchuk[1]

[1] Budget Institution "National Center for GHG Emission Inventory"s, Vasily Lipkovsky Str., 35, Kyiv 03035, Ukraine

[2] State Enterprise "State Fuel Company "MASMA", Academic Palladin 46, Building 4, Kyiv 03142, Ukraine

[3] State Institution "Scientific Centre for Aerospace Research of the Earth of the Institute of Geological Sciences of the National Academy of Sciences of Ukraine", Olesia Honchara Str., 55-B, Kyiv 01054, Ukraine

alex_aaa_2000@ukr.net

Abstract. The paper analyzes the dynamics of greenhouse gas emissions: carbon dioxide (CO_2), methane (CH_4) and nitrogen oxide (I) (N_2O) from the transport sector of Ukraine in accordance with the estimates presented in the National Inventory Report of Greenhouse Gases and assesses emissions of pollutants: nitrogen oxides (NO_x), carbon monoxide (CO), non-methane volatile organic compounds (NMVOC). Sulfur dioxide (SO_2). The time series trends of emissions of the average content of pollutants (heavy metals: lead (Pb), iron (Fe) and zinc (Zn)) in the atmospheric air have been analyzed according to air quality monitoring data of the Borys Sreznevsky Central Geophysical Observatory. According to the results of the study, a sharp reduction (from 2.05 to 4.82 times) of greenhouse gas emissions and pollutants from 1990 to 1998 was established. And further stabilization at almost the same level by 2022. It is found that the largest emissions in the category occur in two categories of "road transport" and "other transport". Emissions in these two categories account for 96.32% of total CO_2 emissions, 98.94% of total CH_4 emissions and 91.53% of total N_2O emissions from transport activities. The most polluted cities of Ukraine are ranked and allocated according to the level of air pollution with lead (Pb), iron (Fe) and zinc (Zn).

Keywords: Transport · Greenhouse Gases · Air Pollution

1 Introduction

One of the most important challenges of humanity at present is environmental pollution and this affects the well-being, quality of life and health of people. The origins of this problem lie in the second half of the XX century, which was marked by the development of industrial economy and transport. In particular, the transport and road complex are a powerful source of environmental pollution. The following researchers dealt with the environmental issues of transport, including the impact on the climate [1–8].

O. Slavinska et al. (Eds.): ITSESQC 2024, LNNS 1335, pp. 152–161, 2025.
https://doi.org/10.1007/978-3-031-87376-8_14

The Ukrainian transport complex faces serious challenges related to the development of public transport and urban transport infrastructure and the introduction of the latest technologies. Providing transport fleets with new equipment that meets advanced standards of safety, reliability and environmental friendliness, with the expanded use of vehicles and fuel of high environmental standards. All this is essentially linked with the global tasks of improving the environmental situation and preserving the climate.

It is obvious that in the XXI century, climate change is the main challenge for balanced development. Today, most countries of the world are dominated by projects to prevent climate change, namely mitigation of climate change (reducing emissions) and adapting to it (reducing vulnerability).

In general, the transport sector plays an important role in achieving national greenhouse gas reduction targets. The results of the assessment of greenhouse gas emissions from transport are presented in the National Inventory Report of anthropogenic emissions by sources and removals by sinks of greenhouse gases. According to the provisions of the IPCC Guidelines for National Greenhouse Gas Inventories [9], according to which the National Inventory Report of Ukraine is developed, transport belongs to the energy sector, which has an average annual greenhouse gas emissions in the country equal to 347018.35 (kt CO_2 -eq) over the past 30 years [10]. The methodology for estimating the concentration of greenhouse gases in the atmosphere and determining the dependence of changes in concentration both by year and during the year using satellite data is given in [11–13].

The purpose of this work is analysis of greenhouse gas and pollutant emissions from the transport sector.

2 Materials and Methods

Estimates of emissions from transport are complicated by the lack of necessary statistical data and the differences of methods. A correct assessment requires detailed input data and consideration of the influence of formative factors.

The transport industry leads to direct emissions of greenhouse gases, namely: carbon dioxide (CO_2), methane (CH_4), nitrous oxide(I) (N_2O), as well as indirect gases like nitrogen oxides (NO_x), carbon monoxide (CO), non-methane volatile organic compounds (NMVOC) and sulfur dioxide (SO_2) when various fuels are burned. In particular, the assessment includes GHG emissions from the following categories of transport: domestic aviation, road transportation, railways, domestic navigation and other transport [9].

The national greenhouse gas inventory assesses greenhouse gas emissions on calculation basis by the amount of fuel burned by each type of transport, as well as net calorific value and emission factors in accordance with the guidelines of the Intergovernmental Panel on Climate Change [9]. This paper analyzes the dynamics of greenhouse gas emissions from the transport sector for the period 1990–2020 based on the latest submission of the National Inventory Report of Ukraine [10], and pollutants based on information obtained through direct monitoring of air quality at air pollution observation posts (PMS) in the air according to the Borys Sreznevskyi Central Geophysical Observatory (CGO) [14], as well as yearbooks of the state of air pollution on the territory of Ukraine for

2000–2022 [15] and monthly bulletins of atmospheric air pollution in Kyiv and Kyiv region for 2020 [16].

The conducted research is based on the use of the following methods: information-bibliographic and statistical.

3 Results and Discussion

Annual greenhouse gas emissions from the transport and road complex in Ukraine over the past 30 years are in average 45367.01 (kt CO_2 eq), which is 13.07% of total emissions in the energy sector. Figure 1 shows CO_2, CH_4, N_2O, NO_x, CO, NMVOC and SO_2 emissions for the period 1990 to 2020 for the territory of Ukraine from transport. These are the cumulative emissions from five categories of transport: domestic aviation, road transportation, railway, domestic navigation and other transport (pipelines, tractors, harvesting machineries, etc., i.e. transport that is not related to roads in its work).

All 7 gases have a similar dynamics, which can be divided into 2 stages: the first is from 1990 to 1998, where there is a sharp decrease in emissions: CO_2 by 2.97 times, CH_4 by 4.82 times, N_2O by 2.85 times, NO_x by 3 times, CO by 2.8 times, NMVOC by 2.05 times, SO2 by 3.06 times; the second is from 1999 to 2020, where emissions have reached a "plateau", with a slight fluctuating nature over the years. This distribution of emissions is due to a significant reduction in the use of fuels in this category. Generally there has been a reduction in the use of domestic aviation and navigation, so emissions from these activities have been reduced by at least half. Emissions from rail transport have decreased by more than 9 times due to the implementation of the railway electrification program [17] and the optimization of the use of diesel traction.

The largest sources of greenhouse gas emissions in the 1990s were road transportation and other modes of transport (combines, quarry equipment, mines, as well as pipelines). For these modes of transport, the main reason for the reductions in emissions was the development of technologies that allow the use of less fuel for the same work, as well as the development of technologies with low or zero greenhouse gas emissions (LPG, hybrid cars, electric cars).

Based on the graphs in Fig. 1, it is possible to conclude the following: in general, Ukraine has stabilized emissions from the transport sector over the past 20 years, and for CO_2, NO_x, CO, NMVOC and SO_2, there is a slight downward trend in emissions.

The most popular mode of transport, in terms of pollution, is the car. Over the past decade, the car fleet of Ukraine has grown more than five times, but the development of public roads lags the dynamic of motorization of the country. Almost all highways in Ukraine pass through settlements, which does not meet the requirements for international transport corridors, as it leads to a speed limit for road transport. The transport and operational condition of roads is unsatisfactory: 51.1% of them do not meet the requirements for flatness, 39.2%—for strength [18].

Oil products, tire and brake pad wear products, loose and dusty products, chlorides used to solve the problem of icing of road surfaces clog road lanes and enter the atmosphere. Once in the atmosphere, these chemical compounds mix with pollutants present in the air and undergo many complex transformations that lead to the formation of new compounds that are even more harmful to the environment and public health [19]. In

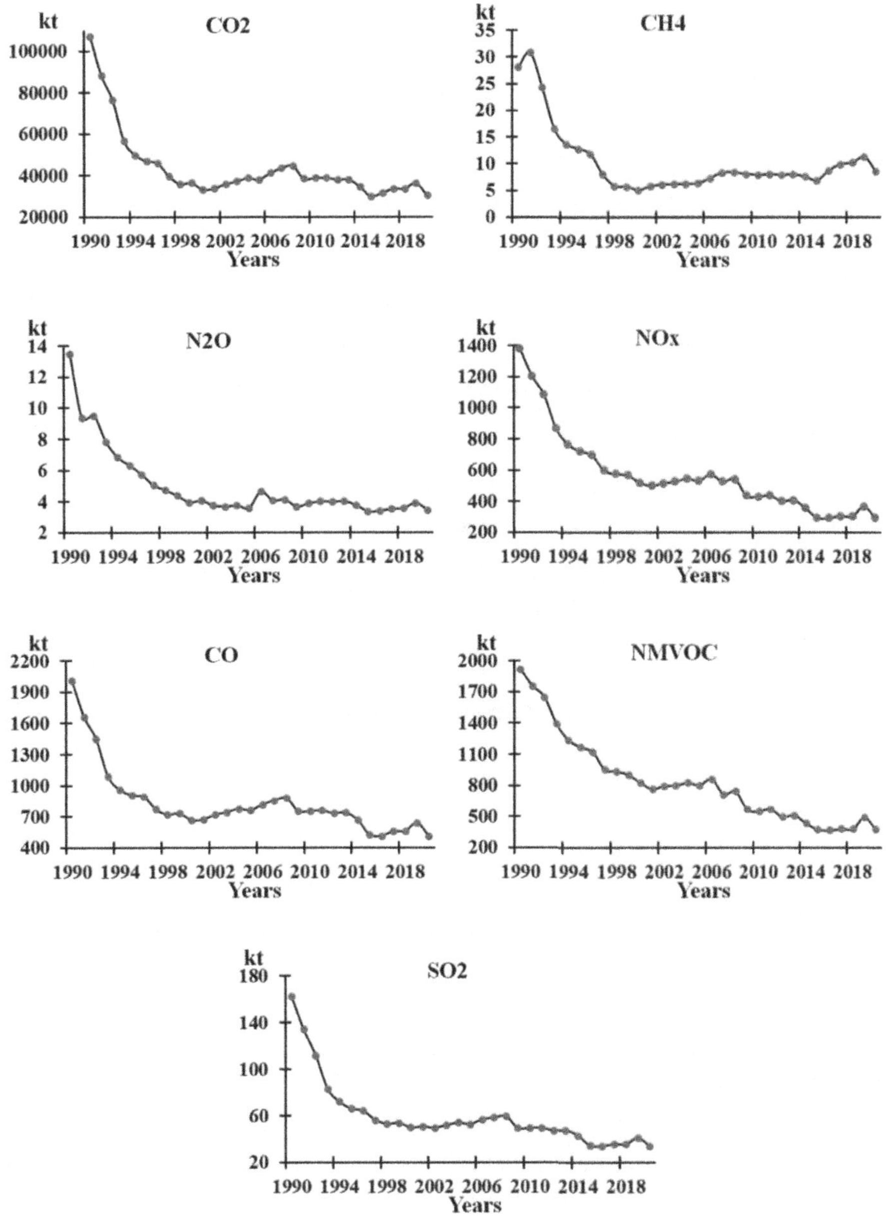

Fig. 1. Dynamics of greenhouse gas emissions from total transport on the territory of Ukraine for the period 1990–2020.

Ukraine, from the categories of "road transportation" and "other transport", the largest emissions of major greenhouse gases occur compared to other categories (see Fig. 2).

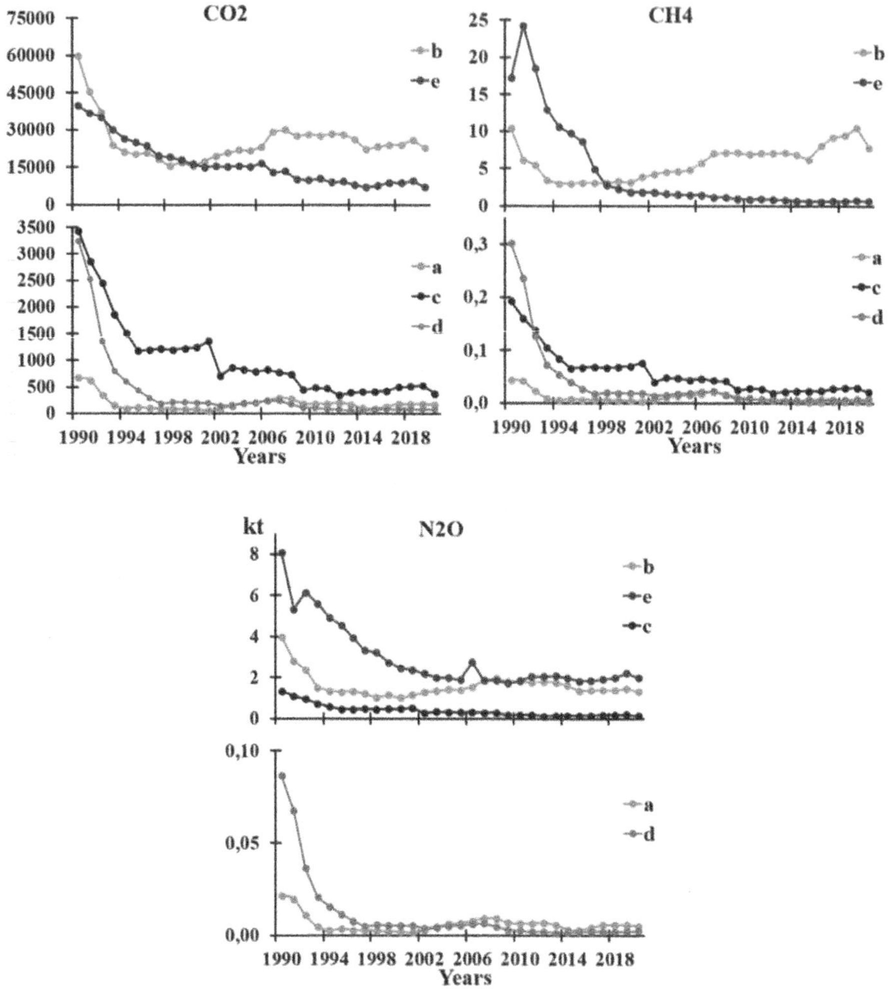

Fig. 2. Greenhouse gas emissions by transport categories from 1990–2020 in Ukraine. Categories of transport: a) domestic aviation, b) road transportation, c) railways, d) domestic navigation, e) other transport.

For the category "road transportation", the average annual values of greenhouse gases are: 25384.84 (kt CO_2), 5.87 (kt CH_4) and 1.61 (kt N_2O), for the category "other transport"—16640.11 (kt CO_2), 4.35 (kt CH_4) and 2.93 (kt N_2O). Thus, emissions in these two categories account for 96.32% of total CO_2 emissions, 98.94% of total CH_4 emissions and 91.53% of total N_2O emissions from transport activities. For the other three categories of transport, greenhouse gas emissions are distributed as follows: for "domestic aviation", the annual averages are: 183.93 (kt CO_2), 0.01 (kt CH_4) and 0.01 (kt N_2O), for "railway": 1019.34 (kt CO_2), 0.06 (kt CH_4) and 0.39 (kt N_2O), for "domestic navigation": 403.45 (kt CO_2), 0.04 (kt CH_4) and 0.01 (kt N_2O). The analysis of greenhouse gas emissions dynamics demonstrates reduction of all gases in all categories since 1990.

Figure 3 demonstrates fourteen air pollution observation posts in the city of Kyiv, which are located on transport routes.

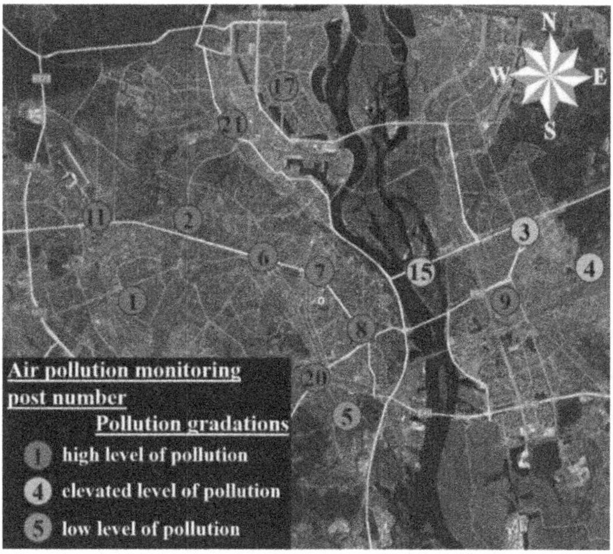

Fig. 3. Scheme of the location of air pollution monitoring posts in the city of Kyiv according to the CGO.

For 2020, the ranking of air pollution observation posts in the city of Kyiv was carried out according to the data of the integral indicator API (air pollution index) by gradations of pollution (Table 1).

The numbers of the posts are indicated in accordance with the numbers used by the Central Geophysical Observatory. According to the level of pollution, there are 3 gradations: low—the API value is less than 5, elevated (API value is from 5 to 7) and high (API value is from 7 to 14). According to the analysis of Table 1, it can be seen that out of 14 air pollution observation posts, the vast majority, namely 10 posts, which is 71.43% of the total number of posts, is in the "high" gradation of pollution. The data from air pollution observation posts record mainly emissions from transport, as they are located on the streets of the city of Kyiv, and not near industrial or other polluting facilities. It is possible to conclude that the main traffic flows of the city of Kyiv are concentrated on a rather limited number of avenues and boulevards, which in turn leads to high values of air pollution on these main transport arteries of the city.

Among the large number of substances that are emitted to the environment, heavy metals are very specific. Once in the environment, heavy metals are included in the biogeochemical cycle and migrate along the links of the trophic chains. Taking into account the cumulative and mutagenic effects of heavy metals, their pollution is a dangerous activity.

Table 1. Ranking of posts of the city of Kyiv by the integral indicator of API (air pollution index) for 2020.

N	Location	API	N	Location	API
20	Demiivska Square	10,88	6	Halytska Square	8,61
21	Sklyarenko Street	10,36	8	Lesia Ukrainka Boulevard	8,07
9	Kaunas Street	10,28	1	Academician Strazheska Street	7,75
7	Bessarabska Square	10,04	3	Pavel Polubotka Street	6,70
11	Beresteiskyi Avenue	9,71	4	Engineer Borodina Street	6,26
17	Obolonsky Avenue	9,63	15	Hydropark	5,32
2	Dovzhenko Street	9,31	5	Nauky Avenue	2,00

Particularly important is the monitoring and establishment of new maximum permissible concentrations for heavy metal contamination of crop production, which is the first link in the trophic chains.

Heavy metals, in particular zinc, copper, lead and cadmium, are very specific in terms of their effect on biological objects. Many of them are necessary for living organisms as trace elements, but due to anthropogenic impact, they can be toxic to biota. Heavy metals are toxic and interfere with the activity of soil microflora. Their concentration in the soil can persist for decades and even centuries [20, 21]. The study analyzed the pollution of cities according to the data of the Central Geophysical Observatory (CGO) (Tables 2, 3 and 4).

Table 2. Ranking of regional centers by the degree of air pollution with lead (Pb) for 2000–2022.

N	City	Pb	N	City	Pb
1	Kharkiv	0,0361	12	Cherkasy	0,0226
2	Kyiv	0,0322	13	Mykolaiv	0,0222
3	Vinnytsia	0,0304	14	Zaporizhzhia	0,0217
4	Sumy	0,0300	15	Khmelnytskyi	0,0217
5	Chernihiv	0,0270	16	Lutsk	0,0213
6	Dnipro	0,0265	17	Odesa	0,0209
7	Ternopil	0,0261	18	Chernivtsi	0,0204
8	Kherson	0,0248	19	Kropyvnytskyi	0,0200
9	Lviv	0,0236	20	Zhytomyr	0,0191
10	Ivano-Frankivsk	0,0235	21	Poltava	0,0191
11	Uzhhorod	0,0235			

Table 3. Ranking of regional centers by the degree of air pollution with iron (Fe) for 2000–2022.

N	City	Fe	N	City	Fe
1	Vinnytsia	8,2013	12	Ternopil	0,7696
2	Ivano-Frankivsk	6,3157	13	Poltava	0,7287
3	Chernihiv	3,6665	14	Uzhhorod	0,7230
4	Lutsk	1,4739	15	Chernivtsi	0,7148
5	Kyiv	1,4178	16	Lviv	0,6309
6	Odesa	1,3426	17	Cherkasy	0,6017
7	Sumy	1,0696	18	Kherson	0,5439
8	Kharkiv	1,0157	19	Zaporizhzhia	0,5400
9	Dnipro	0,8870	20	Zhytomyr	0,4887
10	Mykolaiv	0,8143	21	Khmelnytskyi	0,4752
11	Kropyvnytskyi	0,7696			

Table 4. Ranking of regional centers by the degree of air pollution with zinc (Zn) for 2000–2022.

N	City	Zn	N	City	Zn
1	Vinnytsia	0,3035	12	Uzhhorod	0,0804
2	Kyiv	0,1991	13	Kropyvnytskyi	0,0765
3	Poltava	0,1883	14	Khmelnytskyi	0,0757
4	Cherkasy	0,1683	15	Chernivtsi	0,0700
5	Kharkiv	0,1470	16	Ternopil	0,0670
6	Ivano-Frankivsk	0,1339	17	Kherson	0,0648
7	Lutsk	0,1252	18	Zhytomyr	0,0604
8	Chernihiv	0,1248	19	Zaporizhzhia	0,0604
9	Sumy	0,1135	20	Mykolaiv	0,0591
10	Odesa	0,1045	21	Lviv	0,0577
11	Dnipro	0,0865			

Based on the data obtained in Tables 2, 3 and 4, all cities in Ukraine suffer from air pollution by harmful heavy metals. According to the data obtained, Kharkiv, Kyiv, Sumy, Ivano-Frankivsk and Chernihiv are among the mostly polluted cities of Ukraine. The most polluted city is Vinnytsia, which does not have heavy metallurgy enterprises, so the main polluter is road transport.

4 Conclusions

The study of the analysis of the dynamics of greenhouse gas emissions in the transport sector for the period 1990–2020, for five categories of transport, resulted in distinguishing two periods of multi-year distribution: 1990–1998 and 1999–2020, where in the first period there is a sharp reduction in emissions (from 2.05 to 4.82 times), in the second period a slight fluctuation process was identified, that is, reaching a "plateau", especially in comparison with the first period. In each of these periods, the dynamics of reducing greenhouse gas emissions were identified. It is found that the largest emissions in the category occur in two categories of "road transport" and "other transport". Emissions in these two categories account for 96.32% of total CO_2 emissions, 98.94% of total CH_4 emissions and 91.53% of total N_2O emissions from transport activities.

Based on the study, it can be noted that the transport sector of Ukraine is successfully moving towards meeting the declared goal of reducing greenhouse gas emissions by 2030 [10]. The transport sector of Ukraine cannot become completely climate neutral, because it does not absorb emissions, but only produces them, but it has a great potential to reduce emissions. This can be achieved by implementing measures that the European Union uses or plans to use, namely: increasing the use of bicycles, creating bicycle lanes and car-free zones, reducing the use of cars in cities; rejection of gasoline and diesel cars, permission for electric cars and hydrogen cars, etc.

In the future, solving the challenges of energy supply, greenhouse gas emissions, pollutants in relation to transport can become a driving force in the development of stable transport systems. Particularly such its development that will ensure the fulfillment of socially and economically driven volumes of transportation without harming human health and the environment. The state strategy in the transport sector must necessarily include a section of environmental safety, and monitoring of greenhouse gas emissions, their assessment and forecast in the transport sector can play an important role in achieving national goals for reducing greenhouse gas emissions.

Acknowledgements. We would like to express our gratitude to the Chief Specialist of the Energy Sector of the Budget Institution "National Center for GHG Emission Inventory" Oleksandr Stakhovskyi for providing statistical data.

References

1. Jing, Q.-L., Liu, H.-Z., Yu, W.-Q., He, X.: The impact of public transportation on carbon emissions—from the perspective of energy consumption. Sustainability **14**(10), 6248 (2022). https://doi.org/10.3390/su14106248
2. Din, A.U., et al.: How sustainable transportation can utilize climate change technologies to mitigate climate change. Sustainability **15**(12), 9710 (2023). https://doi.org/10.3390/su15129710
3. Ferrer, A.L.C., Thomé, A.M.T.: Carbon emissions in transportation: a synthesis framework. Sustainability **15**(11), 8475 (2023). https://doi.org/10.3390/su15118475
4. Banister, D., Anderton, K., Bonilla, D., Givoni, M., Schwanen, T.: Transportation and the environment. Annu. Rev. Environ. Resour. **36**(1), 247–270 (2011). https://doi.org/10.1146/annurev-environ-032310-112100

5. Sher, F., et al.: Unprecedented impacts of aviation emissionson global environmental and climate change scenario. Curr. Pollut. Rep. **7**, 549–564 (2021). https://doi.org/10.1007/s40726-021-00206-3
6. Shen, Q., Pan, Y., Feng, Y.: The impacts of high-speed railway on environmental sustainability: quasi-experimental evidence from China. Hum. Soc. Sci. Commun. **10**, 719 (2023). https://doi.org/10.1057/s41599-023-02135-6
7. Smith, R.A.: Railways: how they may contribute to a sustainable future. Proc. Inst. Mech. Eng. Part F: J. Rail Rapid Transit **217**(4), 243–248 (2003). https://doi.org/10.1243/095440903322712847
8. Timmis, A.J., et al.: Environmental impact assessment of aviation emission reduction through the implementation of composite materials. Int. J. Life Cycle Assess. **20**, 233–243 (2015). https://doi.org/10.1007/s11367-014-0824-0
9. IPCC 2008: 2006 IPCC Guidelines for National Greenhouse Gas Inventories—A primer, Prepared by the National Greenhouse Gas Inventories Programme. Institute for Global Environmental Strategies (IGES) (2008)
10. Ukraine's Greenhouse Gas Inventory 1990–2021. Annual National Inventory Report for Submission under the United Nations Framework Convention on Climate Change and the Kyoto Protocol. Ministry of Environmental Protection and Natural Resources of Ukraine, Kyiv (2023)
11. Yelistratova, L., Apostolov, A., Khodorovskyi, A., Tymchyshyn, M.: Monitoring Nitrogen Dioxide (NO2) in environment of Ukraine based on satellite data. Geomat. Environ. Eng. **17**(6), 95–110 (2023). https://doi.org/10.7494/geom.2023.17.6.95
12. Tymchyshyn, M.A., Yelistratova, L.A., Apostolov, A.A., Khodorovskyi, A. Ya., Zakharchuk, I. V.: Satellite monitoring of methane content in the atmosphere on the territory of Ukraine. In: 17th International Conference Monitoring of Geological Processes and Ecological Condition of the Environment, pp. 1–5 (2023). https://doi.org/10.3997/2214-4609.2023520033
13. Tymchyshyn, M.A., Yelistratova, L.A., Apostolov, A.A., Romanciuc, I.F.: Remote detec-tion the CO2 concentration within different land classes cover on the territory of Ukraine. In: Geoinformatics, vol. 21046 (2021). https://doi.org/10.3997/2214-4609.20215521046
14. Boris Sresnevsky Central Geophysical Observatory. http://cgo-sreznevskyi.kyiv.ua/en/
15. Yearbook of the state of atmospheric air pollution on the territory of Ukraine according to the data of the state system of observations of the Hydrometeorological Service. Central Geophysical Observatory, Kyiv (2000–2020)
16. Monthly bulletin of atmospheric air pollution in Kyiv and cities of the Kyiv region for 2020. Central Geophysical Observatory, Kyiv (2020)
17. About the State Electrification Program of railways for 1994–2004. https://zakon.rada.gov.ua/laws/show/115-94-%D0%BF#Text
18. Demchuk, O.A.: Impact of road and rail transport on land pollution by heavy metals within the city of Vinnytsia. Ministry of Education and Science of Ukraine Vinnytsia National Agrarian University (2018)
19. Tánczos, K., Török, Á.: Impact of transportation on environment. Period. Polytech. Transp. Eng. **36**(1–2), 105–110 (2008). https://doi.org/10.3311/pp.tr.2008-1-2.19
20. Petrol fuel quality and its effects on the vehicle technology and the environment. TNO report (2020)
21. Explaining road transport emissions: Publications Office of the European Union. Luxembourg. (2016). https://doi.org/10.2800/71804

Information Security in Smart City Systems: Challenges and Solutions

Andriy Tymoshenko, Serhii Yahodzinskyi, Roman Yaroviy(✉), and Anatolii Kozynets

Private Higher Educational Establishment, European University, 16V. Vernadskoho Akademika Blvd., Kyiv 03115, Ukraine
roman.yaroviy@e-u.edu.ua

Abstract. Modern cities face numerous challenges due to the rapidly increasing strain on infrastructure. The implementation of "Smart City" technologies has emerged as a response to these challenges, utilizing information and communication technologies (ICT) to enhance the efficiency of urban operations, improve citizens' quality of life, and promote sustainable development. This paper examines the key aspects and challenges of information security in Smart City systems, with a focus on the protection of data confidentiality, integrity, and availability. Special attention is given to the experiences of the United States and the European Union, which are leaders in the implementation of smart city technologies. The primary threats to Smart City systems include breaches of data confidentiality, integrity, and availability. The paper provides an analysis of protective methods against these threats, such as encryption, multi-factor authentication, and continuous system monitoring. Issues related to data availability are explored through the lens of vulnerabilities caused by cyberattacks. The presented approaches to addressing these challenges emphasize the need for continuous improvement of security measures. While the integration of cutting-edge technologies can significantly improve residents' quality of life and the efficiency of urban services, it requires substantial investment and attention to cybersecurity concerns.

Keywords: Smart City · Information Security · Cybersecurity · Data Privacy · DDoS Attacks · Blockchain · Artificial Intelligence · Machine Learning · IDS/IPS · Big Data · Cyber Threats · Urban Infrastructure

1 Introduction

Over the past decade, urban population growth has increased significantly around the world [1]. This significantly increased the burden on all services responsible for city management. Many of the technologies that support life in cities have already reached their limits in serving a rapidly growing population. One of the leading trends in technology development in the USA and the European Union is the introduction of highly organized systems known as "Smart Cities" [2]. The concept of Smart City involves the use of information and communication technologies (ICT) to increase the efficiency of

O. Slavinska et al. (Eds.): ITSESQC 2024, LNNS 1335, pp. 162–172, 2025.
https://doi.org/10.1007/978-3-031-87376-8_15

work in cities, improve the quality of life of citizens and promote sustainable development. This includes the integration of various urban systems such as transport, energy supply, waste management, security, etc. The main goal is to create intelligent systems capable of adapting to changing conditions and ensuring sustainable functioning of urban infrastructure [2].

The USA and EU countries are leaders in implementing Smart City technologies. In the US, cities such as New York, San Francisco, and Chicago are actively developing and using smart technologies to manage transportation, energy, and utilities [3, 4].

In the EU, cities such as Barcelona, Amsterdam and Copenhagen are known for their advanced solutions for smart urban resource management and sustainable development [5, 6].

Integrating smart technologies into urban infrastructure requires significant investment and careful planning. However, the main problem remains the provision of information security of these systems [7]. Information security in the context of smart cities involves protecting the confidentiality, integrity and availability of data, which is critical for the sustainable functioning of urban systems. The US and the EU have implemented strict personal data protection standards and regulations, such as GDPR (General Data Protection Regulation) in the EU and CCPA (California Consumer Privacy Act) in the US, which help minimize the risks of data leakage and unauthorized access [8].

At the same time, the introduction of intelligent technologies is associated with several problems, including issues of data protection and security [10]. Information systems of smart cities become potential targets for cyberattacks that can lead to serious consequences, such as disruptions in transportation systems, energy supply, and other critical infrastructure [10].

Therefore, the successful implementation and operation of Smart City systems requires a comprehensive approach to ensuring information security, including the implementation of modern methods of data protection, multi-level authentication and continuous monitoring of the system [11]. The experience of the USA and the EU in this area can be valuable for other countries that seek to improve the efficiency of urban management and ensure the sustainable development of their cities [12].

The main goal of this study is the analysis and research of information security challenges faced by the "Smart City" systems and methods of their elimination.

Information security in Smart City systems is an important topic that attracts the attention of researchers and practitioners in different countries.

In particular, research devoted to the use of artificial intelligence (AI) and machine learning to monitor and quickly respond to cyber threats is of great importance. Works such as the IBM X-Force Threat Intelligence Index (2023) consider the use of AI to automate the process of detecting anomalies in traffic and warning of possible threats.

2 Research Methodology

In order to obtain an objective conclusion regarding the level of threats and the vulnerability of the information systems of the "Smart City", we offer a threat and risk analysis technique that is used to study cyber threats. The methodology explored the main types of attacks, including DDoS, phishing attacks, and SQL Injection. Risks were assessed using data from ENISA [13] and Verizon [14] reports, which include statistics on the frequency and impact of cyber threats.

The reliability of the method is ensured by the use of data from international reports that collect and analyze cyber threats around the world. Threat analysis is necessary to identify potential vulnerabilities and provide an adequate response to new types of attacks.

Availability, integrity, and privacy threats are key challenges for information security in smart cities. DDoS attacks and physical infrastructure attacks threaten the availability of city services, while malware and database attacks can affect the integrity of critical data. Phishing attacks and data breaches are the main threats to privacy, underscoring the need to implement comprehensive protection measures and increase user awareness of the main cyber threats. In Fig. 1. Provided a chart of threat statistics with availability, integrity, and privacy in mind.

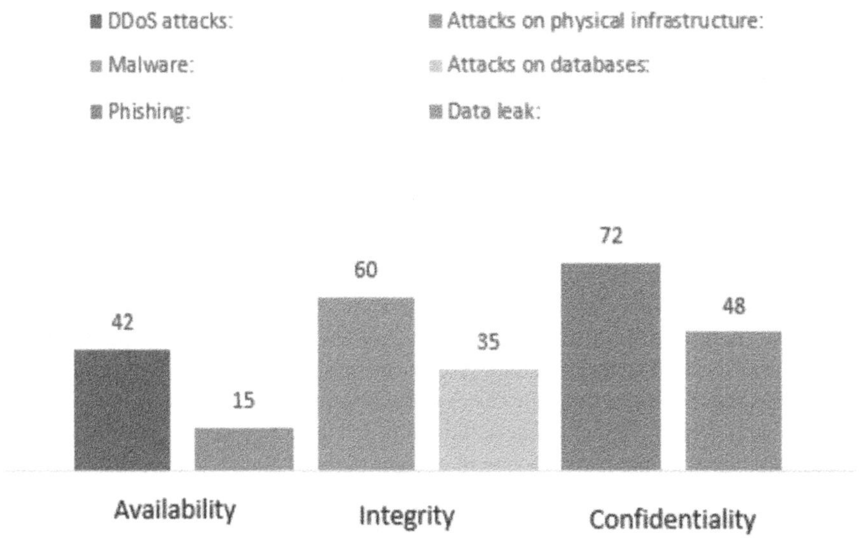

Fig. 1. Threat distribution diagram for smart cities

The study revealed threat statistics related to failures of supporting infrastructure in Smart City systems. These results emphasize the importance of stable operation of critical urban systems and the need for prompt response to failures. Figure 2. Statistics of threats related to failures of the supporting infrastructure are shown, taking into account the average recovery time.

Statistics show that smart cities are actively implementing various protection measures to ensure information security. The use of modern technologies such as AI, ML, blockchain and cloud services, combined with multi-factor authentication, data encryption and intrusion detection systems, allows you to effectively counter cyber threats. User education and the implementation of safety standards also play an important role in creating a safe urban environment. Figure 3. a diagram of the distribution of protective actions on the issue of cyber protection in a smart city is shown.

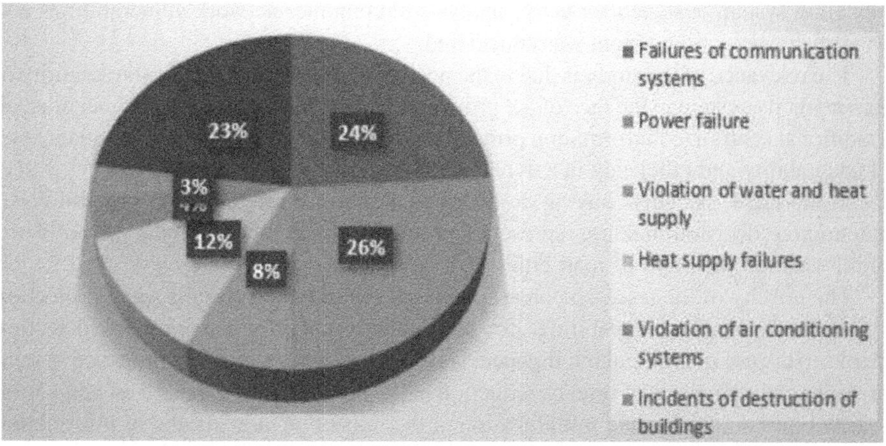

Fig. 2. Statistics on threats related to failures of the supporting infrastructure

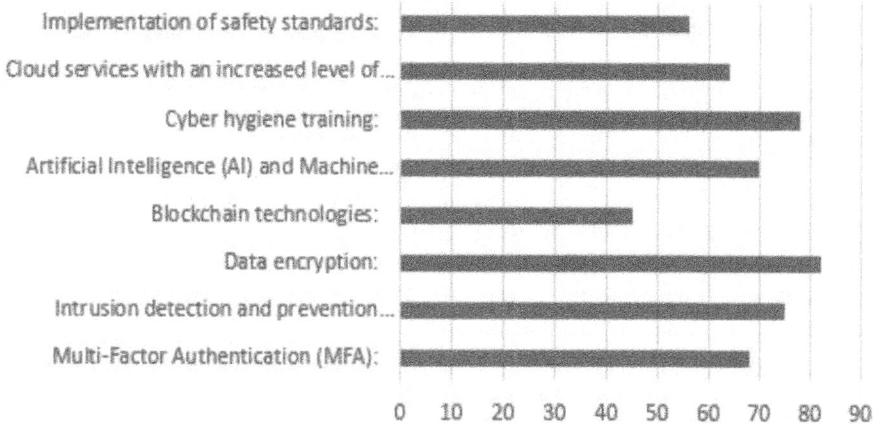

Fig. 3. Diagram of the distribution of protective actions in a smart city

3 Results and Discussion

A promising approach to the construction of complex protection systems in a smart city, which allows overcoming some of the shortcomings, is the use of intelligent information protection systems based on the technology of multi-agent systems. Compared to traditional methods, this approach allows you to significantly increase the effectiveness of cyber protection mechanisms, in particular, their efficiency, adequacy, fault tolerance, resistance to destructive actions, flexibility, etc.

This work is devoted to the development of formal models, implementation architectures of multi-agent protection systems, which serve to analyze vulnerabilities and detect intrusions into computer networks of a smart city. The decision to create an attack modeling system designed for active analysis of computer network vulnerabilities and an intrusion detection system was considered.

The relevance of the study is due to the need to increase the level of cyber security in the smart city system in the face of the growing number and complexity of cyber attacks. Traditional centralized information protection systems do not provide an adequate level of adaptability and efficiency in a dynamically changing threat environment.

Multi-agent systems, having a distributed architecture and the possibility of autonomous decision-making, represent a promising direction for solving the problems of information security of Smart cities.

The novelty of the research consists in the development of a multi-agent protection system with enhanced capabilities for vulnerability analysis and intrusion detection thanks to the use of artificial intelligence. Unlike existing solutions, the proposed system provides active analysis of threats, reduction of the number of false positives, adaptation to new types of attacks, and minimization of the impact on target traffic of information flows.

A general description of the architecture of a typical agent and the cooperation scheme of multi-agent information protection systems.

An intelligent agent is a software or hardware implemented system that has autonomy, a set of "mental properties" and is able to function in a community with other agents.

The architecture of a typical agent includes the following components:

Incoming message buffer—for recording received messages.
Outgoing message buffer—for recording sent messages.
Incoming traffic processor—for processing received messages, parsing, interpreting KQML messages and extracting message content.
Agent dialogue database—stores the attributes of each incoming message (identifiers, message type, its source, etc.).
Meta automaton—controls the semantic processing of incoming messages, directing them to be processed by appropriate automata. In addition, this component manages the parallel execution of various processes within the agent.
Automata—perform basic calculations of the agent, which are determined by its role in the multi-agent system, implementing various scenarios for processing incoming messages.

Differences between agents are reflected in the content of agents' data and knowledge bases. Each agent interacts with other agents, the environment that is perceived and possibly modified by the agents, and the user who interacts with the agents through a user interface. The architecture of a typical agent looks like Fig. 4.

The scheme provides for the cooperation of multi-agent information protection systems. The software code of multi-agent information protection systems, such as the agent-oriented attack modeling system (AAMS) and the multi-agent intrusion detection system (MIDS), is written using the MASDK toolkit, C++, JAVA and XML development tools.

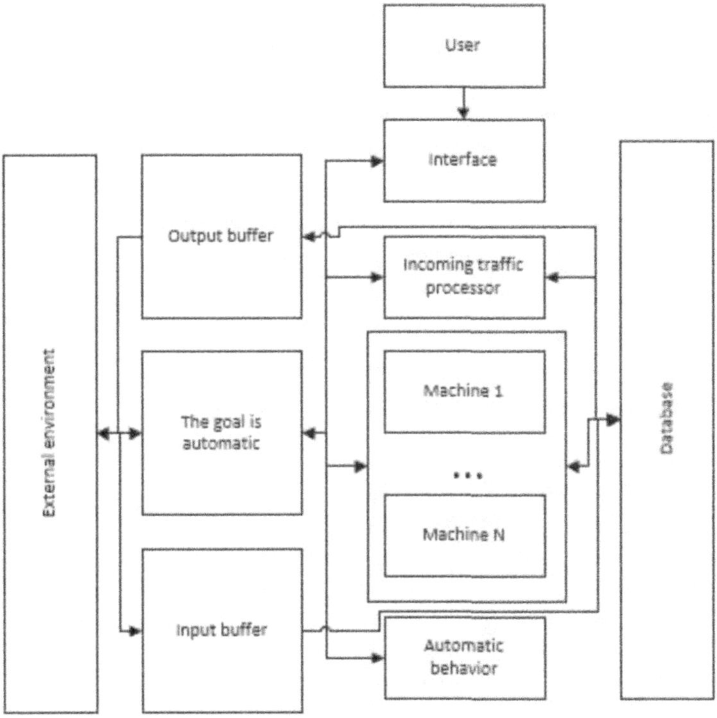

Fig. 4. The architecture of a typical agent

AAMS simulates incoming traffic, i.e. a combination of both normal and abnormal event flows. A stream of abnormal events triggers attacks on a computer network. Inbound abnormal event traffic corresponds to sequences of "single-phase" attacks that use different hosts as the "initiation entry points" of the attacks.

The Ministry of Internal Affairs and Communications is responsible for detecting attacks on the computer network formed by AAMS. This network can include several different segments, and incoming traffic can originate from both inside and outside the protected network. It is assumed that attackers can be located at different points of the network, both outside and inside the protected fragment of the network.

Thus, multi-agent information protection systems provide an opportunity to more effectively detect and prevent threats, thanks to the distributed architecture, the autonomy of agents and their ability to cooperate and adapt in dynamically changing conditions.

When implementing complex coordinated attacks, depending on the established general goal of the attack, a specially designated meta-agent (agent-leader) chooses a general scenario for countering the attack and assigns areas of responsibility to other agents. Agents responsible for individual fragments (steps) of the overall scenario can, in turn, "hire" other agents or implement individual actions independently. For this, special scenarios (plans) of actions and communication protocols are intended.

The scenario is set in the form of a sequence of intentions (sub-goals) and ways to achieve them (actions), which can be described at different levels of detail. Intentions of the lowest level are implemented by specific actions, which are expressed in the form of network packets, operating system commands, or operations with a specific application.

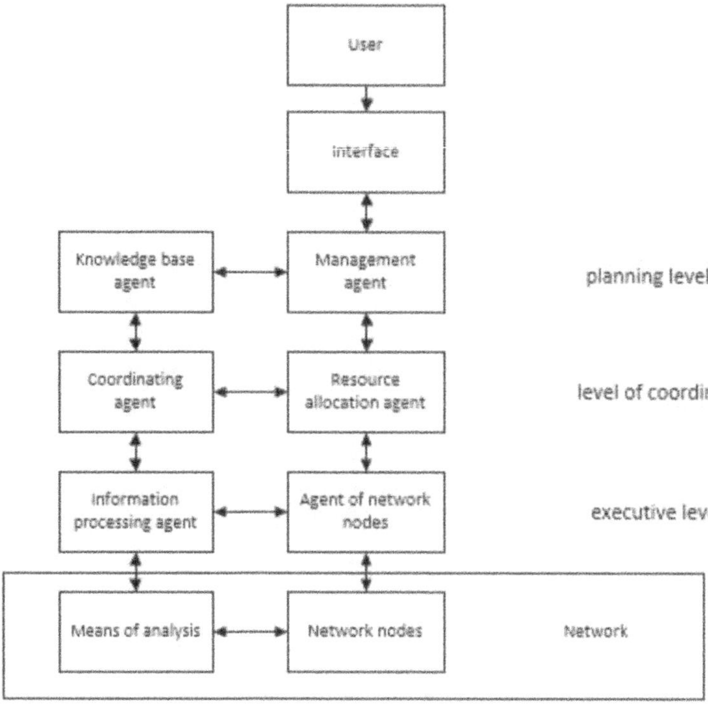

Fig. 5. Interaction of rents in a multi-agent system of computer network security analysis

Figure 5 shows the diagram of the interaction of agents and levels in the multi-agent system of computer network security analysis.

The main tasks of multi-agent systems (MAS) in the protection of the "Smart City" infrastructure.

1. Monitoring and detection of threats in the real time

Multi-agent systems can be used for distributed monitoring of urban networks and systems. Each agent is responsible for a specific part of the infrastructure, such as water, electricity, transport or communication systems. Each agent is able to:

Collect data on system operation.

Detect anomalies or threats based on given parameters and compare them with the historical data.

Respond quickly to anomalies by sending danger signals to other agents or to the central system.

2. Coordination of actions between agents

One of the key features of MAS is the ability of agents to interact with each other, which is important for complex Smart City systems. If a threat is detected, the agent can notify other systems and coordinate actions to contain the threat or minimize its impact.

3. Adaptation to new threats

MAS have the ability to learn and adapt, which is important for cyber security in the face of constantly changing threats. Using machine learning technologies, agents can adjust their threat detection algorithms to be more effective in the future:

Agents can accumulate knowledge about past incidents and threats and use that knowledge to predict new attacks. During a new attack, agents can independently change defense strategies, increasing their effectiveness [15].

4. Automation of Protective Actions

Each agent can perform autonomous protective actions based on predefined scenarios. For example, if an unauthorized access attempt or SQL Injection attack is detected, the agent can automatically:

Block suspicious IP addresses.

Isolate the affected part of the system to prevent the spread of the threat.

Redirect resources to minimize the impact of the attack.

This allows you to significantly reduce the response time to threats and minimize the impact on the infrastructure.

5. Distributed risk management

Multi-agent systems make it possible to implement distributed risk management in the "Smart City". Each agent can conduct risk analysis for their area of responsibility and interact with other agents to coordinate actions. This makes it possible to identify potential threats and implement countermeasures before an attack reaches a critical stage.

6. Simulation and testing of attack scenarios

MAS can be used to simulate attacks and test the defense mechanisms.

This not only allows you to test current defense strategies, but also helps improve the effectiveness of the agent in the event of real attacks.

7. Decentralization of decision-making

Multi-agent systems avoid a central point of failure because decisions are made at the level of each individual agent. This reduces the risk that a single cyber threat can paralyze the entire system, as agents can function independently of each other, handling local problems and making decisions.

The use of artificial intelligence (AI) in multi-agent systems (MAS) significantly increases their efficiency and adaptability, especially in the field of "Smart City" infrastructure protection. AI enables agents to analyze large volumes of data, make informed decisions based on threat analysis, and learn in real time to improve defensive strategies.

AI helps agents detect threats on their own by using machine learning algorithms to analyze anomalies in network traffic or user behavior. Thanks to this: Agents can independently analyze large volumes of data and identify potential threats at an early stage, which increases the chances of successfully preventing cyber attacks.

AI algorithms allow agents to adapt to new types of threats that were not previously known and automatically update their defense strategies.

AI allows agents to predict future threats based on historical data analysis. This makes it possible to create forecasting models: Agents can predict the most likely attack scenarios by analyzing past incidents and changing defense priorities.

AI-powered risk assessment allows agents to quickly assess the threat level and make decisions about protecting the city's critical systems.

Using AI technologies, agents can coordinate their actions more effectively. AI helps agents: Jointly solve complex tasks, in particular, when it is necessary to coordinate the protection of several systems at the same time (power supply, transport, water supply).

Automatically prioritize response to threats based on potential impact on urban systems.

AI in MAS allows agents to learn from previous attacks to act faster and more efficiently in the future: Reinforcement learning allows agents to gain experience during successful or unsuccessful actions to further improve their decisions; this enables the creation of "learning" agents that use knowledge of past threats to detect new types of attacks and respond more quickly to them.

AI allows agents to analyze user behavior, transportation systems, network activity, and other data in real time. This helps: Detect atypical behavior patterns that may indicate a possible attack or intrusion; AI agents can quickly assess whether certain user or device behavior is normal for the system or indicates a threat.

Deep learning algorithms allow agents to process complex data and recognize patterns that a human might miss. This includes: Detection of anomalies in large data flows from sensors and monitoring systems in the city; deep learning increases the accuracy of attack detection and the effectiveness of classification of different types of threats.

AI allows agents not only to detect threats, but also to automatically respond to them without the need for human intervention. This includes: Automatically block suspicious network activity such as SQL Injection attempts, unauthorized access attempts, or anomalous traffic; redirection of resources, for example, switching to a backup power supply if a cyber threat is detected on the main network.

4 Conclusion

Smart city systems face numerous challenges in the field of information security, which is associated with high dependence on information and communication technologies (ICT). These challenges include threats to the privacy, integrity and availability of data, which can have serious implications for critical urban infrastructure such as energy, transport and water supply.

The article discussed key cyber threats to the Smart City, including DDoS attacks, phishing attacks, and SQL Injection. Statistics from sources such as ENISA and Verizon highlight the seriousness of these threats and point to the need for comprehensive protection measures. A method of using multi-agent systems (MAS) with artificial intelligence (AI) was proposed as a promising solution for protecting the infrastructure of the "Smart City". Their use allows for decentralized monitoring, detection of threats in real time, coordination of actions between agents, as well as automation of protective measures.

Intelligent AI agents are able to adapt to new threats, learn from previous experience and increase the effectiveness of their actions when detecting and neutralizing cyber threats. This makes it possible to significantly reduce the response time to attacks and increase the reliability of the "Smart City" systems.

Thus, to ensure the sustainable development of "Smart Cities", it is necessary to implement innovative protection systems, including multi-agent systems using AI, which will ensure flexibility, autonomy and reliability in a dynamically changing cyber environment.

References

1. Komninos, N., Kakderi, C., Panori, A., Tsarchopoulos, P.: Smart city planning from an evolutionary perspective. J. Urban Technol. **26**(2), 3–20 (2019). https://doi.org/10.1080/10630732.2018.1485368
2. Silva, G.M.S.R.: Smart cities and their importance in urban development. Urban Sci. **3**, 48 (2019). https://doi.org/10.3390/urbansci3020048
3. Abualigah, M.A.Z., Alharbi, M.S., Khadher, A.M.: Smart city systems: a review on information security challenges. Inf. Secur. J. Glob. Perspect. **28**, 202–211 (2019). https://doi.org/10.1080/19393555.2019.1566367
4. Chen, J.M.F., Jiang, L., Kim, J.L.: Smart cities: foundations, principles, and application. Comput. Environ. Urban Syst. **74**, 1 (2019). https://doi.org/10.1016/j.compenvurbsys.2018.12.001
5. Verma, R., Malik, B., Kumar, M., Rathi, A., Lal, Y.P.: Security and privacy in smart cities: current trends and future directions. IEEE Trans. Emerg. Top. Comput. **7**, 402–413 (2019). https://doi.org/10.1109/TETC.2017.2764885
6. Rahman, S.A., Williams, A.M.: A study on the impact of GDPR on smart city development. Int. J. Inf. Manag. **47**, 102–107 (2019). https://doi.org/10.1016/j.ijinfomgt.2019.01.004
7. Almazroi, A.H., Bin Taleb, F.M.: Adopting smart cities: a framework for assessing readiness. J. Urban Technol. **26**, 1–16 (2019). https://doi.org/10.1080/10630732.2018.1550093
8. Chatterjee, D., Mishra, K.T., Kalra, E.A., Rathi, M.R.P.: Data Privacy and security issues in smart cities. Int. J. Smart Home. **13**, 1–12 (2019). https://doi.org/10.14257/ijsh.2019.13.3.01
9. General Data Protection Regulation (GDPR). Official Journal of the European Union. European Commission (2019). https://eur-lex.europa.eu
10. Shokry, G., Mahmoud, F.M., Latif, A.J., Khalil, P.S., Saeed, S.G., Mohtaseb, H.A., Taha, M.B.: Security issues in smart cities: a systematic literature review. J. King Saud Univ. Comput. Inf. Sci. **31**, 5–19 (2019). https://doi.org/10.1016/j.jksuci.2017.06.002
11. Schmid, T.C., Dong, P.L., Yang, R.V.: Cybersecurity in smart cities: challenges and opportunities. Future Gener. Comput. Syst. **100**, 108–120 (2019). https://doi.org/10.1016/j.future.2019.02.029
12. Van der Meer, E.S., et al.: The role of artificial intelligence in smart city development. Artif. Intell. Rev. **52**, 163–189 (2019). https://doi.org/10.1007/s10462-018-9622-9
13. ENISA Threat Landscape Report. European Union Agency for Cybersecurity (ENISA) (2023). https://www.enisa.europa.eu/publications

14. Verizon Data Breach Investigations Report (DBIR). Verizon Enterprise Solutions (2023). https://enterprise.verizon.com/resources/reports/dbir
15. Lakhno, V.A., Kasatkin, D.Y., Skliarenko, O.V., Kolodinska, Y.O: Modeling and optimization of discrete evolutionary systems of information security management in a random environment. Machine learning and autonomous systems. Smart Innov. Syst. Technol. **269**, 9–22 (2022)

Smart Cities and Ecological Transport: Challenges, Prospects and Impact on Sustainable Development

Hanna Kharchenko[1](✉) and Volodymyr Kharchenko[2]

[1] Glasgow Caledonian University, G4 0BA Glasgow, Scotland, UK
hanna.kharchenko@gcu.ac.uk
[2] National University of Life and Environmental Science of Ukraine, Kyiv 03189, Ukraine

Abstract. This research examines the challenges, prospects, and impact of integrating ecological transport systems in Ukraine within smart cities, with a focus on sustainable development. Urbanisation has led to significant environmental, economic, and social challenges, including increased traffic congestion, greenhouse gas emissions, and air quality deterioration. Traditional transportation systems often fail to address these issues effectively. The study explores the potential of environmentally friendly transport solutions, such as electric vehicles, hybrid cars, hydrogen-powered vehicles, and non-motorized transport, in mitigating these challenges. The research employs a comprehensive methodological approach, including SWOT analysis, to assess the strengths, weaknesses, opportunities, and threats of ecological transport systems in Ukraine, particularly in the context of ongoing military conflict and economic instability. The findings highlight the environmental benefits of reduced emissions, economic advantages like job creation and lower fuel costs, and social improvements in public health and transport accessibility. However, challenges such as insufficient infrastructure, economic instability, and the need for international support remain. The study concludes that integrating ecological transport into smart cities can significantly contribute to sustainable development, but requires targeted policies, infrastructure investment, and innovative technologies to overcome existing barriers.

Keywords: Smart Cities · Ecological Transport · Electric Vehicles · Sustainable Development · Urbanization · Greenhouse Gas Emissions · Intelligent Transport Systems

O. Slavinska et al. (Eds.): ITSESQC 2024, LNNS 1335, pp. 173–184, 2025.
https://doi.org/10.1007/978-3-031-87376-8_16

1 Introduction

The development of urban agglomerations is accompanied by significant environmental, economic, and social challenges. Urbanisation contributes to increased traffic loads, elevated greenhouse gas emissions, and deteriorating air quality, all of which negatively affect public health and the overall ecological condition of cities. Traditional transportation systems often struggle to address the complexities of urban congestion, limited access for diverse population segments, and high fuel and maintenance costs.

As urban populations and transportation demands continue to grow, there is an increasing need for new transport solutions that are logistically efficient, environmentally sustainable, and economically feasible. In this context, smart cities emerge as a central strategy for sustainable development. Smart cities leverage innovative technologies to optimize the management of urban resources, including transportation systems, aiming to enhance the quality of life for residents while minimising environmental impact.

It is worth noting that environmentally friendly transport systems include various technologies and solutions aimed at reducing harmful emissions, increasing energy efficiency, and improving air quality. These can be electric cars, hybrid cars, hydrogen and biofuel vehicles, as well as bicycle transport and other forms of non-motorized transport. The use of such vehicles makes it possible to reduce the environmental burden on cities, reduce the noise level, and improve the overall environmental situation.

To effectively address urban congestion, energy efficiency, and safety concerns, a hierarchical transportation approach is often recommended [25]. This hierarchy prioritizes high-capacity and space-efficient rail systems, such as trains, metros, and trams, as primary options for intercity and suburban travel. Within this structure, rail-based transit serves as the backbone of urban mobility due to its capacity to move large numbers of people with minimal spatial and environmental impact. Below rail-based transit, public transport options—such as buses and electric shuttles—serve medium-demand routes. At the local level, micro-mobility options, including bicycles and e-scooters, facilitate movement to and from urban nodes and areas not directly accessible by larger transit modes.

The primary purpose of the research is to explore the development and implementation of environmentally friendly transport systems within the context of smart cities. The main objectives are to:

1. Evaluate the effectiveness of various green transport technologies in reducing urban pollution and improving energy efficiency.
2. Analyse the economic impact of adopting sustainable transport solutions, with a focus on cost savings, job creation, and economic growth.
3. Examine the social benefits of environmentally friendly transportation, particularly regarding accessibility, health outcomes, and social equity.

Previous research has extensively explored the environmental impacts of traditional transportation systems [20, 25], highlighting their significant contribution to urban pollution and public health issues. Studies have also examined the economic benefits of transitioning to green transport [1, 21], such as cost reductions in fuel and maintenance, as well as the potential for job creation in the green technology sector. Furthermore, social research has underscored the importance of equitable access to transportation [5, 9, 14], particularly in enhancing the quality of life for marginalized communities.

However, despite these advancements, there are notable gaps in the research concerning the long-term economic impacts of environmentally friendly transport systems in diverse urban contexts. Additionally, the social implications of these systems, particularly their role in promoting inclusivity and equity, have not been thoroughly examined.

Specifically, the study focuses on evaluating the long-term economic and social benefits of green transport solutions, as well as their integration into the broader smart city framework. Through this comprehensive analysis, the study seeks to offer practical recommendations for implementing sustainable transport solutions in urban settings.

2 Literature Review

The development and implementation of environmentally friendly transport systems in smart cities is an important topic of current research that covers various aspects of sustainable development, urbanism, and intelligent transport systems (ITS).

It is worth mentioning the significance of energy consumption per passenger-kilometre (kWh/pass*km) and emissions levels as indicators for comparing the environmental performance of various transport modes. Research by Beaudoin and Lin [6] demonstrates that automated metros and urban ropeways consume significantly less energy per passenger-kilometre than private electric vehicles or buses, due to their high-capacity design, which optimises energy use and minimises congestion in densely populated urban areas. This makes them particularly efficient and sustainable choices for urban transport, as they combine low emissions, high spatial efficiency, and a strong safety record [6].

However, research shows that the development of environmentally friendly transport systems in smart cities requires a more detailed analysis of existing technologies and their impact on urban environments. According to research by Shaheen and Cohen [24], an important aspect is the combination of green transport solutions with transport-sharing models such as car-sharing, ride-sharing, and micro-mobility [24]. They found that such models contribute to a significant reduction in the number of cars on the road and, accordingly, to a reduction in emissions. This is especially important for cities with high population density, where space for new infrastructure is limited. According to studies [7, 10, 14], environmentally friendly transport includes a variety of technologies that aim to reduce harmful emissions and improve energy efficiency. They distinguish the following main directions: electric cars, hybrid cars, water, and biofuel vehicles, as well as bicycle transport and other types of non-motorized transport.

Smart cities use Internet of Things (IoT) technologies, big data, and other innovative solutions to optimize urban transportation systems. Allam and Dhunny [2] indicate that such cities can significantly reduce greenhouse gas emissions by improving traffic flow management and implementing intelligent traffic management systems [2]. Bibri [7] emphasizes that the integration of environmentally friendly vehicles into a smart city strategy is a key factor in achieving sustainable development [7]. One of the key technologies underpinning the concept of sustainable transportation in smart cities is the use of artificial intelligence (AI) and machine learning. Oladimeji et al. [23] consider how these technologies can be used to predict transport demand, optimize public transport routes, and manage EV charging infrastructure. The research notes that the use of AI can significantly increase the efficiency of transport systems, which, in turn, leads to reduced emissions and energy savings [24]. As noted in the research of many scientists, electric cars are one of the most promising technologies for reducing emissions in cities [5, 7, 10, 16]. Hardman et al. [15] investigated the impact of electric vehicles on reducing air pollution and found that they can reduce CO_2 emissions by 30–50% depending on the energy source used to charge them [15]. Yun and Lee [27] point out that the expansion of the network of charging stations is a critical factor for the widespread adoption of electric vehicles [27].

Moreover, hybrid cars, which combine internal combustion engines and electric motors, are an effective intermediate solution on the way to full electrification of transport. Boren [9] emphasize that hybrid vehicles can reduce fuel consumption and emissions [9]. Hydrogen transport, as research by Nikolaidis and Poullikkas [22] shows, has great potential to reduce CO_2 emissions, but requires significant investment in hydrogen production and distribution [22]. Public transport that uses alternative fuels, such as biofuels and electricity, is already being successfully implemented in many cities around the world. González-González et al. [12] note that electric buses and biofuel buses demonstrate high efficiency in reducing emissions and energy consumption [12]. Furthermore, intelligent transport systems (ITS) play a key role in the implementation of environmentally friendly transport solutions. Jia Wang [14] examine how ITS can optimize traffic flow, reduce congestion, and improve the overall efficiency of transportation systems [14]. The integration of ITS into urban transport networks allows not only to increase the efficiency of transport flows but also to reduce energy consumption and emissions of harmful substances.

It should be noted that the implementation of environmentally friendly transport systems also has significant social and economic benefits. Cervero and Guerra [11] emphasize the importance of integrating transport and land use to create sustainable urban environments [11]. Moreover, reducing fuel costs, improving air quality, and reducing noise pollution contribute to improving the quality of life of the urban population [5, 10, 16]. In addition, investments in the production and maintenance of environmentally friendly transportation create new jobs and stimulate economic development [17, 18, 20].

The literature review shows that the development and implementation of environmentally friendly transport systems in smart cities is a multifaceted problem that requires a comprehensive approach and the integration of various technologies and solutions. Despite the significant progress in this field, there are certain challenges, such as the need to expand the infrastructure for charging electric vehicles, ensuring the availability of hydrogen technologies, and the integration of intelligent transport systems. Further research and innovation in this area are necessary to ensure the sustainable development of cities and improve the quality of life of their residents.

3 Methodology

To achieve the goal of the research, which consists of the analysis of the implementation of environmentally friendly transport systems in the conditions of smart cities and the assessment of their impact on the environmental situation, economic indicators, and social aspects of the life of the urban population, a comprehensive methodological approach was applied.

To determine the energy efficiency of each mode, energy consumption per passenger-kilometre (kWh/pass*km) across all transport alternatives was analysed, including automated metros, urban ropeways, electric vehicles, and public buses. This metric provides insights into each mode's energy efficiency by measuring the energy required to transport a single passenger over a kilometre. Automated metros and urban ropeways, known for their high passenger capacities and optimised energy use, are expected to exhibit lower kWh/pass x km values, positioning them as efficient urban transit options.

In the next step, the strategic SWOT analysis was used to assess the internal environment and external factors affecting the development of environmentally friendly transport systems in Ukraine. This analysis made it possible to identify the strengths and weaknesses of ecological transport, as well as the opportunities and threats arising in the conditions of military conflict and economic instability.

Synthetic analysis, which involves the integration of the results of quantitative and qualitative analyses to form a comprehensive understanding of the impact of ecological transport on the sustainable development of smart cities. This allows you to summarize the conclusions, identify the main trends, and develop practical recommendations for further development.

This systematic approach provides a thorough analysis of the topic, enabling a comprehensive assessment of the challenges, opportunities, and impact of ecological transportation on the sustainable development of cities.

4 Results and Discussion

The rapid urbanization of modern cities raises multifaceted challenges that include increased traffic congestion, elevated greenhouse gas emissions, and deteriorating air quality. Traditional transportation systems often fall short of addressing these issues, leading to significant environmental and social consequences. In cities like Kyiv and Lviv, the reliance on conventional vehicles has exacerbated pollution levels, contributing substantially to the overall urban emission footprint. For instance, the transport sector is responsible for approximately 15% of total CO_2 emissions in Ukraine, with a significant portion arising from high-density traffic areas [13, 19]. It's important to highlight that electric vehicles are the most widespread type of environmentally friendly transport. They work exclusively on electricity and have a zero level of emissions of harmful substances during operation.

According to [28], Ukraine demonstrates positive dynamics in the introduction of environmentally friendly vehicles from 2020 to 2023. The number of electric cars on Ukrainian roads has increased by more than 300%, from 11,000 in 2020 to 35,000 in 2023. In addition, more than 8,000 hybrid cars were imported, and the number of public transport using alternative fuels (including biofuels and electric buses) increased by 40% in the same period. Such indicators indicate a significant interest in environmentally friendly transport and the potential for its further development in Ukraine [28]. Notably, the main advantages of electric cars include high energy efficiency, reduced dependence on fossil fuels, and reduced greenhouse gas emissions. However, for their widespread implementation, it is necessary to create an extensive network of charging stations.

It is worth noting that in five months of 2024, a record 20.8 thousand electric cars were imported into Ukraine (Fig. 1). Despite all the negative factors, including the blackout, the demand for electric cars has grown.

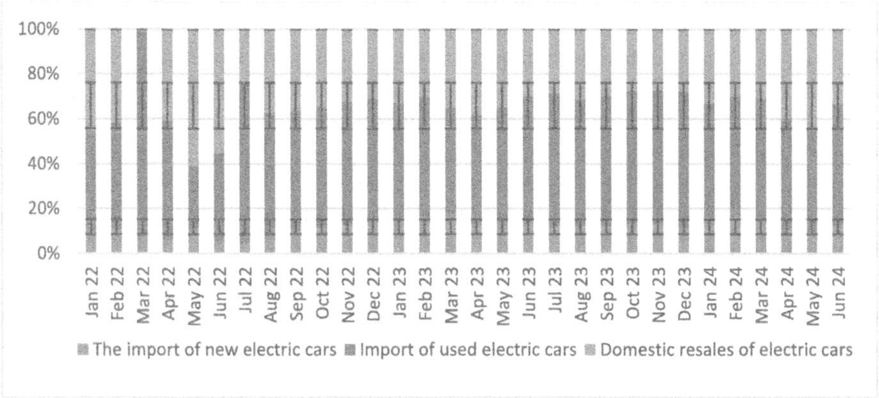

Fig. 1. Dynamics of the electric cars market [28].

It is worth saying, that the number of electric cars is growing rapidly every year (Fig. 2). In the analysed period of 2024, 20,785 electric cars were imported, which is 2.1 times more than in the same period of 2023, when only 9,736 electric cars were imported [27].

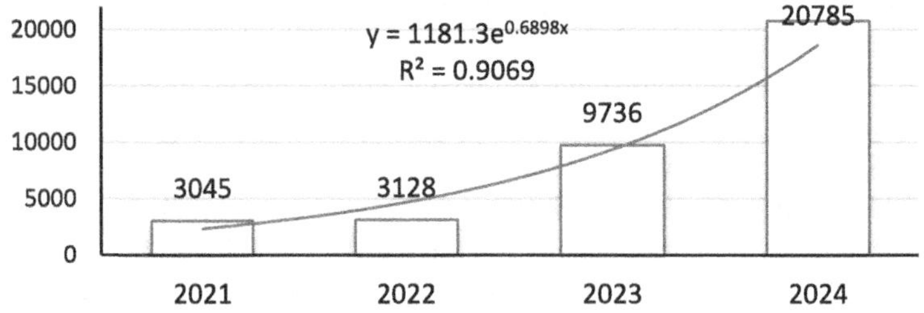

Fig. 2. The dynamics of the number of imported new electric cars [28].

In general, electric cars account for 12.9% of the total number of imported cars. Among electric cars, the Tesla brand has the highest demand in the domestic market with 4,742 cars. Volkswagen is in second place: with 3,194 electric cars, followed by

Nissan with 3,034 electric cars. The most popular models were Nissan Leaf, Tesla Model 3, and Volkswagen ID.4 [28].

The leader among regions in the import of electric cars to Ukraine is the Lviv Region, accounting for 13.5% of the total imports. This is followed closely by Kyiv, which represents 13.1% of the electric cars market, and the Odesa Region, contributing 9.3% (Fig. 3) [19, 28].

The significant uptake in these regions reflects both local governmental support for green technologies and growing consumer interest. In Lviv, the local government has implemented various incentives, including tax reductions and subsidies for electric car purchases, which have spurred higher adoption rates. Kyiv's strong performance is attributed to its large population and extensive infrastructure development, including the expansion of electric car charging stations.

It's worth pointing out that electric cars play a key role in the development of the smart city concept, where technologies are used to improve the quality of life, reduce the impact on the environment, and increase the efficiency of management processes. A smart city integrates advanced technologies to optimize transportation systems, energy networks, and infrastructure.

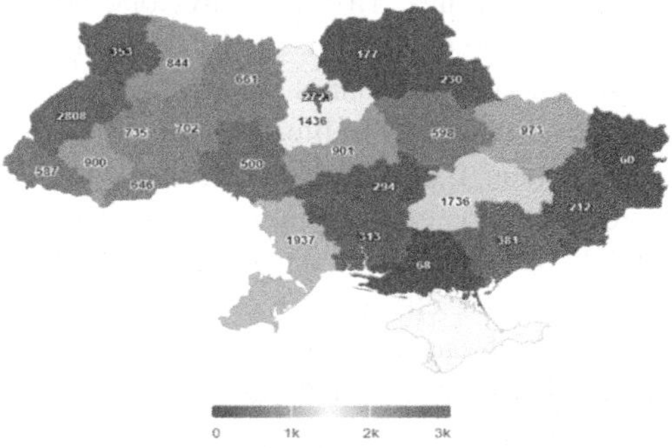

Fig. 3. Geography of import of electric cars to Ukraine [28].

In addition, electric cars can interact with intelligent transport systems, including traffic monitoring and management systems. It facilitates reduce congestion and improve traffic management by automatically adjusting traffic signals and providing information on the most efficient routes for drivers.

Expanding the network of charging stations is critical for the successful introduction of electric cars. By 2023, there were about 1,500 charging stations in Ukraine, which is insufficient to meet the growing demand [19, 28]. A strategic plan for the development of the infrastructure of charging stations is necessary condition for stimulating the use of electric cars.

The electric vehicles using is also driving the development of new technologies for energy storage and distribution, such as high-capacity batteries that can be used as part of an energy storage system for smart grids. This allows to optimise of energy consumption and reduces the load on the energy infrastructure.

Additionally, the introduction of special lanes for electric vehicles can help reduce traffic jams and stimulate the use of environmentally friendly transport. Such lanes allow electric cars to avoid traffic jams and reach their destination faster, which is an additional advantage for their owners.

In recent years, initiatives to allocate special lanes for electric vehicles have been implemented in Ukraine. In 2020, experimental projects on the allocation of separate lanes for electric vehicles on main highways began in Kyiv. By the end of 2023, such lanes have been created on more than 30 km of roads in the city [13].

Using of special lanes for electric cars in Kyiv showed that owners of such vehicles spend an average of 20% less time on trips than users of conventional cars. This has significantly increased the attractiveness of electric cars among the population. In addition, studies have shown that such lanes contributed to reducing the overall level of congestion on the city's main thoroughfares by 10–15% during peak hours [13].

Public transport running on biofuel is an important step towards reducing emissions in cities. Biofuels made from plant materials or waste reduce the carbon footprint of transportation. In Ukraine, the introduction of biofuel buses has already shown positive results in reducing emissions and improving air quality.

Non-motorized transportation such as bicycles and electric scooters is becoming increasingly popular in urban settings. They not only contribute to reducing the level of pollution but also improve the physical health of residents. Cities with developed infrastructure for cyclists, such as bike lanes and rental stations, have seen reduced congestion and improved overall quality of life.

According to Beaudoin and Lawell [6] underground and unban ropeways typically indicate the lowest energy consumption per passenger-kilometre 0.05–0.15 kWh/pass*km and 0.02–0.1 kWh/pass*km accordingly due to optimised operations, continuous flow of passengers, and efficient use of energy through electric propulsion. Their infrastructure supports high-capacity travel with minimal stop-and-go energy loss.

While buses are a flexible transportation option, they usually have lower passenger capacities compared to metros and ropeways, which can lead to higher energy consumption per pass-km, especially in urban areas with heavy traffic. It is important to note that although electric buses, consumed from 0.2–0.4 kWh/pass*km, present a more sustainable option compared to diesel buses, their efficiency can vary depending on passenger load and route conditions [6].

While EVs offer zero emissions, their kWh/pass* km can be higher than that of public transit modes, especially in urban settings with lower passenger density. Electric Vehicles commonly consume between 0.15–0.25 kWh/pass*km, depending on the model, battery type, and driving conditions.

It should be noted that the development of the ecological transport system in Ukraine is closely related to its internal and external environment. Therefore, we will apply SWOT analysis, a method of strategic analysis that aims to determine the strengths and weaknesses of ecological transport (table 1).

The war in Ukraine significantly affected the context of introducing ecological transport. While there are significant strengths and opportunities, the challenges associated with military action pose serious obstacles to sustainable development. Despite this, there is a chance for a post-war modernisation of transport infrastructure with an emphasis

Table 1. SWOT-analysis of the ecological transport system of Ukraine.

Strengths	Weaknesses
Support of the state and international organizations Ukraine receives support from the government and international organizations (EU, World Bank) for the implementation of ecological transport, including aid during wartime *Development of the IT sector* Ukraine's strong IT sector creates opportunities for the development of smart transport systems and the introduction of innovations in the transport infrastructure, which is especially important in wartime conditions *Increasing environmental awareness* Growing awareness about environmental problems, especially against the background of war, where environmental and energy issues are gaining new relevance *Adaptation of infrastructure to challenges* Ukraine's ability to adapt transport infrastructure to wartime conditions, including reconstruction and modernization during conflict	*Insufficient funding* Limited resources due to the war, which significantly complicates the financing of projects for the development of ecological transport *Unsatisfactory infrastructure* Significant damage to transport infrastructure due to hostilities, limiting opportunities for the introduction of new technologies *Population migration* Forced migration within the country and abroad creates an additional burden on urban infrastructure and changes transport needs
Opportunities	**Threats**
Investment in innovation The possibility of attracting international investment and assistance for the recovery and development of ecological transport, including innovations in the conditions of post-war recovery *Development of renewable energy* The development of the renewable energy sector can support the electrification of transport and reduce dependence on energy imports, especially important in times of war *Implementation of new legislative initiatives* Implementation of legislative initiatives aimed at supporting ecological transport in the conditions of recovery and reconstruction of the country after the war *Reconstruction and modernisation* Complex reconstruction and modernisation of infrastructure taking into account the latest environmental standards and technologies	*Prolonged military aggression* Continuation of the war may lead to further destruction of infrastructure, which will significantly complicate the implementation of ecological transport *Economic instability* War increases economic instability, which can reduce investments in green transport and delay the development of projects *Dependence on external resources* High dependence on the import of technologies and energy resources can create additional challenges for the sustainable development of ecological transport *Environmental and Humanitarian Crises* War creates additional environmental and humanitarian crises that divert resources and attention from the development of sustainable transport initiatives

on environmental solutions, which could become an important factor in the country's recovery.

It is worth noting that the implementation of ecological transport in Ukraine has a significant potential for promoting the sustainable development of the country, taking into account its environmental, social and economic challenges. According to analytical data and expert assessments, this impact can be considered through three key aspects: environmental, social and economic.

The main environmental advantage of ecological transport is a considerable reduction in emissions of greenhouse gases and pollutants [3, 5]. The transport sector in Ukraine is responsible for about 15% of total CO_2 emissions, with a significant portion of these emissions occurring in large cities with heavy traffic [19]. Switching to electric and alternative fuel vehicles can reduce these emissions by 30–50% depending on the type of technology and energy source. For example, according to research, the use of electric buses in large cities of Ukraine, such as Kyiv and Lviv, has already contributed to a 20–25% reduction in air pollution compared to the use of diesel buses [19].

In addition, the development of intelligent transport systems (ITS) and traffic optimization with the help of big data and Internet of Things (IoT) technologies can reduce congestion, which will also contribute to the reduction of emissions of harmful substances [2, 21].

At the same time, ecological transport has a positive effect on the health and quality of life of the population. Reducing emissions and improving air quality lead to a reduction in pollution-related diseases such as respiratory and cardiovascular diseases. Studies show that in cities where ecological transport has already been implemented, the health indicators of the population have improved by 10–15% [26].

In addition, ecological transport promotes social integration. For example, the expansion of the network of bicycle paths and pedestrian zones in cities makes them more accessible and attractive to different population groups. It also promotes the development of a healthy lifestyle culture among the urban population.

Ecological transport can become a driving force of economic growth in Ukraine. Investments in infrastructure, such as charging stations for electric vehicles and biofuel production, create new jobs and support the development of related industries. According to research, every billion hryvnias invested in the development of ecological transport creates about 15–20 thousand new jobs in such sectors as construction, IT, production, and maintenance of transport [28].

In addition, reducing fuel costs and improving the efficiency of the transport system can help reduce government spending on energy imports. This, in turn, strengthens the country's energy independence and improves economic stability.

5 Conclusion

The Ukrainian shift towards ecological transportation holds significant promise for supporting sustainable urban development, improving environmental quality, and fostering public health. The transport sector, which contributes approximately 15% of Ukraine's CO_2 emissions, has already seen reductions in emissions by 20–25% in cities like Kyiv and Lviv, thanks to the introduction of electric buses. Expanding the use of electric and biofuel-powered public transport, alongside the growing presence of electric vehicles (EVs)—with imports increasing over 300% from 2020 to 2023-demonstrates both the demand for and impact of these green solutions on air quality and urban livability.

Beyond environmental gains, ecological transport initiatives offer considerable social and economic benefits. Improved air quality and reduced noise pollution contribute to better health outcomes, with cities that have implemented green transportation reporting a 10–15% improvement in public health indicators. Additionally, expanded access to clean transportation options, such as the growing network of bike paths and pedestrian zones, encourages social inclusion and a more active lifestyle among urban populations. Economically, every billion hryvnias invested in ecological transport infrastructure, such as EV charging stations and biofuel production, is projected to create 15,000–20,000 jobs across IT, construction, and maintenance sectors, further strengthening Ukraine's economy.

However, to achieve maximum effect, challenges related to war, economic instability and lack of infrastructure must be overcome. Attracting international support, developing renewable energy sources and creating favourable conditions for investors are critical factors for the successful implementation of ecological transport in Ukraine.

Integrating ecological transport into Ukraine's smart city initiatives presents a pathway to enhanced urban resilience, sustainability, and economic independence. This transformation not only advances environmental health and quality of life for citizens but also opens doors for innovation in transport and energy management. In these ways, the ecological transport sector can be pivotal to Ukraine's post-war recovery, demonstrating a commitment to sustainable development and setting a benchmark for future city planning.

References

1. Aldenius, M., Mullen, C., Pettersson-Löfstedt, F.: Electric buses in England and Sweden – overcoming barriers to introduction. Transport. Res. Part D Transp. Environ. **104**, 103204 (2022)
2. Allam, Z., Dhunny, Z.A.: On big data, artificial intelligence and smart cities. Cities **89**, 80–91 (2018)
3. Almalki, F.A., Alsamhi, S.H., Sahal, R., et al.: Green IoT for eco-friendly and sustainable smart cities: future directions and opportunities. Mobile New App. **28**, 178–202 (2023)
4. Alsamhi, S.H., Ma, O., Ansari, M.S., Meng, Q.: Greening Internet of things for greener and smarter cities: a survey and future prospects. Telecommun. Syst. **72**(4), 609–632 (2019)
5. Bamwesigye, D., Hlavackova, P.: Analysis of sustainable transport for smart cities. Sustainability. 11 (2019)
6. Beaudoin, J., Lawell, C.L.: The effects of public transit supply on the demand for automobile travel. J. Environ. Econ. Manag. **88**, 447–467 (2018)
7. Bibri, S.: Data-driven smart sustainable cities: a conceptual framework for urban intelligence functions and related processes, systems, and sciences. Advances in the leading paradigms of urbanism and their amalgamation, 143–173 (2020)
8. Boren, S., Ny, H.: A strategic sustainability analysis of electric vehicles in EU today and towards 2050. Int. J. Environ. Chem. Ecol. Geol. Geophys. Eng. **10**(3), 207–215 (2016)
9. Boren, S.: Electric buses' sustainability effects, noise, energy use, and costs. Int. J. Sustain. Transp. **14**(12), 956–971 (2020)
10. Ceder, A.: Urban mobility and public transport: Future perspectives and review. Int. J. Urban Sci. **25**(4), 455–479 (2020)
11. Cervero, R., Guerra, E., Ai, S.: Beyond mobility. The sustainable urban development reader (2022)

12. González-González, E., Nogués, S., Stead, D.: Automated vehicles and the city of tomorrow: a backcasting approach. Cities **94**, 153–160 (2019)
13. Government of Ukraine: Initiative on Special Lanes for Electric Vehicles in Kyiv. Ministry of Infrastructure Report, Kyiv (2023)
14. Jia Wang, S., Moriarty, P.: Energy savings from smart cities: a critical analysis. Energy Procedia. **158**, 3271–3276 (2019)
15. Hardman, S., Shiu, E., Steinberger-Wilckens, R.: Comparing high-end and low-end early adopters of battery electric vehicles. Transp. Res. Part A: Policy Pract. **121**(88), 40–56 (2016)
16. Ismagilova, E., Hughes, L., Rana, N., Dwivedi, Y.: Role of smart cities in creating sustainable cities and communities: a systematic Literature Review. ICT unbounded, social impact of bright ICT adoption. TDIT 2019. IFIP advances in information and communication technology, 558 (2019)
17. Klopp, M.J., Petretta, D.L: The urban sustainable development goal: Indicators, complexity and the politics of measuring cities. Cities. 63, 92–97, (2017)
18. Lucas, K., Philips, I., Mulley, C., MaMattioli, L.: Is transport poverty socially or environmentally driven? Comparing the travel behaviours of two low-income populations living in central and peripheral locations in the same city. Transp. Res. Part A: Policy Pract. **116**, 622–634 (2018)
19. Impact of Electric Vehicles on CO_2 Emissions in Urban Areas. Environmental Report, Ministry of Environmental Protection and Natural Resources of Ukraine, Kyiv (2023)
20. Mingwei, L.: Huijuan, Z: Contribution of low-carbon transport policy to the improvement of urban traffic ecological environment. Nat. Environ. Pollut. Technol. **20**, 363–370 (2021)
21. Nikitas, A., Michalakopoulou, K., Njoya, E.T., Karampatzakis, D.: Artificial intelligence, transport and the smart city: definitions and dimensions of a new mobility era. Sustainability. **12**, 2789 (2020)
22. Nikolaidis, P., Poullikkas, A.: A comparative overview of hydrogen production processes. Renew. Sustain. Energy Rev. **67**, 597–611 (2017)
23. Oladimeji, D., Gupta, K., Kose, N.A., Gundogan, K., Ge, L., Liang, F.: Smart transportation: an overview of technologies and applications. Sensors. **23**, 3880 (2023)
24. Shaheen, S., Cohen, A.: Planning for shared mobility (2018). https://escholarship.org/content/qt0dk3h89p/qt0dk3h89p.pdf
25. World Bank Reports on Urban Transport, https://www.worldbank.org/en/topic/transport
26. World Health organisation. Air pollution, https://www.who.int/health-topics/air-pollution/publications#tab=tab_1
27. Yun, Y., Lee, M.: Smart city 4.0 from the perspective of open innovation. J. Open Innov. Technol. Market Compl. **5**(4), 92 (2019)
28. 2.1 times more electric cars were imported to Ukraine in 2024, https://opendatabot.ua/analytics/electrocars-2024-5

Modelling the Distribution of Passenger Traffic on Marching Routes with a Combined Mode of Movement

Oksana Hulchak[1], Yevheniia Shapenko[1](✉), Iryna Tkachenko[2], and Oleksandr Bilonoh[1]

[1] National Transport University, Mykhaila Omelyanovych-Pavlenka St. 1, Kyiv 01010, Ukraine
e.shapenko@ntu.edu.ua

[2] O.M. Beketov National University of Urban Economy in Kharkiv, Marshal Bazhanov 17, Kharkiv 61002, Ukraine

Abstract. The paper establishes the theoretical prerequisites for studying the methods of organising traffic on urban and suburban routes. Different approaches to planning the route network are considered, including optimisation of traffic schedules, balancing the load on different types of transport and ensuring the convenience of transfers. Particular attention is paid to the analysis of passenger traffic on the studied routes. Based on a detailed analysis of urban passenger transport systems, key performance criteria such as efficiency, environmental friendliness, travel time and comfort were identified. The conducted research allowed to evaluate the motivational factors influencing the choice of transport connection by passengers, which made it possible to build a model of passenger flow distribution on routes with a combined mode of movement. The results of the study can be used not only to optimise existing routes, but also to create new transport models for cities, taking into account future trends in infrastructure development and increased demand for public transport. The study also takes into account the impact of modern technologies on passenger transport management and the improvement of the quality of transport services. The recommendations provided can also be used to develop strategies for the development of large cities, namely the introduction of modern information technologies for traffic management, the development of electric transport and the introduction of smart transport systems that automate data collection and traffic control.

Keywords: Organization of Traffic · Urban Passenger Transportation · Distribution of Passenger Traffic · Optimization of Routes · Intelligent Transport Systems · Transport Modeling

1 Introduction

Modern cities are facing increasingly complex challenges in the field of transport infrastructure due to the constant growth in demand for passenger transport. The organisation of efficient public transport is one of the key problems that requires a comprehensive

O. Slavinska et al. (Eds.): ITSESQC 2024, LNNS 1335, pp. 185–195, 2025.
https://doi.org/10.1007/978-3-031-87376-8_17

approach to solving. An important aspect of this process is passenger traffic forecasting, which allows not only to plan routes but also to ensure an appropriate level of passenger service by adapting the transport system to their needs.

The study of passenger flow modelling methods and the analysis of motivational factors influencing the choice of a route are becoming necessary for building optimal transport solutions. This article discusses the main approaches to organising traffic on urban and suburban routes, analyses the effectiveness of various methods of passenger flow management, and proposes models that can be used to improve the operation of transport systems.

The results of the study can be useful for planning new routes and optimising existing transport networks, which, in turn, will help to improve the quality of transportation and the overall transport situation in cities.

1.1 Problem Definition

Route discovery and planning play an important role in the management of real-time data and geo-data-based social services. In this light, it is necessary to conduct a survey of passengers and analyse the existing literature on route discovery and planning. Researchers [2, 3] provide a detailed description for each representative study of route search and planning. This survey summarises the results of existing studies on route discovery and route planning, thus revealing some new insights that may be useful to researchers and software developers in the field of spatial data management and geographic information systems.

The study was based on field observations and the method of expert assessments. The respondents were passengers of different age categories in accordance with their quantitative ratio. In specially designed questionnaires, passengers were asked to answer a number of questions that determine the importance of criteria when choosing a type of transport.

1.2 Research Gaps and Relevance

An overview of recommendations for motivating the choice of travel route is given in the work of scientists [1]. The authors found that recommendations for choosing routes and their application are widely used, directly affecting the travel habits of citizens. The development of smart and efficient travel routes based on big data has become a central issue in research on route recommendations.

In studies performed prior to 2012, it was argued that the choice of travel mode was virtually independent of the cost of travel. Now, with different levels of service, calculations without taking into account the cost of travel can cause significant errors. Therefore, it is necessary to verify the conclusions of previous studies in order to use them in new conditions. Given the object of the study—a single route—only a part of the journey is subject to study—the choice of the type of connection. In this regard, the subject of the study was to establish the list and importance of factors that determine the choice of passengers' mode of transport in a market economy.

1.3 Literature Review

The last few decades have seen a number of studies aimed at developing efficient route discovery and planning methods. In the article, Ke and Xuan [2] review existing research on route discovery and planning for urban transport. In particular, the route discovery problem can be broadly defined as the search for objective routes and trajectories over a set of routes and trajectories, respectively, based on query information. Whereas the route planning problem can be defined as the discovery of one or more new routes from the road network based on the travel requirements provided by users. Existing relevant studies by Yifeng et al. [7] review the literature on trajectory data management, Xin et al. [8] conducted intelligent analysis of trajectory data, Sheng et al. [9] and Jean Damascene and Sabine [10] trajectory data analytics and processing of space-time queries.

1.4 Research Gaps

This study has certain unresolved problems in the scientific literature, namely the lack of generally accepted standards for assessing the efficiency of transport routes using a combined mode of movement. Artificial intelligence and big data technologies are not integrated into passenger flow modelling, especially for combined routes. The impact of buses with a combined mode of movement on social groups and the environment is also not yet sufficiently studied.

2 Research Methodology

2.1 Research Approach

The study was based on field observations and the method of expert assessments. The respondents were passengers of different age categories in accordance with their quantitative ratio. In specially designed questionnaires, passengers were asked to answer a number of questions that determine the importance of criteria when choosing a type of transport.

2.2 Data Collection Process

The analysis of the results of the 2024 study shows that the most significant factors influencing passengers' choice of transport mode are the time spent on the journey, the fare, the level of transport service, the number of transfers, the interval of movement and the filling of vehicles. The change in the importance of the factors of travel time and cost recorded in the 2017 and 2018 surveys is explained by the fact that the average interval of vehicles on the route network decreased from 4.1 min to 2.6 min due to the additional involvement of about 3,000 route taxis in Kyiv. In addition, the respondents' opinion was significantly influenced by the campaign related to the preparation for the increase in public transport fares.

2.3 Data for the Study

The surveys were conducted in 2017, then after the increase in urban transport tariffs in 2018 and again in 2024. According to recommendations [2, 4], the group of respondents for each survey should consist of:

Formula 1: Calculating the required number of respondents:

$$n = \frac{t^2 \cdot \sigma^2 \cdot N}{\Delta^2 \cdot N + t^2 \cdot \sigma^2} \tag{1}$$

where

t— an indicator of reliability;
σ^2— between—group variance;
N— the size of the general population;
Δ— sampling error.

As a result of the calculations, the recommended sample size should consist of 385 people. In the course of the research, 400 people were interviewed. The questionnaire data was processed according to the methodology described in studies [2, 4].

The consistency of the respondents' assessments was determined using the coefficient of variation of the opinions of the entire team.

Formula 2. Coefficient of variation of respondents' opinions:

$$v = \frac{\sqrt{D}}{M} \tag{2}$$

where

D— *the variance,*
M— *mathematical expectation,*

$$M = \frac{\sum_{i=1}^{n} a_i}{n}, \tag{3}$$

$$D = \frac{1}{n-1}\sum_{i=1}^{n}(a_i - M)^2, \tag{4}$$

Formula 5. Concordance coefficient

$$W = \frac{\sqrt{\sum_{i=1}^{n}(a_i - M)^2}}{\sqrt{\sum_{i=1}^{n} a_i^2} + M\sqrt{n}}, \tag{5}$$

Since the value of $v < 0.3$ for each group of assessments, the degree of consistency of experts' opinions can be considered sufficient [4]. This conclusion is confirmed by the obtained values of the coefficient W ($0.1 < W < 0.2$), which indicates an average degree of consistency [4]. In addition, the reliability of the results obtained is confirmed by the results of sociological studies by other authors [5] and the processing of experimental data [4].

2.4 Statistical Methods

The last few decades have been devoted to a number of studies aimed at developing effective methods for route discovery and planning. In [5], experts review existing research on route finding and planning for urban transport. In particular, the problem of route search can be generally defined as the search for objective routes and trajectories over a set of routes and trajectories, respectively, based on query information. Scientists [8] proposed the structure of an integral index of efficiency, which helps to choose the best approach and includes the main economic, technological and social indicators. Whereas the route planning problem can be defined as the discovery of one or more new routes from the road network based on the travel requirements provided by users. Existing relevant studies review the literature on trajectory data management [7], trajectory data mining [8], trajectory data analytics [9], and spatio-temporal query processing [10].

The analysis of the results of the 2024 study shows that the most significant factors influencing passengers' choice of the type of transport are the time spent on the trip, the transport tariff, the level of transport service, the number of transfers, the interval of movement and the filling of vehicles. The change in the importance of the factors of travel time and cost recorded in the 2017 and 2018 surveys is explained by the fact that the average interval of vehicles on the route network decreased from 4.1 min to 2.6 min due to the additional involvement of about 3,000 route taxis in Kyiv. In addition, the respondents' opinion was significantly influenced by the campaign related to the preparation for the increase in public transport fares.

3 Results and Discussion

3.1 Modelling the Distribution of Passenger Traffic on Routes

Let's consider the most important criteria and analyse their impact on the choice of a movement route (Table 1).

Taking into account the analysis of the results obtained and the studies of other authors [4, 5], it can be argued that one of the main criteria that determine the choice of route is the total travel time. However, its assessment by passengers differs depending on the level of transport service. Thus, the results of sample observations show that when travelling on public transport routes, passengers overestimate the actual travel time by 15–20%, and on taxi routes—underestimate it by up to 35–40%. This proves that passengers' subjective perception of the passage of time should be taken into account when modelling passenger flows, while in modern models, passengers' choice of a route is based on real time consumption.

The level of tariff has a significant impact on the passenger's choice of the type of transport and the volume of traffic. In order to study this dependence, a relevant sociological study was conducted in 2024. Processing the results with the help of the Microsoft Excel application 'Solution Finder' based on the least squares method [16] allowed us to establish an empirical relationship (correlation coefficient R = 0.957)

$$P = \left[0{,}87 + 0{,}13\exp\left(3{,}4\left(\frac{\tau_п}{\tau_б} - 1\right)^{5-2{,}8(\tau_п/\tau_б)}\right)\right]^{-1}, \quad \tau_п \geq \tau_б \tag{6}$$

Table 1. The most important criteria and their significance when choosing a relocation route.

Criterion name	Weighted average score (2017 p.)	Weighted average score (2018)	Weighted average score (2024 p)
Total costs of of time spent on travelling	7,53	6,82	7,60
Number of transfers	7,31	6,11	6,69
Interval of movement	6,81	5,38	5,99
Adherence to the schedule	5,01	4,92	6,21
Quality of service provision	4,88	4,78	6,79
Monetary value	4,87	5,71	7,71

where P—the proportion of passengers who will not refuse to travel if the fare is increased;

$\tau_б$ i $\tau_п$—respectively, the basic and increased transportation tariff.

After the tariff increase in 2018, field observations confirmed the results of the previous study. At the same time, the volume of traffic has changed by 15%, with 10% more passengers using high-speed services only when absolutely necessary. The reason why 54% of passengers opt for high-speed services with higher fares is a lack of time, and 25%—because of the lack of or overcrowding of mass transit vehicles. The increase in tariffs caused a change in the average age of users of rapid transit from 30 to 35 years (Figs. 1, 2, and 3). Many methods have been proposed to determine the level of transport service [2, 4, 11–13]. The analysis of these methods has established that the level of transport service is determined by the levels of road safety, vehicle occupancy, time spent on the movement of passengers and the regularity of traffic. In design calculations, it is extremely difficult to take into account such characteristics as traffic regularity and safety, which depend on driver discipline and road conditions. Assuming that the introduction of different traffic modes on the route does not change the values of these characteristics, the expected result is that they do not affect the passenger's choice of transport mode.

Thus, to assess the level of transport service, ceteris paribus, it is advisable to use the time spent by the passenger on moving and filling the vehicle. The results of the study and the conclusions of many authors [14, 15] confirm that when a passenger chooses a travel route, it is necessary to take into account the interval of vehicle movement. Based on the analysis of the results obtained, taking into account the absence of transfers during route trips, the main characteristics that a passenger is guided by when choosing a travel route are the total time spent on travel, the interval of movement, the tariff for transportation and the filling of the cabin.

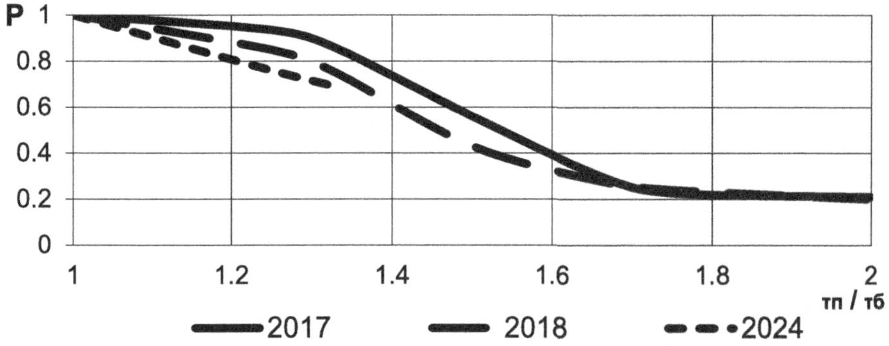

Fig. 1. Impact of a fare increase (τp) on the change in the share of passengers (P) who will not abandon the trip

The decision to use a mode of transport is made by many passengers based on a subjective assessment of various factors, such as the time spent waiting and the duration of the trip, the fare and the occupancy of the vehicle, etc. The presence of a set of criteria and the subjectivity of decision-making by many participants in the transport system determine the probabilistic nature of the choice. The theoretical basis for modelling the choice process should be the results of research in the field of psychophysics, the subject of which is the study of the relationship between physical and psychological processes. The analytical expression of this relationship is called the basic psychophysical law. For a long time, several expressions of the basic psychophysical law have been proposed [16]. In practice, a passenger chooses the type of transport based on a comparison of changes in the value of a certain factor or their combination. From this point of view, the models of G. Fechner and S. Stevens are suitable for use in engineering calculations, since, firstly, they provide a comparison of the values of factors for competitive options, secondly, they have simple analytical solutions, and thirdly, and most importantly, they have empirical confirmation. The accumulated experience shows that the Stevens model is preferable. It is supported by the results of studies [16–19], which prove the expediency of using a power function to determine the distribution of passengers among alternative routes.

Thus, individual factors fi (i = 1, 2,…, n) that influence the choice of a mode of transport are considered as some physical stimuli. A change in their magnitude causes a proportional change in the response, which in physical terms is the proportion of passengers choosing a particular route. That is, the share of passengers (P) is described by the differential equation

$$\frac{\mathrm{dP}_i}{P_i} = a_i \cdot \frac{\mathrm{df}_i}{f_i}, \tag{7}$$

where

a_i— the proportionality coefficient.

Integration of (1) allows us to obtain the dependence

$$P_i = C \cdot f_i^{a_i}, \tag{8}$$

where

C— a constant determined by additional conditions.

Consideration of m alternative routes according to the criterion fi requires the following condition

$$\sum_{j=1}^{m} P_{ij} = 1 \tag{9}$$

From (8) and (9) we obtain

$$C = \left[\sum_{j=1}^{m} f_{ij}^{a_i}\right]^{-1} \tag{10}$$

Then

$$P_{ij} = f_{ij}^{a_i}\left[\sum_{j=1}^{m} f_{ij}^{a_i}\right]^{-1} \tag{11}$$

Taking into account the fact that each route is characterised by several factors and it is advisable to describe the generalised characteristic in the form of their product, we obtain

$$X_j = \prod_{i=1}^{n}\left[f_{ij}^{a_i}\left[\sum_{j=1}^{m} f_{ij}^{a_i}\right]^{-1}\right] \tag{12}$$

The proportion of passengers who will choose j route based on its generalised characteristics out of m alternative routes, taking into account condition (9), is determined by

$$P_j = \frac{\prod_{i=1}^{n}\left[f_{ij}^{a_i}\left[\sum_{j=1}^{m} f_{ij}^{a_i}\right]^{-1}\right]}{\sum_{j=1}^{m}\left[\prod_{i=1}^{n}\left[f_{ij}^{a_i}\left[\sum_{j=1}^{m} f_{ij}^{a_i}\right]^{-1}\right]\right]} \tag{13}$$

Or

$$P_k = \left[\sum_{j=1}^{m}\prod_{i=1}^{n}\left(\frac{f_{ij}}{f_{ik}}\right)^{a_i}\right]^{-1}, 1 \le \mathrm{k} \le \mathrm{m} \tag{14}$$

The practice of using the combined mode of movement on urban routes shows a predominant combination of the two forms. Then, expression (14) for determining the share of passenger traffic that will use the non-stop mode has the following form:

$$P_{з} = \left[1 + \prod_{i=1}^{n}\left(\frac{f_{iд}}{f_{iз}}\right)^{a_i}\right]^{-1} \tag{15}$$

where fiд and fiз are the significant factors, respectively, of the additional and normal traffic modes.

Based on the analysis of the results of a sociological survey, the article identifies significant factors for the distribution of route correspondence. These include the total time spent on travel (t), the cost of a trip (τ), the interval of vehicle movement (i) and the degree of occupancy of their interior, the numerical estimate of which is the coefficient of dynamic use of passenger capacity (γ). With this in mind, expression (15) has the form:

$$P_з = \left[1 + \left(\frac{t_\partial}{t_з}\right)^a \left(\frac{\tau_\partial}{\tau_з}\right)^b \left(\frac{i_\partial}{i_з}\right)^c \left(\frac{\gamma_\partial}{\gamma_з}\right)^d\right]^{-1} \tag{16}$$

where a, b, c, d—empirical coefficients.

The values of the empirical coefficients in expression (16) are determined on the basis of the results of field observations and the solution of the system of linear equations.

$$\sum_{j=1}^{n}(a \times ln\left(\frac{t_{дj}}{t_{зj}}\right) + b \times ln\left(\frac{\tau_{дj}}{\tau_{зj}}\right) + c \times ln\left(\frac{i_{дj}}{i_{зj}}\right) + d \times ln\left(\frac{y_{дj}}{y_{зj}}\right) = \\ = \sum_{j=1}^{n}\left(\frac{1}{P_{зj}} - 1\right) \tag{17}$$

As a result of the calculations, the values of the coefficients were established:

a = −2.11; b = −2.73; c = −1.17; d = −0.38, provided that the conditions $t_\partial < t_з$, $\tau_\partial > \tau_з$, $i_\partial < i_з$, $\gamma_\partial < \gamma_з$.are met. The degree of conformity was assessed based on the correlation level, the coefficient of which was R = 0.971.

For an urban route using conventional and taxi modes of transport, the distribution of traffic volumes between them is determined on the basis of expression (16), taking into account the signs of the coefficients

$$\begin{cases} P_3 = \left[1 + \left(\frac{t_3}{t_m}\right)^{2,11} \left(\frac{\tau_3}{\tau_m}\right)^{2,73} \left(\frac{i_3}{i_m}\right)^{1,17} \left(\frac{\gamma_3}{\gamma_m}\right)^{0,38}\right]^{-1} \\ P_m = \left[1 + \left(\frac{t_m}{t_3}\right)^{2,11} \left(\frac{\tau_m}{\tau_3}\right)^{2,73} \left(\frac{i_m}{i_3}\right)^{1,17} \left(\frac{\gamma_m}{\gamma_3}\right)^{0,38}\right]^{-1} \end{cases} \tag{18}$$

The organisation of urban passenger transport involves, on the one hand, the ability of passengers to choose routes of movement, and on the other hand, to form passenger flows by changing the characteristics of urban transport routes, which determines the importance of practical application of the results of the study of motivation for passenger choice. Changing conditions of economic activity have determined the need to clarify the results of previous studies on the factors influencing passengers' choice of travel route.

3.2 Discussion of Results

Experimental studies (2017, 2018, 2024), carried out by the method of expert assessments, have established that the determining factors, in terms of their influence on the passenger's choice of the route, are the cost of the trip, the duration of the trip, the interval of movement and the filling of the bus interior. Thus, the necessity of taking into account the cost of the trip in motivating passenger choice is proved. The reliability of the conclusions is confirmed by stable results over five years, a reasonable sample size of respondents and an average degree of consistency, which corresponds to a value of the concordance coefficient of less than 0.2.

Based on the results of these studies, a model of discrete passenger choice of travel route is proposed. The basic psychophysical law in the formulation of S. Stevens is taken as the theoretical basis for defining the model.

3.3 Limitations of the Study and Prospects for Future Research

Future research in this area will aim to improve the efficiency and flexibility of the transport system, ensuring convenient movement of citizens with minimal time and resources. [8, 16].

4 Conclusion

The analysis of urban passenger transport studies shows different and sometimes opposite motivations of the main transport participants (passengers, carriers, local authorities), which is determined by the level of transport service, transport performance and cost of transportation. Therefore, it is recommended to adopt these characteristics to assess the efficiency of urban passenger transport and to use them in the construction of urban transport models.

References

1. Shiming, Z., Zhipeng, L., Li, Y., Fei, T.: A survey of route recommendations: Methods, applications, and opportunities. Inf. Fusion. **108**(3), 102413 (2024)
2. Ke, L., Xuan, R., XiaoBing, P., Lisi, C., Siqi, F.: Route search and planning: a survey. Big Data Research Volume. **26**, 100246 (2021)
3. Naumov, V., Bekmagambetova, L., Bitileuova, Z., Zhanbirov, Z., Taran, I.: Mixed Fuzzy-Logic and Game-Theoretical Approach to Justify Vehicle Models for Servicing the Public Bus Line. Communications. Sci. Lett. Univ. Zilina. **24**(1), 26–34
4. Shuo, S., Danhuai, G., Jiajun, L., Ji-Rong, W.: Prediction-based unobstructed route planning. Neurocomputing **213**, 147–154 (2016)
5. Ke, L., Xuan, R., XiaoBing, P., Lisi, C., Siqi, F.: Route search and planning: a survey. Big Data Research. **26**, 100246 (2021)
6. Taran, I., Litvin, V.: Determination of rational parameters for urban bus route with combined operating mode. Transp. Probl. **13**(4), 158–171
7. Yifeng, Z., Xuefeng, C., Xin, C., Shengchao, Q.: Optimal route search with the coverage of users' preferences. 24th International Joint Conference on Artificial Intelligence (IJCAI) (2015)
8. Xin, C., Lisi, C., Gao, C., Xiaokui, X.: Keyword-aware optimal route search. Proceedings of the VLDB Endowment (PVLDB) **5**(11), 1136–1147 (2012)
9. Sheng, W., Zhifeng, B.J., Shane, C., Gao, C.: A survey on trajectory data management, analytics, and learning. ACM Computing Surv. **54**(2), 1–36 (2021)

10. Jean, D.M., Sabine, T.: Trajectory data mining: a review of methods and applications. J. Spat. Inf. Sci. **13**, 61–99 (2016)
11. Hongjie, L., Tengfei, Y., Xinhuan, Z., Hongzhe, X.: Quantum entanglement and self-attention neural networks: an investigation into passengers and stops characteristics for optimal bus stop localization. Inf. Fusion. **112**, 102527 (2024)
12. Haoge X., Yong C., Chuanjia L., Xiqun (Michael) C.: Space-Time adaptive network for origin-destination passenger demand prediction. Transp. Res. Part C Emerg. Technol. **167**, 104842 (2024)
13. Yongjie, Y., Jinlei, Z., Lixing, Y., Yang, Y., Xiaohong, L., Ziyou, G.: Short-term passenger flow prediction for multi-traffic modes: a transformer and residual network based multi-task learning method. Inf. Sci. **642**, 119144 (2023)
14. Wenzhe, S., Jan-Dirk, S., Koji, F.: Estimating the route-level passenger demand profile from bus dwell times. Transp. Res. Part C Emerg. Technol. **130**, 103273 (2021)
15. Xiaowei, H., Zhenyang, Q., Wei, G.: Optimization of electric bus dispatching interval considering stochastic traffic conditions. Transportation Letters (2024)
16. Joaquin de Cea, E.F.: Transit assignment for congested public transport systems: an equilibrium model. Transp. Sci. **27**, 133–147 (1993)
17. Xin, W., Bingfeng, S., Meng, X., Fang, Z., Rui, J.: A passenger flow spatial–temporal distribution model for a passenger transit hub considering node queuing. Transp. Res. Part C Emerg. Technol. 163 (2024)
18. Antonio, P., Antonino, V.: Network design and vehicle routing problems in road transport systems: Integrating models and algorithms. Transp. Eng. 16 (2024)
19. Tongfei, L., Yao, G., Jie, X., Min, X., Jianjun, W., Huijun, S.: Ridesharing user equilibrium model without the en-route transfer: An OD-based link-node formulation. Transp. Res. Part E: Log. Transp. Rev. 187 (2024)

Intelligent Transport Technologies and Smart Logistics

Solving the Fuzzy Traveling Salesman Problem Using Genetic Algorithm with Clustering by Ward's Method

Eugene Ivohin(✉) and Konstantin Yushtin

Taras Shevchenko National University of Kiyv, Volodymyrska Street, 64/13, Kyiv 01601, Ukraine
ivohin@univ.kiev.ua

Abstract. The article describes a method for solving the fuzzy traveling salesman problem on a given transportation network based on a two-stage approach. In the fuzzy traveling salesman problem, the task is to find the shortest route with a fuzzy specified travel time between individual cities of the network. The travel time along the network is given in the form of fuzzy trapezoid numbers. A method for converting fuzzy numbers into a special form is proposed, and operations on such numbers are considered. The route is formed in two stages: at the first stage, the cities of the network are grouped into clusters, after which the optimal solution to the traveling salesman problem is found for each cluster using a genetic algorithm. A modification of the genetic algorithm scheme is used based on improving the mutation operation and increasing the diversity of populations. Two variants of the cluster formation topology with a limited number of available paths and a fully connected topology are considered. Numerous experiments are conducted, the results of which confirm the constructiveness of using the preliminary clustering methodology for a large number of cities in the network. Pre-clustering improves the overall route duration in the traveling salesman problem compared to solutions obtained by using a genetic algorithm without using clusters.

Keywords: Genetic Algorithms · Travelling Salesman Problem · Clustering · Ward's Method

1 Introduction

Today, the problem of finding the shortest route between arbitrary points of transport networks is very relevant: the volumes and needs of transport and logistics services are constantly growing. Their main goal is to find the most effective route from the point of view of some criterion for servicing the maximum number of consumers. It is clear that an unsuccessful choice of delivery route introduces additional resource costs into the servicing process, the volumes of which must be reduced as much as possible.

It is worth noting that there is a set of factors that determine the duration of the route: for example, time of day, weather conditions, and even the "load" of passing a particular section. This means that the time of passing each stage of the route will differ depending

O. Slavinska et al. (Eds.): ITSESQC 2024, LNNS 1335, pp. 199–210, 2025.
https://doi.org/10.1007/978-3-031-87376-8_18

on various subjective and objective conditions, which are factors influencing the total duration of the trip.

1.1 Problem Statement

The Traveling Salesman Problem (TSP) is a classical problem of combinatorial optimization [1], where it is necessary to find the shortest possible route that passes through a given list of cities, visiting each city exactly once and returning to the starting city. Solving the TSP has wide applications in logistics, route planning, manufacturing, and many other fields [2].

One of the main applications of the TSP is the optimization of supply routes. Solving the TSP helps to minimize delivery time and reduce transportation costs, which is critical for companies operating in the logistics and retail sectors. The traveling salesman problem is used in the optimization of manufacturing operations, including those involving robots, where the task involves minimizing the tool's travel paths during production processes. The traveling salesman problem is still considered one of the most challenging problems to solve, classified as NP-complete.

The traveling salesman problem is a well-known combinatorial problem. As the objective function in the TSP, the route length or travel duration is most often used. Given the subjectivity of estimates for the duration of travel along the stages of the route, let's consider the formulation of the traveling salesman problem with fuzzily defined travel durations on the transportation network [3]. In this case, it is necessary to find a cyclic permutation of the city numbers that the salesman must visit, such that the time costs are minimized while considering the constraint that each point is visited no more than once. The mathematical formulation of the fuzzy traveling salesman problem can be stated as follows: it is necessary to minimize the objective function, taking into account the specified method of comparing fuzzy numbers

$$\sum_{i=1}^{n}\sum_{j=1}^{n}\tilde{t}_{ij}x_{ij}, \tag{1}$$

where the time costs of movement between points are represented in the form of a matrix $\tilde{T} = \{\tilde{t}_{ij}\}$, $i, j = \overline{1, n}$ with elements as fuzzy numbers [4], and the possible routes between cities are determined by a matrix X, subject to the fulfillment of the following constraints:

$$\sum_{i=1}^{n} x_{ij} = 1 \text{ for any } j = 1, 2, \ldots, n,$$

$$\sum_{j=1}^{n} x_{ij} = 1 \text{ for any } i = 1, 2, \ldots, n, \tag{2}$$

$$x_{ij} = 0 \text{ or } 1 \text{ for any } i, j = 1, 2, \ldots, n,$$

$$i - j + nx_{ij} \leq n - 1, 1 \leq i \neq j \leq n.$$

Constraints in the form (2) define the rules for the traveler's movement between points on the route: from each city the traveler must leave for only one other city and must arrive at each new city only from one other. The last inequality ensures the absence of local cycles in the route and guarantees that the traveler passes all the cities of the network in one closed cycle.

1.2 Review of Solving Methods

There are many methods for solving the problem, which can be divided into two groups: exact and approximate (heuristic, metaheuristic). Exact methods guarantee finding an optimal solution, but require a lot of computing resources and time to solve the problem. This group includes, for example, the exhaustive search method, the dynamic programming method and mathematical programming methods (branches and bounds, cuts, etc.) [5]. Heuristic methods (nearest neighbor, insertion, greedy) allow you to find a solution close to the optimal one, but do not guarantee its accuracy. In real life, transport networks consist of a large number of nodes, conditions can change quickly over time, so the speed of finding optimal routes is a critical factor. That is why researchers very often use approximate metaheuristic methods, in which it is difficult to provide optimality guarantees, such as genetic algorithm (GA), particle swarm method, ant colony method, simulated annealing (RS) method and others. Such approximate algorithms allow you to quickly get a solution within the given constraints. The choice of method depends on the size of the problem and the available time for calculation.

The most significant results in solving the TSP have been obtained using approximate methods. A typical approach to finding solutions in optimization problems using approximate methods is based on an iterative process aimed at improving the results. With such methods, a better solution within the corresponding neighborhood should be found during each iteration. If such a solution is obtained, it becomes the current solution, and a new iteration begins. This process continues until the improvement of the objective function decreases to nearly zero or the specified number of iterations is completed. It is evident that such methods are focused on finding only local optima, and the location of the found optimum can significantly depend on the choice of the initial calculation point. The global optimum can only be found by chance. To increase the probability of finding the global optimum, multiple experiments with different initial points are conducted, which significantly increases the time required to obtain the final result.

In this regard, the development of algorithms that do not have the aforementioned drawback is of interest. Genetic algorithms belong to such methods. The applying of genetic algorithms is based on the use of populations of possible solutions evolving through crossbreeding, mutation and selection. GA is a stochastic heuristic optimization method, the main idea of which is associated with the implementation of the evolutionary development of species [6]. The primary mechanism of evolution is natural selection, the essence of which is that more adapted individuals have a higher chance of survival and reproduction, and, accordingly, produce more offspring compared to less adapted individuals. Through the transfer of genetic information, offspring inherit the main qualities of their parents. In biology, chromosomes, formed from a set of genes, are responsible for preserving an individual's genetic information. During the reproduction of organisms, the fusion of two parental cells occurs, and their interaction forms the new genetic material of the offspring. The main interaction mechanism is crossover (also known as recombination). During crossover, the DNA of the ancestors is divided into two parts, which then exchange halves. During inheritance, mutations can occur due to various influences, resulting in changes to some genes in one of the parent's cells. The altered genes are passed on to the offspring, giving it new properties. If the new properties are beneficial, they are likely to be retained in the gene pool, leading to a significant increase in the adaptability of the species.

Genetic algorithms are used in a wide range of fields, including optimization, machine learning, engineering design, finance, biology, and artificial intelligence [7]. However, GAs are not universal methods for solving all optimization problems [8]. Such algorithms are best suited for problems characterized by complexity (the problem must be difficult to solve using traditional methods), a powerful set of feasible solutions (there are a large number of possible solutions to the problem), and an evaluation function (there is a criteria function that allows evaluating the quality of each possible solution). In the case of the TSP, we have all the properties, and the power of the set of possible solution options is (n-1)!, where n is the number of cities. Therefore, GAs can be an effective way to solve it.

To improve the GA scheme, a modification is often used based on one of the methods for improving mutation and increasing population diversity, which belongs to the PNP (Pick Near Point) group of algorithms [9]. In this case, some new mutations are created by slightly altering one or more parameters of the existing best solution, which allows for improved exploration of the solution space near current locally optimal solutions, maintains diversity in the population to avoid premature convergence, and addresses existing algorithmic conflicts.

Additional efficiency of genetic algorithms is provided by the scheme of using preliminary grouping of nodes in a cluster in the presence of points of common (close) localization or segment formations in the network. Preliminary formation of the structure of the transport network by means of a set of clusters allows to significantly accelerate the speed of GA operation.

2 Main Results

2.1 Definitions and Basic Assumptions

This study proposes an approach to solving the TSP using a GA based on the sequential execution of two stages. In the first stage, the nodes (cities) of the transportation network are grouped into so-called clusters, and in the second stage, the GA is used to solve the traveling salesman problem within each cluster. The solutions to the traveling salesman problem within the clusters represent subproblems of the initial TSP, where the optimal routes are obtained by minimizing the travel durations using the GA.

For clustering the initial data (the set of cities in the network), it is proposed to apply Ward's method, which ensures the shortest inter-cluster route. The application of Ward's method allows to create spherical or nearly spherical clusters to represent the real grouping, especially in case of compact arrangement of nodes. This method can be combined with other clustering methods to provide more accurate grouping result. According to the classical Ward's method, the distance between clusters is taken as the normalized increase in the sum of squared distances of objects to the cluster centers resulting from their merger [10]. Unlike other methods of cluster analysis, Ward's method uses analysis of variance techniques to evaluate distances between clusters. At each step of the algorithm, two arbitrarily chosen clusters are merged, provided that the increase in the objective function, in the form of the within-cluster sum of squared distances, is minimized. Ward's method allows to reduce the influence of possible artifacts, aims at the sequential merging of closely located clusters and "seeks" to create optimally sized clusters when the number of clusters is initially unknown.

Let us return to the formulation of the problem (1), (2). The duration of travel between cities in the network depends significantly on a set of objective and subjective factors that influence the assessment of the time it takes to overcome individual stages of the route. We formalize the subjective uncertainty of a traveling salesman in assessing the travel time in the network based on fuzzy numbers. We assume that a traveling salesman has some time to travel each section (the time of possible delay), which gradually becomes undesirable with an increase in the value. Therefore the fuzzy travel durations $\tilde{t}_{ij}$ between arbitrary cities $i, j = \overline{1, n}$ will be represented as trapezoidal fuzzy numbers.

Definition: A trapezoidal fuzzy number $\tilde{A}$ [11] is an ordered quadruple of real numbers (a_1, a_2, a_3, a_4), $a_1 \le a_2 \le a_3 \le a_4$, for which the membership function $\mu_{\tilde{A}}(x)$ is defined as follows:

$$\mu_{\tilde{A}}(x) = \begin{cases} \frac{x-a_1}{a_2-a_1}, \; for \;\; a_1 \le x \le a_2 \\ 1, \; for \, a_2 \le x \le a_3 \\ \frac{a_4-x}{a_4-a_3}, \; for \, a_3 \le x \le a_4 \end{cases} \tag{3}$$

If the representation of a trapezoidal fuzzy number is approached using a Gaussian distribution with the corresponding characteristics, then in the generalized case, the trapezoidal fuzzy number can be represented in a slightly different form:

$$\tilde{A} = (a_1, a_2, a_3, a_4) = ([a_2, a_3], \alpha, \beta) = (m, w, \alpha, \beta) \tag{4}$$

where the midpoint $m = \frac{a_2+a_3}{2}$ and the half-width of the plateau $w = \frac{a_3-a_2}{2}$ are used, and the coefficients $\alpha = a_2 - a_1$ and $\beta = a_4 - a_3$ define the left and right distribution of the fuzzy number $\tilde{A} = (a_1, a_2, a_3, a_4)$ respectively.

To operate with fuzzy numbers, it is necessary to define operations based on the description provided above. The midpoint is taken as the simple arithmetic mean of the plateau boundaries, and the left and right distributions are considered according to the lattice rule, where for arbitrary real numbers a, b, we set $a \cap b = \min\{a, b\}$ and $a \cup b = \max\{a, b\}$.

Then, for arbitrary trapezoidal fuzzy numbers $\tilde{A} = \left(m(\tilde{A}), w(\tilde{A}), \alpha_1, \beta_1\right)$ and $\tilde{B} = \left(m(\tilde{B}), w(\tilde{B}), \alpha_2, \beta_2\right)$, the operations of addition, subtraction, multiplication, and division can be defined, which in the general case we denote by the symbol $\circ$ as follows:

$$\begin{aligned}\tilde{A} \circ \tilde{B} &= \left(m(\tilde{A}) \circ m(\tilde{B}), w(\tilde{A}) \cup w(\tilde{B}), \alpha_1 \cup \alpha_2, \beta_1 \cup \beta_2\right) = \\ &= \left(m(\tilde{A}) \circ m(\tilde{B}), \max\left(w(\tilde{A}), w(\tilde{B})\right), \max(\alpha_1, \alpha_2), \max(\beta_1, \beta_2)\right)\end{aligned} \quad (5)$$

Finally, we obtain

$$\begin{aligned}\tilde{A} + \tilde{B} &= \left(m(\tilde{A}) + m(\tilde{B}), w(\tilde{A}) \cup w(\tilde{B}), \alpha_1 \cup \alpha_2, \beta_1 \cup \beta_2\right) \\ &= \left(m(\tilde{A}) + m(\tilde{B}), \max\left(w(\tilde{A}), w(\tilde{B})\right), \max(\alpha_1, \alpha_2), \max(\beta_1, \beta_2)\right) \\ \tilde{A} - \tilde{B} &= \left(m(\tilde{A}) - m(\tilde{B}), w(\tilde{A}) \cup w(\tilde{B}), \alpha_1 \cup \alpha_2, \beta_1 \cup \beta_2\right) \\ &= \left(m(\tilde{A}) - m(\tilde{B}), \max\left(w(\tilde{A}), w(\tilde{B})\right), \max(\alpha_1, \alpha_2), \max(\beta_1, \beta_2)\right) \\ \tilde{A} \times \tilde{B} &= \left(m(\tilde{A}) \times m(\tilde{B}), w(\tilde{A}) \cup w(\tilde{B}), \alpha_1 \cup \alpha_2, \beta_1 \cup \beta_2\right) \\ &= \left(m(\tilde{A}) \times m(\tilde{B}), \max\left(w(\tilde{A}), w(\tilde{B})\right), \max(\alpha_1, \alpha_2), \max(\beta_1, \beta_2)\right) \\ \tilde{A} \div \tilde{B} &= \left(m(\tilde{A}) \div m(\tilde{B}), w(\tilde{A}) \cup w(\tilde{B}), \alpha_1 \cup \alpha_2, \beta_1 \cup \beta_2\right) \\ &= \left(m(\tilde{A}) \div m(\tilde{B}), \max\left(w(\tilde{A}), w(\tilde{B})\right), \max(\alpha_1, \alpha_2), \max(\beta_1, \beta_2)\right)\end{aligned}$$

To perform the operations of comparison and ranking of fuzzy numbers, we will use the method based on the median average value. In other words, if for each $\tilde{A} = (a_1, a_2, a_3, a_4) \in F(R)$ a ranking function $\Re : F(R) \to R$ with a median average value in the form of $\Re(\tilde{A}) = \left[\left(\frac{a_2+a_3}{2}\right) + \left(\frac{\beta-\alpha}{4}\right)\right]$ is defined, then for any two trapezoidal fuzzy numbers $\tilde{A} = (a_1, a_2, a_3, a_4)$ and $\tilde{B} = (b_1, b_2, b_3, b_4)$, we have the following possible comparison options:

- $\tilde{A} \succ \tilde{B}$ if and only if $\Re(\tilde{A}) > \Re(\tilde{B})$;
- $\tilde{A} \prec \tilde{B}$ if and only if $\Re(\tilde{A}) < \Re(\tilde{B})$;
- $\tilde{A} \approx \tilde{B}$ if and only if $\Re(\tilde{A}) = \Re(\tilde{B})$.

The processes of handling fuzzy numbers involve a defuzzification stage—converting a fuzzy result into a precise (numerical) value. This is an important step in the methodology of applying the fuzzy approach, especially in tasks of fuzzy control and fuzzy business logic, where it is necessary to transform fuzzy solutions into specific events or numerical values. There are various defuzzification methods, among which the most common are the Center of Gravity (CoG) or centroid method, the mean of maxima method, and the maximum method. To compare the research results, we will use the Center of Gravity method. In this method, the defuzzification point is calculated as the center of gravity of the fuzzy set. For a discrete set, the formula is as follows [12]:

$$CoG = \frac{\sum_{i=1}^{n} \mu(x_i) \cdot x_i}{\sum_{i=1}^{n} \mu(x_i)}, \quad (6)$$

where x_i are the points in the result space, and $\mu(x_i)$ is the membership degree of each point.

2.2 Review of Numerical Experiments

Let us consider the results of solving the traveling salesman problem using the scheme provided in [13] (Fig. 1).

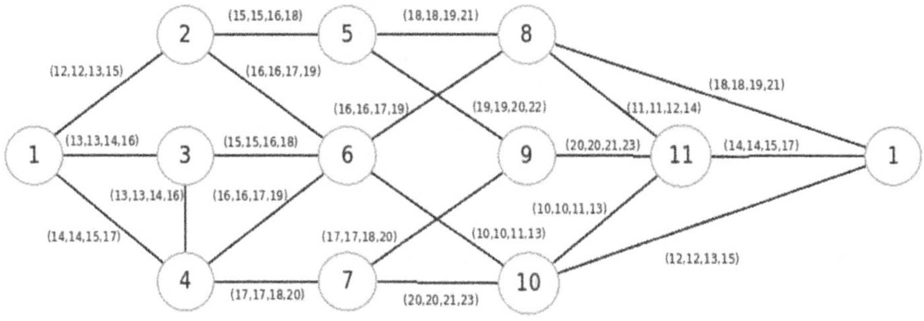

Fig. 1. The example of the traveling salesman problem with fuzzy travel durations between nodes.

For conducting numerical calculations based on the software implementation, the diagonal elements $\tilde{t}_{ii}$ of the matrix $\tilde{T}$ should be set to large positive numbers in order to obtain the values $x_{ii} = 0$ for all $i = 1, 2, ..., n$ in the solution. In the subsequent discussion, numerical calculations for the corresponding crisp traveling salesman problem are provided to verify the results of solving the fuzzy problem.

As in any other small network, an arbitrary exact procedure for finding the optimal route can be used to solve the traveling salesman problem. However, using fuzzy travel time in the network does not allow us to obtain an optimal solution. Therefore, to find an approximate solution, we will consider a genetic algorithm with a modification to improve mutation and increase population diversity in the form of PNP algorithm as an optimization method. In the process of forming mutations, a new solution is created by slightly changing one or more parameters of the existing solution. This can help in exploring the solution space near current locally optimal solutions, and allows maintaining diversity in the population to avoid premature convergence of the genetic algorithm. We will conduct a preliminary grouping of network nodes in a cluster using the Ward method. To carry out clustering, we will consider paths with travel time between nodes on the transport network as grouping objects. In this example, at the initial stage, we have 20 clusters.

Let's define the similarity metric for different paths. This can be, for example, the number of common nodes between two paths or the inverted geographical distance between them. For this task, the paths are defined by the fuzzy travel durations between points. When solving the proposed clustering problem, two metrics can be considered the most effective for calculating the intra-cluster distance in the context of path grouping, both based on the use of the time distance between arbitrary pairs of paths in the cluster. These metrics are the sum of pairwise distances and the maximum distance. For a cluster C with paths $p_1, p_2, \ldots, p_n$, the intra-cluster travel duration D is determined as:

$$D(C) = \sum_{i=1}^{n-1} \sum_{j=i+1}^{n} d\left(p_i, p_j\right), \tag{7}$$

or as a maximum time distance

$$D_{\max}(C) = \max_{i \neq j}\{d(p_i, p_j)\}, \tag{8}$$

where $d(p_i, p_j)$ is the time between arbitrary paths p_i and p_j, which can be calculated using any numerical metric (e.g., Euclidean distance or Hamming distance).

For calculating the inter-cluster travel durations, different approaches will be used depending on the task requirements and data characteristics. Among the most commonly used approaches are the calculations of the average, minimum, and maximum time distance between pairs of elements from two clusters. These methods define the ways to compute the distance between clusters, which are accordingly known as the nearest neighbor method $d(A, B) = \min(d(a, b) : a \in A, b \in B)$, the farthest neighbor method $d(A, B) = \max(d(a, b) : a \in A, b \in B)$, and the average distance method, where the latter is calculated as the mean of all squared pairwise distances between the elements of the two clusters $d(A, B) = \frac{1}{|A| \times |B|} \sum_{a \in A} \sum_{b \in B} d^2(a, b)$.

It is important to note that the choice of a specific method for calculating distances between clusters significantly depends on the specifics of the task. For example, it is known that the nearest neighbor method can lead to a chaining effect, where clusters become elongated, while the farthest neighbor method typically forms more compact and well-separated clusters. The average distance method provides a balance between these methods, making it an effective choice for many applied tasks. Therefore, in our case, we will use the average method for calculating the distance between clusters.

2.3 Genetic Algorithm and Numerical Results

Thus, for solving the TSP, a two-stage algorithmic procedure is proposed, which involves grouping the cities in the route into clusters, finding the optimal traveling salesman routes within each cluster using GA, and subsequently merging the clusters to form the final solution to the traveling salesman problem on the given transportation network:

1. Form all possible nodes of the TSP that participate in constructing the route.
2. Determine the optimal number of clusters to which the TSP nodes will be grouped using Ward's method.
3. Identify the nearest nodes for inter-cluster paths.
4. Iteratively rearrange the two extreme nodes in each cluster using the optimal inter-cluster distances with the nearest neighboring clusters.
5. Define genetic parameters such as crossover probability, mutation probability, population size, and the number of GA iterations.
6. Create an initial population of paths within each cluster for the selected two extreme nodes in the cluster from step 4.
7. Apply the genetic selection operator within each cluster.
8. Run the genetic crossover operator for each cluster.
9. Execute the genetic mutation operator for each cluster.
10. Repeat steps 8 and 9 according to the number of iterations, using the results of step 9 as the initial population for step 8.
11. Determine the proximity between clusters to identify the route for sequential joining.

12. Obtain k clusters and determine the duration of the resulting route for the initial traveling salesman problem.

The results of the initial stages of the clustering process for the given example are presented in Table 1.

Table 1. Initial stages of the clustering process for the given example

Step	Cluster merging	Crisp metric	Fuzzy metric (CoG)
1	6–10 and 10–11	150	$\frac{363}{2}$
2	1–2 and 1–3	$\frac{985}{4}$	$\frac{1149}{4}$
3	6–10-11 and 11–8	$\frac{2023}{6}$	$\frac{1217}{3}$
4	2-1-3 and 3–4	$\frac{3083}{6}$	$\frac{1793}{3}$

Subsequently, a jump in the inter-cluster metric is observed, and the clustering process can be stopped. After the first stage, we have two clusters with four nodes each (2-1-3-4 and 8-11-10-6) and clusters with three separate nodes (5, 7, and 9).

The clustering state is as follows (Fig. 2):

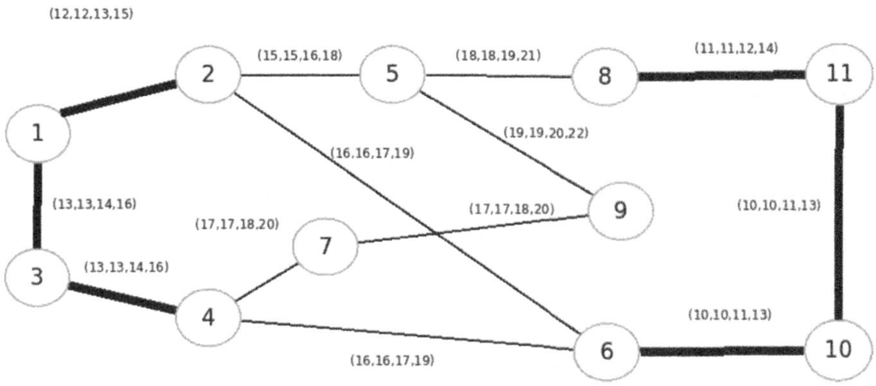

Fig. 2. Clustering after the first stage.

As a result of further optimization, the final solution is obtained (Fig. 3).

The optimal solution to the fuzzy traveling salesman problem (1), (2) is determined by the following path $1 \rightarrow 2 \rightarrow 6 \rightarrow 10 \rightarrow 11 \rightarrow 8 \rightarrow 5 \rightarrow 9 \rightarrow 7 \rightarrow 4 \rightarrow 3 \rightarrow 1$.

For the corresponding crisp problem, the best-known result is 156 units [12]. In the case of the fuzzy traveling salesman problem, the duration of the route is represented by a fuzzy number (156, 156, 167, 189), which allows us to conclude that one of the approximate variants of the fuzzy optimal solution has been obtained.

Next, we will consider the results of numerical calculations regarding the use of the two-stage approach for the traveling salesman problem in cases with different numbers

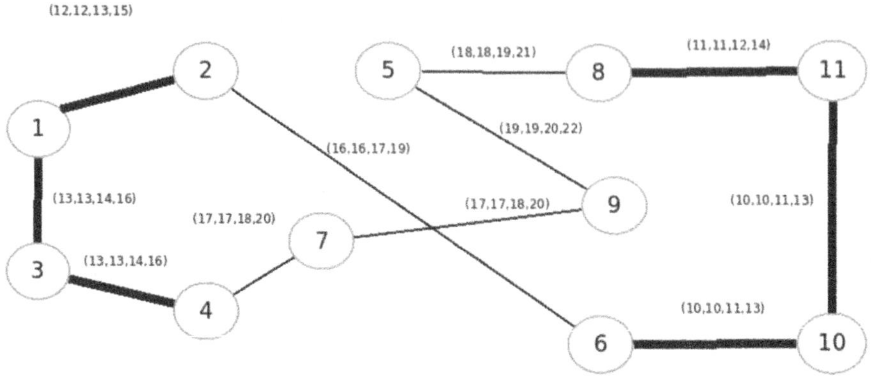

Fig. 3. Final solution to the traveling salesman problem.

of cities, assuming that the cities are randomly generated on a 200x200 plane and the probability of a direct path between cities is 50%.

Table 2 presents comparative results obtained using GA without PNP (average deterioration compared to the optimal route duration) and using the two-stage method without PNP (Table 2), as well as results obtained using GA with PNP (average deterioration compared to the optimal route duration) and using the two-stage method (Table 3).

Table 2. Comparison of numerical calculation results (GA without PNP).

Number of cities	Classic GA without PNP, worse by, %		Two-stage GA without PNP, worse by, %	
	100%	50%	100%	50%
19	2.71%	2.49%	2.36%	2.24%
20	2.91%	4.92%	2.54%	4.35%
21	3.19%	5.11%	2.81%	4.56%
22	3.33%	5.22%	2.91%	4.62%
23	3.55%	5.51%	3.13%	4.88%

Based on the results of the numerical experiment, as shown in Tables 2 and 3, it can be observed that the use of the two-stage method based on the genetic algorithm for solving the traveling salesman problem allows for relatively better results than using the genetic algorithm alone.

The combination of the cluster method and the genetic algorithm provides the opportunity to achieve savings in time and resources, while maintaining high accuracy of results. The use of preliminary clustering of nodes of the transport network reduces the size of the search space and provides high-quality initial solutions that are effectively optimized by the genetic algorithm within each cluster. This combination is especially noticeable in the case of solving traveling salesman problems in large transport networks,

Table 3. Comparison of numerical calculation results (GA with PNP)

Number of cities	Classic GA without PNP, worse by, %		Two-stage GA without PNP, worse by, %	
	100%	50%	100%	50%
19	2.48%	2.43%	2.24%	2.17%
20	2.83%	4.81%	2.53%	4.26%
21	3.09%	4.94%	2.77%	4.35%
22	3.21%	5.05%	2.90%	4.46%
23	3.41%	5.25%	3.05%	4.66%

where it is important to ensure an optimal balance between accuracy and computational efficiency.

3 Conclusions

For the fuzzy traveling salesman problem model, based on the fuzzy representation of travel durations between nodes in the transportation network, a two-stage scheme for finding the optimal traveling salesman route is proposed. The essence of this approach involves applying a clustering procedure for the nodes using Ward's method at the first stage, followed by the use of a genetic algorithm at the second stage to find the most accurate approximate solution by solving the resulting subproblems within all clusters. For the calculations, a formalization of fuzzy travel durations in the network is proposed, including methods for approximating fuzzy values, arithmetic operations, and methods for ordering fuzzy numbers. A method for transforming trapezoidal fuzzy numbers based on a Gaussian-like approach is also proposed. The proposed methodology is illustrated with examples, and a comparison of the results obtained using the proposed approach and those obtained using the direct application of the genetic algorithm is provided.

References

1. Bienstock, D., Zambelli, G.: Special Issue: Integer Programming and Combinatorial Optimization (IPCO)// Mathematical Program. **192** (2022). https://doi.org/10.1007/s10107-022-01782-y
2. Traub, V.: Approximation algorithms for traveling salesman problems. Cambridge University Press (2024)
3. Ivohin, E.V., Gavrylenko, V.V., Ivohina, K.E.: On the influence of fuzzy perception of the time passage speed on the solutions of optimization planning problems. Artifical intelligence **1**(95), 93–103 (2023). https://doi.org/10.15407/jai2023.01.093
4. Syropoulos, A., Grammenos, T.: A modern introduction to fuzzy mathematics. Wiley (2020). https://doi.org/10.1002/9781119445326
5. Jeter, M.W.: Mathematical Programming. Routledge (2020)
6. Cicirello, V.A.: Evolutionary computation: theories, techniques, and applications. Appl. Sci. (2022). https://doi.org/10.3390/books978-3-7258-2124-2

7. Chambers, L.D.: Practical Handbook of Genetic Algorithms. CRC Press (2019)
8. Topuria, N., Kikvidze O.: Application of genetic algorithm in common optimization problems. Int. Ann. Sci. **8**(1), 17–21 (2019). https://doi.org/10.21467/ias.8.1.17-21
9. Deep, K., Mebrahtu, H.: Variant of partially mapped crossover for the travelling salesman problems. Int. J. Combinatorial Optimi. Prob. Inf. **3**(1), 47–69 (2012)
10. Großwendt, A., Röglin, H., Schmidt, M.: Analysis of ward's method. ACM-SIAM Symposium on Discrete Algorithms (2019). https://doi.org/10.1137/1.9781611975482.182
11. Reddy, G.M.: Fuzzy ahp with trapezoidal fuzzy numbers and topsis. LAMBERT Academic Publishing (2020)
12. Voskoglou, M.G.: Fuzzy sets, fuzzy logic and their applications. Mathematics (2020). https://doi.org/10.3390/books978-3-03928-521-1
13. Ivohin, E.V., Gavrylenko, V.V., Ivohina, K.E.: On the recursive algorithm for solving the traveling salesman problem on the basis of the data flow optimization method. Radioelectron. Comput. Sci. Control **3**, 141–147 (2023). https://doi.org/10.15588/1607-3274-2023-3-14

Intelligent Technologies for Optimising the Workplaces of Tractor Drivers: Safety and Comfort in Modern Machine and Tractor Units

Maksym Krainiuk(✉) and Oleh Shcherbak

Kharkiv National Automobile and Highway University, 25 Yaroslav Mudryi Street, Kharkiv 61002, Ukraine
olenakrainiuk@khadi.kharkov.ua

Abstract. This study is devoted to identifying changes in the working conditions of tractor drivers in workplace computerization and developing proposals for updating sanitary standards for tractors and agricultural machinery. The paper analyzes modern technologies, such as the Internet of Things (IoT), artificial intelligence (AI), biometric sensors, and energy-efficient systems, which can regulate the microclimate in tractor cabs. Based on a review of the literature and practical examples, it is determined that introducing these technologies significantly changes operators' working conditions, increasing their comfort and safety, but also requires the adaptation of existing regulations. The study emphasizes the need to revise sanitary requirements to consider new factors, such as process automation and the impact of intelligent systems on the health of operators. The results can be used to improve the regulatory framework in the field of agricultural labour, which will contribute to increasing the efficiency of agricultural production and protecting workers' health.

Keywords: Tractor Operator · Workplace Computerization · Microclimate Control · Agricultural Machinery

1 Introduction

1.1 Problem and Relevance of the Study

Digitalisation and labour safety in machine and tractor systems, the introduction of computer control systems for various types of equipment, and the use of satellite and local systems require a new approach to the organisation of the operator's workplace.

For example, according to statistics, about 250 injuries are recorded in Ukraine during the operation of agricultural machinery alone per year, which is 4–5% of the total, but among fatalities, this level reaches 10–11%.

The duration of temporary disability per victim is more than 30 days, which indicates the seriousness of the injuries [1, 2]. Statistics may not consider minor injuries, which suggests the possibility of hidden injuries. The State Labour Service reports such cases.

O. Slavinska et al. (Eds.): ITSESQC 2024, LNNS 1335, pp. 211–222, 2025.
https://doi.org/10.1007/978-3-031-87376-8_19

During the workplace assessment, it was found that tractor drivers are exposed to a range of adverse factors, such as physical, chemical, biological, ergonomic and psychosocial, which can seriously impact health and safety.

The issue of ensuring optimal microclimatic conditions in the cabs of machine and tractor units is critical to improving operator comfort, safety and productivity. In particular, given the wide variability in working conditions, seasonal temperature fluctuations and challenging operating conditions, traditional climate control systems are often unable to effectively adapt to individual user needs and provide a stable microclimate. This can lead to reduced labour productivity, increased fatigue, and even dangerous situations for the health of operators.

The study's relevance is determined by the need to introduce the latest technologies, such as the Internet of Things (IoT), artificial intelligence (AI), biometric sensors, energy-efficient systems and nanotechnology, to create intelligent microclimate control systems. Given the rapid development of technology and the growing demands for comfort and safety in various industries, introducing these innovations will significantly improve operators' working conditions, increase labour efficiency and reduce health risks. Research in this area will contribute to developing new approaches to climate control in the cabs of machine and tractor units, which is essential for the development of modern engineering and the productivity of various sectors of the economy.

1.2 Objective of the Study

The study aims to determine the impact of workplace computerisation on the working conditions of tractor drivers and develop recommendations for improving sanitary standards and requirements for designing machine-tractor units to ensure safe and comfortable working conditions. The study's objectives are as follows: (1) Analysis of modern technologies for controlling the microclimate in the cabs of machine and tractor units and their impact on the comfort and safety of operators. (2) Determination of possible changes in the working conditions of tractor drivers due to the introduction of computerised control systems.

2 Review of Current Literature

Computerisation and automation of processes in machine and tractor systems are becoming leading trends that help improve labour efficiency, task accuracy, and employee comfort. The introduction of modern information and communication technologies in machine cabs not only optimises operations but also significantly changes the operator's working conditions, reducing physical exertion and increasing the level of automation of routine processes [3, 4].

The introduction of computerised systems into the cabs of machine and tractor units can have both positive and negative effects on the health of operators [5, 6]. While it reduces the need for physical labour, thereby reducing the risk of injury and overstrain, it's equally important to be aware of the potential negative effects. Prolonged work with computerised systems can lead to new challenges, such as increased cognitive load and the risk of developing occupational diseases associated with the use of displays. Understanding these potential negative effects is crucial for promoting operator safety.

Modern research indicates the need to revise and update the sanitary norms and standards that regulate the working conditions of machine and tractor system operators. Previous studies [7, 8] highlighted the importance of developing new recommendations that consider modern technological changes and their impact on workers' health. In particular, this applies to the microclimate parameters in cabs, workplace ergonomics and equipment requirements.

The use of sensor technologies and artificial intelligence in managing the microclimate and other machine systems can significantly improve operators' working conditions. Papers [9, 10] explore the capabilities of such systems to automatically adjust the working environment to the operator's individual needs. These technologies provide optimal comfort and safety and can also reduce the impact of adverse external factors.

An analysis of current literature shows that innovations in computerising machine and tractor systems have great potential for further development. In particular, works [11, 12] consider the possibility of introducing intelligent systems that can not only adapt to external conditions but also actively contribute to maintaining the health of operators, for example, by monitoring biometric data and providing recommendations on work and rest.

A review of the current literature indicates researchers have a significant interest in computerising workplaces in machine and tractor units and its impact on working conditions. Despite considerable progress, issues related to the adaptation of sanitary standards and ensuring employee safety remain relevant and require further research.

3 Discussion and Results

3.1 Adapting Sanitary Standards and Ensuring Driver Safety in Modern Machine and Tractor Units

It is interesting to note that the development of technology and the introduction of new types of equipment, such as on-board computers, have significantly changed the working conditions of tractor drivers. However, the question of how these changes affect occupational health and safety requirements requires further analysis.

Currently, there are several regulations governing the safety of tractor drivers, such as DSP 3.3.2.041–99 'Sanitary Rules for the Equipment and Arrangement of Tractors and Agricultural Machinery' and the rules for labour protection in agriculture, approved by Order of the Ministry of Social Policy of Ukraine No. 1240 of 29 August 2018. However, none of these documents considers the new realities associated with integrating on-board computers into machine and tractor systems.

The presence of on-board computers on machine-tractor units adds new types of workload to the driver, particularly the sensory workload associated with the need to continuously monitor numerous machine parameters and the intellectual workload, including information processing and decision-making. Such workloads can lead to fatigue, stress, and even errors during work.

In this context, it is essential to consider that modern working conditions for tractor drivers require additional occupational safety measures. It is necessary to update existing regulations, add new requirements for working with onboard computers, and develop recommendations to reduce the risks associated with new workload types.

Introducing technologies that reduce harmful factors (vibration, noise, dust, etc.) certainly helps improve working conditions. However, it is essential to remember the new challenges faced by machine operators working with modern machinery and to ensure that their occupational health and safety are adequate.

3.2 Using Smart Technologies to Monitor Working Conditions

Modern machine and tractor units have increasingly sophisticated and robust systems to ensure operator comfort and safety. One of the critical aspects of this comfort is the cab climate, which directly impacts the driver's performance and health. Integrated sensor systems for climate control are one of the most effective solutions for ensuring optimal working conditions.

Integrated sensor systems can continuously monitor critical parameters of the cab and work area climate, such as temperature, humidity, carbon dioxide (CO_2) and air quality. These systems consist of sensors placed in different cab parts to provide a detailed picture of the current conditions. Let's look at potential solutions using intelligent systems to improve working conditions and monitor them.

Sensor systems for monitoring the microclimate in the cab can automatically adjust the ventilation, heating or cooling parameters, providing comfortable working conditions for the driver (Fig. 1).

1. Temperature control can be carried out by temperature sensors, which automatically adjust the operation of heaters or air conditioners to maintain optimal working conditions in the event of an increase or decrease in temperature to an unacceptable level.
2. Humidity sensors provide humidity control, and the system can automatically adjust the humidity level by activating dehumidifiers or humidifiers.
3. Monitoring carbon dioxide (CO_2) levels is mandatory, as high levels of CO_2 in the cab can cause drowsiness and reduced concentration, which increases the risk of accidents. CO_2 sensors monitor the concentration of this gas, and the ventilation system automatically provides fresh air to maintain safe CO_2 levels.
4. Air quality is monitored by additional sensors that can measure other harmful impurities in the air, such as dust or toxic substances. The system activates air filtration and updates its composition if an elevated pollution level is detected.
5. Our integrated sensor systems provide automatic microclimate control, ensuring your comfort and convenience. These systems not only collect information about current conditions but also automatically adjust the operation of ventilation, heating, or cooling systems based on the data received. This means you can concentrate on your work without being distracted by uncomfortable conditions, increasing your productivity and reducing stress levels.

Introducing such systems in modern machine and tractor units is an essential step towards raising the standards of safety and comfort at the workplace of the tractor driver. This helps protect employees' health but also helps reduce the impact of adverse conditions on their productivity and the quality of work performed.

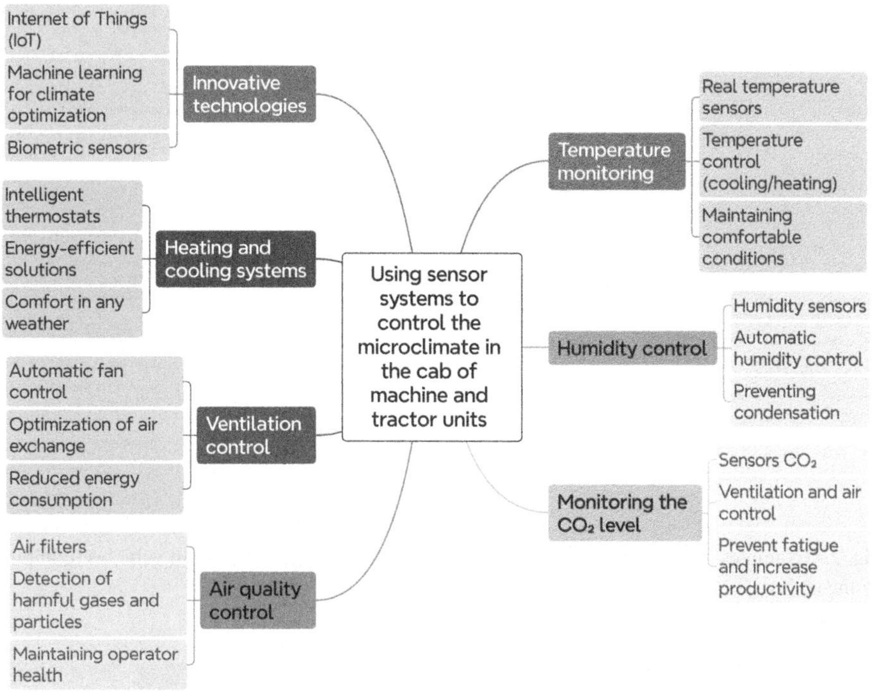

Fig. 1. Use of sensor systems to control the microclimate in the cab of machine and tractor units (compiled by the authors).

3.3 Analysing the Use of Sensor Systems for Cabin Climate Control

Modern manufacturers of machine and transport units are actively introducing innovative sensor systems to control the microclimate in cabs, ensuring increased comfort and efficiency for operators. Let's look at specific examples of companies and technologies that use modern information systems to provide comfort and safety. Figure 2 illustrates examples of the use of sensor systems to control the microclimate in cabs of various types of transport, including automotive, aviation, construction, utility, railway and other machinery and transport equipment. It demonstrates how leading companies such as Tesla, BMW, John Deere, Boeing and others are implementing intelligent systems to monitor and control microclimate parameters such as temperature, humidity, CO_2 levels and ventilation. This underlines the importance of technology in providing a comfortable and safe environment for operators and passengers.

John Deere [13] is one of the leaders in introducing innovative technologies into its machinery (Fig. 3). It is implementing the ComfortGard™ system:

- It is equipped with multiple sensors that monitor temperature, humidity and air quality in the cab.
- Automatic climate control regulates the air conditioning and heater to ensure constant comfort regardless of the outside conditions.

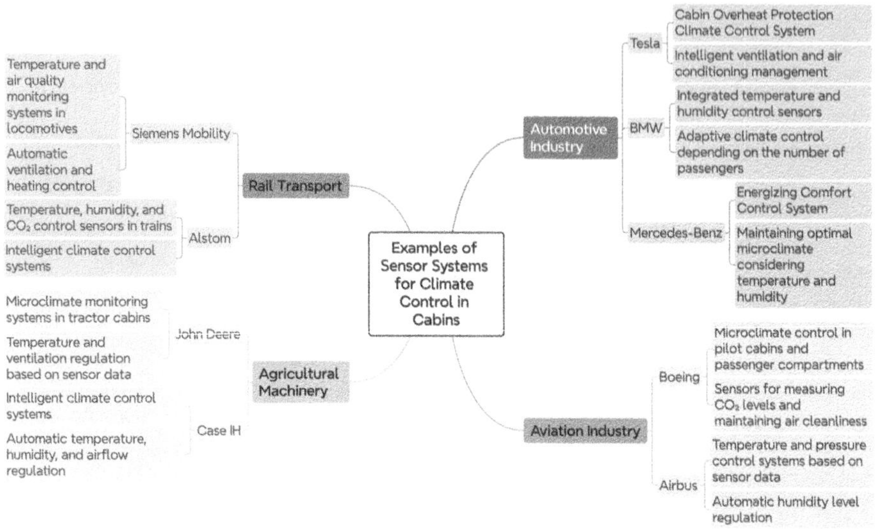

Fig. 2. Examples of the use of sensor systems to control the microclimate in vehicles (compiled by the authors).

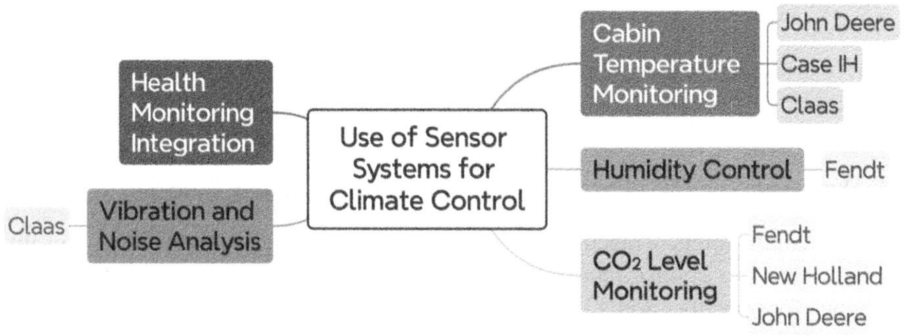

Fig. 3. Examples of the use of sensor systems to control the microclimate in the cab of transport vehicles (compiled by the authors).

- The high-efficiency air filtration system protects the operator from dust, allergens and other pollutants, which is especially important when working in the field.

It can also be integrated with intelligent systems:

- The microclimate systems are integrated with on-board computers, allowing the operator to adjust the settings via an intuitive interface. The 8R series tractors are equipped with temperature, humidity and CO_2 sensors. These sensors automatically adjust the climate control settings to create a comfortable environment for the operator.
- It is possible to save personalised settings for different operators.
- The JDLink system allows you to remotely monitor the condition of the equipment and the microclimate in the cab via mobile applications.

CLAAS [14] is known for its high-tech solutions (Fig. 3). The machines' cabs are equipped with a state-of-the-art CLAAS Climate Control system consisting of sensors for monitoring temperature, humidity, and air quality, automatically adjusting the microclimate for optimal operator performance. The CLAAS PREMIUM CAB system includes:

- Solar sensors are used to adjust the cooling intensity based on insolation.
- An automatic carbon dioxide control system that ensures a constant supply of fresh air and prevents CO_2 build-up, which increases operator concentration and performance.
- A highly effective noise and vibration isolation system that works in tandem with sensors to minimise noise and vibration exposure.

New Holland [15] is actively implementing advanced technologies to improve operator comfort and safety by introducing the Hori-zon™ Cab system (Fig. 3), which includes:

- A set of sensors for monitoring the microclimate to ensure automatic temperature and humidity control.

The use of air quality sensors that detect the presence of dust and other contaminants activates the appropriate filtration systems.

- An integrated control system that allows you to remotely monitor and adjust microclimate parameters via mobile applications or an onboard computer.

Fendt is one of the leading manufacturers of machine transport units, which is actively introducing innovative technologies to improve the comfort and efficiency of operators (Fig. 3). Fendt uses energy-efficient climate control systems in its machines, including heat recovery technologies and intelligent compressors, which reduce energy consumption and maintain an optimal cabin climate. Thanks to intelligent control algorithms, Fendt climate control systems adapt to the current working conditions, ensuring a comfortable environment for the operator regardless of external factors. This makes Fendt products not only efficient but also environmentally responsible.

Many other manufacturers of machine and transport units cooperate with specialised technology companies to implement advanced solutions. For example:

- Bosch develops high-precision sensor modules and control systems for climate control used in various types of transport.
- Honeywell provides comprehensive air quality monitoring and management solutions, including sensors for CO_2, VOC (volatile organic compounds) and other parameters.
- Siemens offers integrated automation and control systems that can be adapted for use in the cabs of various vehicles to ensure an optimal microclimate.

3.4 Modern Technologies for Microclimate Control in Machine and Tractor Units

Modern technologies for controlling the microclimate in the cabs of machine and tractor units significantly increase the comfort and safety of operators. The use of the latest developments, such as the Internet of Things (IoT), artificial intelligence (AI), biometric sensors, energy-efficient systems and nanotechnology, allows us to create intelligent climate control systems that adapt to individual user needs and provide a stable working environment. Let's analyse these technologies in more detail in terms of their potential use to ensure the safety and comfort of tractor operators in modern machine and tractor units (Fig. 4):

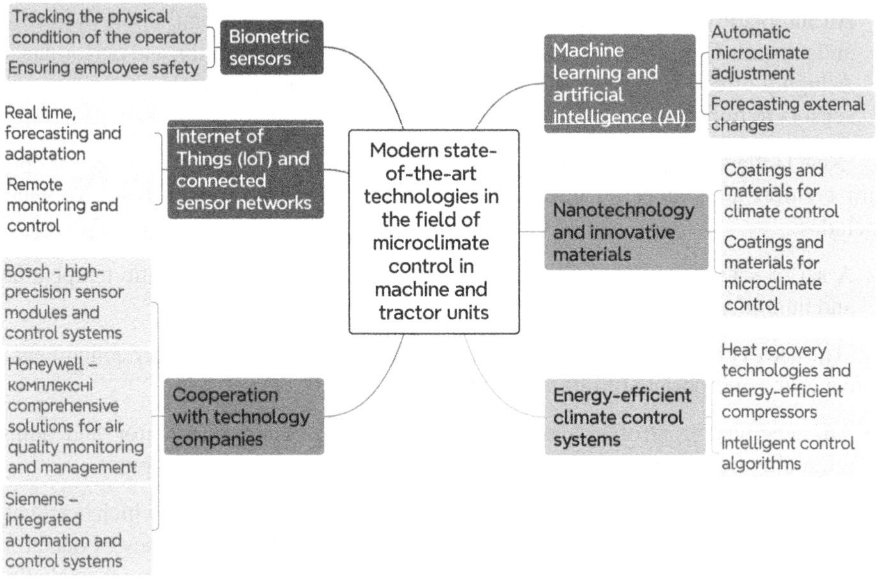

Fig. 4. Modern technologies for microclimate control in machine and tractor units (compiled by the authors).

1. Internet of Things (IoT) and connected sensor networks
   - Real-time operation, forecasting and adaptation. Sensor systems integrated with the IoT allow for real-time monitoring and control of the microclimate in the tractor cab. The data collected from these sensors includes microclimate parameters and air quality data for the working area. Thanks to the connection to cloud services, this data is processed and analysed, which allows you to predict possible changes and avoid unfavourable conditions.
   - There is a possibility of remote monitoring and control. Modern IoT systems can control the microclimate remotely via cloud platforms. For example, managers of transport companies can remotely monitor the conditions in the cabs of various vehicles, assess the working conditions of operators and respond quickly to changes in external factors or internal microclimate parameters. This significantly increases the safety and productivity of employees.
2. Machine learning and artificial intelligence (Ai) can perform:
   - Automatic microclimate adjustment. AI allows the microclimate system to adapt to the operator's individual needs based on historical data analysis. For example,

AI can learn user preferences for temperature, light level, and ventilation intensity and automatically adjust these parameters to ensure maximum comfort during work.

– Predicting external changes. AI is also used to proactively predict changes in external conditions, such as weather changes or seasonal temperature fluctuations. Based on these predictions, the climate control system can adjust the appropriate parameters in advance, maintaining optimal conditions regardless of external influences, thereby ensuring a secure and stable working environment for the operator.

3. Biometric sensors will hide:

   – Tracking the physical condition of the operator. Biometric sensors, whether built into the seat or integrated into wearable devices, monitor the physical condition of the operator. They measure heart rate, stress level, body temperature, signs of fatigue, and other indicators. This allows the microclimate in the cab to be automatically adjusted to create a more comfortable working environment and to provide signals to the operator and manager about the need for rest if stress or fatigue levels exceed acceptable limits.
   – Ensuring employee safety. A climate control system that is equipped with biometric sensors can also respond to critical changes in the operator's physical condition, providing a protective layer. For example, if the sensors detect high levels of fatigue or stress, the system can automatically reduce the tractor's work intensity, increase ventilation, or even turn off equipment to ensure the worker's safety, thereby enhancing the sense of protection for the operator.

4. Energy-Efficient climate control systems:

   – Energy-efficient climate control systems include using heat recovery technologies that reuse the heat generated by the tractor's systems to heat the cab. This reduces energy consumption and ensures a stable microclimate. Energy-efficient compressors also play an essential role in maintaining optimum temperature conditions with minimal energy consumption.
   – Intelligent control algorithms for climate control systems optimise their operation, reducing energy consumption and increasing efficiency. They consider current conditions, such as outdoor temperature, solar activity, and humidity, and adjust the system accordingly. This ensures comfortable conditions with minimal energy consumption.

5. Nanotechnology and innovative materials:

   – The use of nanotechnology in the production of materials for tractor cabs allows the use of special coatings that reflect heat or provide better insulation. This helps maintain a stable microclimate in the cab regardless of external conditions. For example, nanocoatings on the cab windows can reflect the sun's rays, preventing the interior from overheating.
   – Antibacterial and anti-allergenic materials. Using innovative materials with antibacterial and anti-allergenic properties in air filtration systems improves the air quality in the cab. This is important for the operator's health, especially when

working in confined spaces for long periods. This filtration system reduces the risk of allergic reactions and provides a cleaner and safer working environment.

3.5 Structural and Functional Scheme for Ensuring Optimal Working Conditions for Operators of Modern Machine and Tractor Units

To better illustrate the composition and functionality of the monitoring system for tractor cabin microclimate and safety, we provide both structural and functional diagrams. These diagrams clarify the interactions between system components, from data collection through sensors to the execution of environmental adjustments within the cabin. The structural diagram (Fig. 5) outlines the system's key components and their configuration, while the functional diagram (Fig. 6) details the workflow and processes involved in achieving optimal working conditions for operators.

Structural Schema of Monitoring System	Sensors	Temperature Sensors
		Humidity Sensors
		CO_2 Sensors
		Air Quality Sensors
	Central Controller	Receives data from sensors
		Processes data and makes adjustments
	Actuators	Air Conditioner and Heater
		Ventilation Systems
		Filtration Blocks
	Control Algorithm	Automatic climate adjustment based on sensor data

Fig. 5. Structural scheme for ensuring optimal working conditions for operators of modern machine and tractor units

The structural and functional diagrams above demonstrate the integration of sensor networks, control algorithms, and actuators, creating a dynamic system capable of maintaining a stable and safe environment within the tractor cabin. By automatically adjusting to external and internal conditions, this system significantly enhances operator comfort, reduces fatigue, and contributes to overall productivity.

These frameworks emphasize the importance of continuous monitoring and intelligent adaptation in modern agricultural machinery, highlighting how innovative technologies can improve the safety and efficiency of machine operators' workplaces.

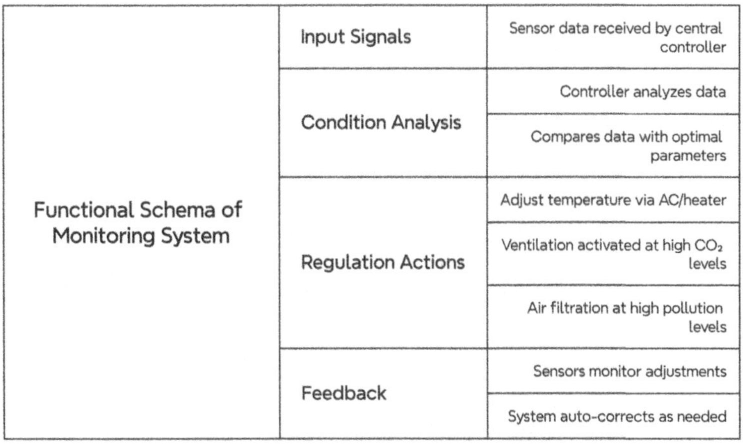

Fig. 6. Functional scheme for ensuring optimal working conditions for operators of modern machine and tractor units

4 Concusion

The use of modern sensor systems to control the microclimate in the cabs of machines is becoming a standard. Introducing such technologies helps improve operators' comfort and safety, increase their productivity and reduce health risks. The further development and integration of the latest technologies, such as IoT, AI and biometric sensors, opens up even more opportunities to optimise working conditions and increase the efficiency of operations in various areas.

A study aimed at identifying possible changes in operators' working conditions during the computerisation of workplaces confirmed the relevance of analysing and revising existing safety and health standards. The use of modern technologies in managing machine-transport units affects the working conditions and health of operators, which requires adapting the regulatory framework to new working conditions.

Based on a literature review and analysis of practical examples, the main aspects that require attention are the automation of work processes, the impact of sensor systems on operator comfort, and the need for changes in sanitary standards. For instance, the computerisation of transport systems involves the use of autonomous vehicles and smart logistics, which opens up new opportunities for improving efficiency and operator comfort, but at the same time requires more careful monitoring of the impact of these technologies on employee health.

The proposed recommendations can serve as a basis for future research in this area and for revising existing occupational safety standards. This will help improve working conditions, ultimately increasing the efficiency of various production processes and ensuring workers' health protection.

References

1. State Statistics Service of Ukraine, 2020–2022, https://www.ukrstat.gov.ua
2. Prevention of hidden injuries at agricultural enterprises. Labour protection and fire safety, 19 April 2023 http://surl.li/swubb
3. Benos, L., Bochtis, D.D.: An analysis of safety and health issues in agriculture towards work automation. Information and Communication Technologies for Agriculture—Theme IV: Actions, pp. 95–117. Springer International Publishing, Cham (2022)
4. Lunner-Kolstrup, C., Hörndahl, T., Karttunen, J.P.: Farm operators' experiences of advanced technology and automation in Swedish agriculture: a pilot study. Agromed. **23**, 215–226 (2018). https://doi.org/10.1080/1059924X.2018.1458670
5. Shirokov, Y., et al.: Analysis of changes in working conditions of a tractor driver during computerization of control systems of a tractor unit. Sci. Herit. **84**(1), 62–67 (2022)
6. Caffaro, F., et al.: Machinery-related perceived risks and safety attitudes in senior Swedish farmers. J. Agromedicine **23**(1), 78–91 (2018). https://doi.org/10.1080/1059924X.2017.1384420
7. Krainiuk, M., Krainiuk, O., Buts, Y., Barbashyn, V.: Substantiation of changes in the certification of the tractor driver's workplace in modern machine-tractor units. Munic. Econ. Cities. **3**(184), 209–215 (2024). https://doi.org/10.33042/2522-1809-2024-3-184-209-215
8. Yu, K.M., Krainiuk, O.V., Sherbak, O.V.: Hygienic assessment of the working conditions of the tractor driver's workplace in modern machine-tractor units. Modern Research in Science and Education. Proceedings of the 10th International Scientific and Practical Conference, pp. 263–273. BoScience Publisher, Chicago (2024)
9. Subeesh, A., Mehta, C.R.: Automation and digitization of agriculture using artificial intelligence and internet of things. Artif. Intell. Agric. **5**, 278–291 (2021). https://doi.org/10.1016/j.aiia.2021.11.004
10. Mana, A.A., et al.: Survey review on artificial intelligence and embedded systems for agriculture safety: a proposed IoT agro-meteorology system for local farmers in Morocco. Smart Embedded Systems and Applications, 211–242. CRC Press (2022)
11. Hota, S., Tewari, V.K., Chandel, A.K.: Workload assessment of tractor operations with ergonomic transducers and machine learning techniques. Sensors. **23**(3), 1408 (2023). https://doi.org/10.3390/s23031408
12. Grandi, F., et al.: A transdisciplinary digital approach for tractor's human-centred design. Int. J. Comput. Integr. Manuf. **33**(4), 377–397 (2020). https://doi.org/10.1080/0951192X.2019.1599441
13. Shearer, S.A., Pitla, S.K., Luck, J.D.: Trends in the automation of agricultural field machinery. Proceedings of the 21st Annual Meeting of the Club of Bologna, Italy (2010)
14. Hackfort, S.: Unlocking sustainability? The power of corporate lock-ins and how they shape digital agriculture in Germany. J. Rural. Stud. **101**, 103065 (2023). https://doi.org/10.1016/j.jrurstud.2023.103065
15. New Holland: official page of the company. https://agriculture.newholland.com/en-ie/europe

Using Modern Document Information Systems in the Management of Transport Companies

Halyna Sichkarenko[1], Vira Gamaliia[2], and Artem Zabuga[3](✉)

[1] National Transport University, Mykhaila Omelianovycha-Pavlenka St. 1, Kyiv 01010, Ukraine
[2] State University of Infrastructure and Technologies, Kyrylivska St. 9, Kyiv 02000, Ukraine
[3] National Technical University of Ukraine "Igor Sikorsky Kyiv Polytechnic Institute", Prosp. Peremohy 37, Kyiv 03056, Ukraine
azabuga@ukr.net

Abstract. The rapid development of technologies has revolutionized the field of management of transport companies. Digitization is an indicator of their success in the market. Document and information systems are the cornerstone of the digital transformation of companies. As world experience shows, digital transformations, in particular in combination with artificial intelligence, significantly affect all work processes, generate new skills, competencies, knowledge management technologies, organizational capabilities, etc. The report defines the place and role of document and information systems and technologies (DIST) in the digitalization of transport companies. The leading information technologies have been identified from the position of their role in the management of documentation processes in companies. The real advantages and disadvantages of using artificial intelligence in working with documents are revealed. Widespread problems of DIST implementation and ways to eliminate them in Ukrainian transport companies are classified. The authors conclude that it is inappropriate to reduce the essence of document's, information systems' and technologies' digitization only to the technical and informational aspects of their functioning, without taking into account the impact on corporate strategy. It has been proven that the latest DIST contribute to increasing the efficiency of the company's production activities through operational collection and a rational mechanism for processing and systematizing information. It can also help for shortening the response time to external changes, optimizing the enterprise organizational structure, reducing the amount of operational and management costs, forming an optimal development strategy due to the access of all managers to a single information base, etc.

Keywords: Document and Information Systems · Information Technologies · Management of Transport Companies · Digitization · Data Security

O. Slavinska et al. (Eds.): ITSESQC 2024, LNNS 1335, pp. 223–235, 2025.
https://doi.org/10.1007/978-3-031-87376-8_20

1 Introduction

The implementation and using intelligent transport systems involves digitalization of all processes. The digital transformation of companies begins with the implementation of document information systems and technologies (DIST). The latter allow automating the processes of creating, transmitting, receiving, classifying, verifying, saving, searching, using and protecting transport data in all sources of information. They act as a leading part of the general information management system of the company, a basis for digitization of business processes and all technological, engineering, research activities etc.

Scientific studies devoted to the process of digitalization of companies, where DIST plays a decisive role, mainly consider individual components of this phenomenon and have many debatable issues. In Ukraine, there is a lack of research that would go beyond a purely technical and technological point of view, consider the use of DIST in the management of companies, and determine which practical developments in this area can be adapted for development.

The purpose of the article is a critical analysis and generalization of scientific approaches to the key aspects of the using document and information systems and technologies in the context of the digital transformation of transport companies of Ukraine from a practical point of view, as well as the identification of real problems during the implementation of DIST in the management system and ways to eliminate them.

2 Research Methodology

The research methodology was the methods of analysis and synthesis, systematization, comparison—for consideration of existing DIST; abstract-logical—to generalize the experience of their application; statistical and sociological—at clarifying the advantages and disadvantages of popular Ukrainian services and AI tools for working with document information, etc.

3 Results and Discussion

The digital transformation of companies begins with the introduction of document and information systems and technologies. A popular article by famous scientists [9] is called: "Document management system—the way to digital transformation".

The authors' analysis of publications on the Google Academy platform for the query "document management system" revealed 1,630 items (since 2001). The peak of publications came in 2023 (229 against 178 in 2021), which proves the relevance of the study. It was found that the conceptual-categorical apparatus is not fully formed; there are active discussions among researchers [4, 5, 8, 13, p. 11; 15]. Mostly, the concept of digital transformation of companies is of abstract and technological nature [22, p. 118]. We join the authors who believe that such approach does not provide a complete picture of the process in today's conditions. They suggest calling digital transformation as a process when new or existing models and operations are created or supported the transformation of organizational structures, resources or relationships with internal and external actors [15]. There is a combined impact of several digital innovations, leading to the emergence of new actors, structures, practices, values and opinions that change the rules of the game in organizations, industries or spheres [7, p. 52]. So, in new approaches to the digitalization of document circulation in companies, their socio-humanitarian aspect is emphasized. It includes skills, organizational capabilities, value orientations of employees, competencies, social practices and structures. With the using AI, new, in particular, ethical issues arise. The authors emphasize the growing influence of the social and humanitarian factor on the implementation of DIST in transport companies.

The market of software products of Ukraine offers a variety of services created according to the requirements of current legislation. Each company can choose them by itself. For example, since 2019, transport companies of Ukraine have been implementing the e-CMR (consignment note) project, which aims to replace paper waybills with electronic ones. The start of mandatory use of e-CMR was determined on August 1, 2022. This is a state initiative to simplify the process of transportation, increase its quality and transparency, reduce costs for business, and bring Ukrainian legislation closer to European standards. PTAH is a popular platform that unites all participants in the document circulation and provides a single standard for the exchange, verification, protection and transportation of legally significant documents signed by EDS (electronic digital signature) and QES (qualified electronic signature). Free services include Paperless, where users can exchange accounting and legal documents: contracts, invoices, reports of completed works, expense invoices, etc. Among the paid services, the most popular in Ukraine are M.E.Doc (from the Ukrzvit company), VCHASNO, Document.Online, Art-Office, Comarch EDI, FossDoc, ASCON and about 10 other products [19]. As a rule, the services are available around the clock and can provide qualified technical assistance, maintain a sufficient level of security in terms of confidentiality and data protection, and also offer to customize the functionality of the service to the individual needs of the client. They mostly have a flexible variable pricing system and a free "starter" tariff. According to our data, the VCHASNO and Document.Online systems are the most popular. The latest document information systems offer increasingly reliable security measures, including encryption, user authentication and access control.

It should be noted that product advertisements on official websites may differ significantly from reality. With this in mind, the authors offer a table of advantages and disadvantages of popular services that can help with their selection and implementation (Table 1).

Therefore, the main disadvantages of document technologies, which are present in all types of systems, are additional funding, time spent on implementation, adaptation, updating, maintenance, technical support and training of personnel, and dependence on the Internet. Threats to security and confidentiality of information remain the biggest drawback.

Technologies with artificial intelligence are of special importance in working with documents. Thus, NLP (natural language processing) algorithms with AI can automatically translate, analyze and understand the content of documents, generate text similar to human speech, distinguish emotions and moods expressed in documents. AI can analyze documents for hidden information, identify and redact certain data to ensure privacy, make it easier to understand key points without reading the entire document, create concise summaries, and more. AI search engines can understand the context and

Table 1. Popular Ukrainian document management services: advantages and disadvantages.

Services / site	Advantages	Disadvantages
M.E.Doc https://medoc.ua/	1. A full-fledged EDM (electronic document management) service has been implemented. It is possible to sign primary documents and tax invoice and send them to the counterparty 2. A built-in system of tips when filling out reports 3. It is possible to create many company and business profiles 4. Support for the network version of the system, including in terminal mode 5. Automatic calculation and filling of fields from the enterprise map 6. The software consists of numerous plug-in modules and can be customized for any client	1. Software overload, slowness 2. To work with a number of keys, you need to purchase a license 3. In the SaaS mode of the service, it is possible to sign reports only in the Windows system through crypto provider 4. It can work with documents only with the keys of the Ukraine Central Security Agency 5. High requirements for technical support
VCHASNO https://vchasno.ua/	1. 24/7 access to documents from any device 2. 30 s to search for a document – quick access to the cloud archive 3. Intuitive interface 4. Protection of confidentiality of information 5. Any number of employees can work autonomously 6. Support on Facebook-group, availability of videos, test point of sale for training	1. Rather difficult to use 2. High dependence on technologies that may become obsolete 3. May require significant investment of time to implement technology and training staff 4. Weak technical support, delay in updating 5. For small and medium businesses

(*continued*)

Table 1. (*continued*)

Services / site	Advantages	Disadvantages
Art-Report plus https://art-zvit.com.ua/	1. The possibility of working with several enterprises at the same time, drawing up consolidated reporting 2. It is possible to create many profiles of companies and individual entrepreneurs 3. Automatic calculation and filling of fields from the enterprise map 4. Handy tips from auxiliary guides 5. Ability to work with several reports simultaneously from one screen	1. Contains a limited list of reports 2. Old interface, poor reporting information 3. There is no network version of work, the possibility of working with a single database from many computers
A single window (EDZV) https://opz.org.ua/index.php	1. The software is completely free 2. Allows you to create and sign reports 3. Ability to sign and register tax invoices 4. Unified format of the transport message	1. No express software support, delay in updating 2. Difficulty with installation, requires many additional components to work 3. The functions of superimposing QES, sending reports, and exchanging electronic tax invoices are not fully implemented

(*continued*)

Table 1. *(continued)*

Services / site	Advantages	Disadvantages
PTAH https://edi.com.ua/uk	1. Allows you to combine different accounting systems for instant transfer and exchange of documents signed by QES 2. The possibility of EDS integration into any accounting system 3. Speed of work, mobile solution for drivers 4. Easy access and quick search of documents through convenient filters 5. Protection against unauthorized access, separation of rights and multi-factor authentication	1. It takes time and effort to implement 2. Requires additional costs for integration with existing systems 3. Needs staff training 4. May be limited in functionality for certain sectors 5. Requires constant updating and support
KeepinCRM https://keepincrm.com/	1. Flexible interface 2. Constantly updated, can be modified as needed 3. Not overloaded with functionality, simple and clear 4. Features turn on and off as needed so they don't get in the way 5.Implementation control and prompt response 6. Ability to conduct all communication with clients within the system without using other software	1. Difficult to learn 2. Not all tasks are done quickly 3. No integration with Meta 4. Inconvenient searching 5. For small and medium businesses

Source Created by the authors based on the analysis of sites, reviews of product users [1, 6, 14, 16, 21] and analysis of publications [12, 17, 19, 20] intent of queries, providing more accurate and relevant search results in the document repository. Customer relationship management (CRM) systems are implemented to analyze customer data and determine the best engagement strategies [18]. AI systems can continuously learn and improve their performance.

However, at present, the achievements of AI in working with document information are used by a small part of transport companies. Many people do not understand its main components and implementation strategies. This is evidenced by a study on artificial intelligence and information management, which was conducted by AvePoint together with the Association for Intelligent Information Management and the Center for Intelligent Political Leadership in 2023 [2]. The report summarizes the opinions of more than 750 business leaders in 10 sectors of the economy from 16 countries. Although 88% of respondents say they have an information management strategy, only just over half of them have basic measures in place such as archiving, retention and lifecycle management policies (56%). Meanwhile, companies with innovative information management strategies are 1.5 times more likely to benefit from AI than others. Advanced companies see the main benefits in increasing efficiency and productivity (74%), improving the decision-making process (67%) and optimizing resources (65%) [2].

Already, 89% of top-managers recognize that a high level of data quality is critical to success. According to Gartner forecasts, by 2025, 10% of all generated data will come with the help of artificial intelligence. Companies will need to focus even more on information management to ensure the high quality and reliability of their data [3].

Therefore, the using AI in working with documents not only increases its efficiency and reduces the number of manual intervention and errors, but also optimizes the sequence of operations, speeds up the processing of documents and enables companies to make informed decisions based on operational information. However, it also carries certain threats. A serious problem currently remains the lack of qualified specialists who can effectively implement and manage AI. In 2024, AvePoint conducted research and analyzed the perspectives of more than 750 global business leaders on real-world applications of artificial intelligence and information management. According to the results, 95% of companies encountered obstacles during the implementation of AI, primarily due to the quality and organization of internal data, misprioritization of their data, etc. [2] (Table 2).

Despite all fears and understandable difficulties in using DIST, they can be reduced. The implementation of the latter together with AI can significantly increase the efficiency of management; decisions made and make production processes more adaptable to changes in the business environment. Competent planning, employee training, taking into account the features of a specific organization at choosing technological solutions, as well as security issues, remain important. We will provide an overview of several ways of integrating AI into information systems, in particular, through a secondary analysis of survey materials of 49 leading Ukrainian Internet marketing experts regarding the platforms and tools they directly use [10] (Table 3).

So, a common problem with the use of AI in companies is hallucinations, false or fabricated information provided at the request of users. Hallucinating AI refers to the generation of output that may seem reliable but is actually wrong. But according to P. Gerasimov, SEO Team Lead at Innovation Group, artificial intelligence hallucinations

Table 2. Common problems of the introduction of document and information systems and technologies in companies and ways to eliminate them.

Problem	Description	Ways of elimination
Initial investment	Purchase of equipment such as computers, printers, scanners, servers, etc	Making a detailed budget, explore the possibilities of free programs
Training and recruitment costs	Untrained staff and users of new systems can lead to errors and working inefficiencies	Training staff and users should be carried out prior to the implementation of DIST with ongoing support thereafter
System failure and insufficient technical support	The process may stop completely, resulting in data loss or corruption	Providing reliable support, including vendor support and internal IT capabilities
Incorrect distribution of work among users	There may be a delay in the work that interrupts the entire work process	Engaging users early in the transition period
Adaptation	System changes, including the company's existing work processes, which may not integrate well with the newer ones	Ensuring compatibility with DMS and plan integration during the system design phase
Threats to security and confidentiality of information	New systems may have vulnerabilities or may not comply with data protection regulations	Conducting a thorough safety assessment and ensuring compliance with applicable regulations
Performance issues	Newer systems may have operational issues, such as slow response times, which reduce performance	Testing and optimizing performance before full implementation

Source Generated by the authors based on [9, 10, 12, 14, 16]

are less of a problem than errors caused by the human factor. The more detailed the task is formulated, the better the result is obtained [10]. Obviously, it is necessary not only to use AI, but better to cooperate with it, which depends on the level of personnel training.

Companies are starting to implement cloud technologies in order to increase business flexibility and reduce IT infrastructure costs. According to Microsoft Office 365, 82% of small and medium-sized enterprises were able to reduce costs after switching to cloud technologies, and 70% began to reinvest the saved money back into the business [11]. Experts call SaaS the #1 technology for investing in the digital transformation of business processes. However, the lack of interest of the IT staff in the development of technological SaaS, due to possible salary reductions and layoffs, certain psychological barriers, and fear of information leakage may be obstacles to its implementation. In addition, the activities of potential customers are not sufficiently transparent: double

Table 3. Advantages and disadvantages of common AI tools for working with texts among Ukrainian companies.

AI platforms	Advantages	Disadvantages
ChatGPT used by 82% of Ukrainian business practitioners	– tested many times in different scenarios, which made it possible to optimize and create many useful plugins – it's easy to brainstorm, work out different versions of texts – an independent service that integrates into various search engines – can be used for free	– too hallucinatory – has primitive writing (incorrect formulating is possible) – constant problems due to inaccurate writing of prompts – database valid until September 2023 – not adapted to marketing tasks
Google Bard	– the possibility to using the search system for processing new materials – due to the availability of Internet access, you can search and use up-to-date information, – gives several answer options to choose from	– does not understand long instructions, so it is better to give prompts step by step – often invents false information, gets confused in data and makes mistakes – prompts created for ChatGPT will work here not so effectively
Jasper.AI used by 14% of survey participants	– there are already written prompts – automatically generates content, checks for errors, uses natural language processing – adapted to the performance of marketing tasks – better manages with naming, writing texts for banners and generating product cards – provides access to the Jasper Art neural network capable of generating simple images on demand	– generates content that does not correspond to reality or does not correspond to the application – high price – it takes many generation attempts to get the perfect result you need

(continued)

accounting, unlicensed software, etc. The native SaaS market is not yet sufficiently developed; it lacks interesting options, offers, guaranteed quality [23].

The experience of working with document and information systems of the State Agency for Infrastructure Restoration and Development of Ukraine deserves attention and dissemination. After the full-scale invasion, a digital ecosystem for accountable management of restoration DREAM was created and made publicly available. It is unique

Table 3. (*continued*)

AI platforms	Advantages	Disadvantages
Microsoft Bing Chat used by 14% of the interviewed experts	– actually GPT-4, but with other basic settings and a more user-friendly graphical interface – applies GPT-4 to compare Bing chatbot responses to baseline data and evaluate model performance	– weak sequential memory – does not save initial settings until the last request, needs regular reminder – often gives inaccurate answers to the same question
Notion AI used by about 10% of respondents	– great functionality – improved prompts and modules for solving typical problems – tools for working with text – high speed of work	– lack of own application – paid interface
OpenAI actively used by 6% of respondents	– own applications – optimal for large business – enriches technology through feedback from human testers – applies GPT-4 to compare Bing chatbot responses to baseline data and evaluate model performance	– can invent false information, gets confused in data and makes mistakes

Source Generated by the authors based on [10]

not only for Ukraine but also for the world. This state electronic ecosystem provides a single digital route for all projects for the restoration of critical infrastructure and long-term reconstruction and modernization of Ukraine. DREAM collects, organizes and publishes open data at all stages of restoration projects (the number of which is currently 8201) in real time, implementing the highest standards of transparency and accountability. Anyone, anywhere, can monitor the effectiveness of project implementation and use this information to reduce risks, create accurate reporting and improve the overall effectiveness of projects. The State Enterprise "National Institute of Infrastructure Development" will test the fully Ukrainian product Task Manager "E-Road"—a platform for building information and analytical, reporting and monitoring systems. Its advantages: systematization and organization of work processes of employees of "DerzhdorNDI" and Ukravtodor; the ability to prioritize tasks and projects; control of deadlines and monitoring the effectiveness of task performance; prevention of corruption risks at the planning and execution stage of tasks. This allows developing road construction in accordance with world standards, opening up new opportunities for the market of contracting organizations and international cooperation.

4 Conclusion

The conducted analysis revealed an ambiguous picture of the results of DIST implementation in companies and their satisfaction with popular platforms and services, as well as slow pace of implementation of innovative systems and technologies, which has a limited positive impact. It remains relevant to improve the theoretical toolkit, to analyze the influence of the latest DISTs on the emergence of new actors, structures, practices, corporate culture and values.

The authors of the article come to the conclusion that digitalization of DIST in company management is a dynamic process of implementation and using digital innovations that change organizational and cultural structures and lead to the emergence of new practices. Cloud technologies and artificial intelligence are in low demand among native companies in working with document information due to insufficient understanding of components and implementation strategies, lack of interesting offers, guaranteed quality. Meanwhile, advanced companies see the benefits of AI technologies in improving the decision-making process, work efficiency and resource optimization. The authors classified the general advantages and disadvantages of popular Ukrainian services, based on user feedback, survey results, and expert assessments. This can help with their selection and implementation. It can be stated that the main disadvantages of technologies for working with documents are financial and time costs for implementation, maintenance and training of personnel, the need for updating and technical support, etc. The biggest drawback of DIST remains threats to the security and preservation of confidentiality of information, as well as the lack of qualified specialists, especially able to work with AI technologies. Hallucinations, false or fabricated information provided at the request of users are common disadvantages of using AI. From the point of view of the interviewed experts, AI hallucinations are less of a problem than errors caused by humans without the appropriate level of training and experience. Companies spend most of their time and energy not on technical difficulties, but on psychological adaptation of people to the new work format.

It has been proven that DISTs increase the efficiency of any company's activities through the operational collection, processing and systematization of information, reduction of response time to external changes, optimization of the organization of the enterprise. Prompt obtaining of options for solving management tasks, increase in the level of labor productivity, reduction of the volume of operational and management costs, formation of an optimal development strategy in conditions of competition are also important advantages of these technologies. The experience of working with document and information systems of the State Agency for Infrastructure Restoration and Development of Ukraine and the State Enterprise "National Institute of Infrastructure Development" deserves attention and dissemination for effective using.

The research issue is of urgent interest to both practitioners and researchers. Further development in this direction will help to optimize the management of companies, including in transport, to improve the results of their work, which are focused on end users.

References

1. A community for sharing experience between users of UkrBlank, UkrWarehouse and UkrZarplata, https://www.softbalance.com.ua/forum/index.php
2. AI and Information Management Report. Discover the Data Problem That's Stalling AI Success (2024). https://www.avepoint.com/ebooks/ai-and-information-management-report-2024
3. Blackburn, A.: New Study Reveals Crucial Role of Information Management in AI Success (2024). https://info.aiim.org/aiim-blog/new-study-reveals-crucial-role-of-information-management-in-ai-success
4. Chan, J., Logan, S., Moses, L.B.: Rules in information sharing for security. Criminol. Crim. Just. (2020). https://doi.org/10.1177/1748895820960199
5. Chesterman, C. W.: Communications for Complex Systems Governance. Complex System Governance. Cham, 315–338 (2022). https://doi.org/10.1007/978-3-030-93852–9
6. EVO company, Vchasno Kasa project reviews. First independent review site in Ukraine, https://www.otzyvua.net/uk/kompaniya-evo-proekt-vchasno-kasa
7. Hinings, B., Gegenhuber, T., Greenwood, R.: Digital innovation and transformation: An institutional perspective. Inf. Organ. **28**, 52–61 (2018). https://doi.org/10.1016/j.infoandorg.2018.02.004
8. Jordan, S., Zabukovšek, S.: Organizational Maturity and Sustainability Orientation Influence on DMS Life Cycle–Case Analysis. Sustainability. 15, 4308 (2023). https://doi.org/ https://doi.org/10.3390/su15054308
9. Jordan, S., Zabukovšek, S., Šišovska Klančnik, I.: Document Management System – A Way to Digital Transformation. Naše Gospodarstvo/Our Economy. **68**(2), 43–54 (2022). https://doi.org/10.2478/ngoe-2022-0010
10. Kirik N: The best AI tools for working with texts (2023). http://surl.li/urevs
11. Confidentiality and SaaS-solutions. Scallium (2021). https://ua.scallium.pro/cyber-security-saas
12. Markevich, K.: Digitization: advantages and ways to overcome challenges (2021). https://razumkov.org.ua/statti/tsyfrovizatsiia-perevagy-ta-shliakhy-podolannia-vyklykiv
13. Obikhod, S.: Implementation of information and communication technologies into the business process management system of domestic enterprises in the context of the development of the digital economy. Econ. Manage. Admin. **4**(98), 10–17 (2021). https://doi.org/10.26642/ema-2021-4(98)-10-17
14. OPZ.ORG.UA. Forum, https://opz.org.ua/forum/index.php
15. Plekhanov, D., Franke, H., Netland, T.H.: Digital transformation: A review and research agenda. Eur. Manag. J. **41**(6), 821–844 (2023). https://doi.org/10.1016/j.emj.2022.09.007
16. Reviews. KeepinCRM, https://keepincrm.com/ https://keepincrm.com/feedback
17. Savko, I.: Which program to choose for filing reports electronically, http://surl.li/uresh
18. Singh, R., Gildhiyal, P.: An artificial intelligence-enhanced methodology for automating global document management systems operations. Conference: 2023 IEEE International Conference on Integrated Circuits and Communication Systems (2023). https://https://doi.org/10.1109/icicacs57338.2023.10099523
19. The most popular electron document management services. Legal newspaper online (2020). https://yur-gazeta.com/publications/practice/inshe/naypopulyarnishi-servisi-elektronnogo-dokumentoobigu.html
20. Top 10 best CRM systems for Ukraine (2024). http://www.livebusiness.com.ua/tools/crm/
21. Ukraine's first independent review site, https://www.otzyvua.net/uk/

22. Vial, G.: Understanding digital transformation: A review and a research agenda. J. Strat. Inf. Syst. **28**, 118–144 (2019). https://doi.org/10.1016/j.jsis.2019.01.003
23. What is Software as a Service (SaaS)? Clouds from across the ocean. "Business Technologies Online" company, https://cbto.com.ua/library/saas

AI-Based Solutions for the Safe Transport of Dangerous Goods in the Context of Ukraine's European Integration

Olena Krainiuk[1(✉)], Yuriy Buts[2], Artur Kagramanian[2], Vitalii Barbashyn[3], and Mykola Yatsiuk[3]

[1] Kharkiv National Automobile and Highway University, Yaroslav Mudryi Str., 25, Kharkiv 61002, Ukraine
olenakrainiuk@khadi.kharkov.ua
[2] Ukrainian State University of Railway Transport, Feuerbach Square 7, Kharkiv 61050, Ukraine
[3] O. M. Beketov National University of Urban Economy in Kharkiv, Marshal Bazhanov Str., 17, Kharkiv 61002, Ukraine

Abstract. The purpose of the study was to analyse the possibilities of using artificial intelligence (AI) to improve the safety of dangerous goods transportation in Ukraine, considering the European integration process. The study assessed the current state of use of AI technologies, analysed the compliance of Ukrainian standards with international requirements, and identified areas for further research in this area.

The study results showed that AI could significantly improve transportation safety using hazardous materials by optimising routes, predicting emergencies, automatically monitoring the technical condition of transport, and introducing autonomous vehicles. Examples of successful AI implementation in the EU and the USA, as well as in Ukraine, where pilot projects are already underway, are identified.

The study's conclusions emphasise the need to adapt international practices to improve transport safety in Ukraine. It is recommended to use AI more actively in logistics processes, integrate technologies by European standards, and continue to develop innovative solutions in cooperation with international partners.

Keywords: Artificial Intelligence · Dangerous Goods · Transport Safety · Eurointegration · Route Optimisation · Risk Monitoring

1 Introduction

Dangerous goods, such as chemicals, explosives and radioactive materials, can cause significant damage to the environment, human health and infrastructure in the event of an accident. Even minor disturbances can lead to severe consequences, including soil and water contamination, loss of life, and substantial economic losses due to cleanup and recovery efforts. Improving transport control and safety methods is important to mitigate these risks.

O. Slavinska et al. (Eds.): ITSESQC 2024, LNNS 1335, pp. 236–247, 2025.
https://doi.org/10.1007/978-3-031-87376-8_21

There have been several serious accidents involving dangerous goods in Ukraine. For example, in 2020, a petrol tanker overturned in the Odesa region, causing fuel leakage and environmental damage. In 2019, a truck carrying ammonia was involved in an accident in Mykolaiv region, prompting evacuations and rescue efforts. An accident on the Kyiv-Chop highway in 2021, where a truck carrying chemicals collided with another vehicle, led to the road's closure for decontamination. These incidents confirm the importance of strict control over transporting dangerous goods.

Artificial intelligence (AI) is essential in ensuring the safety of dangerous goods transported by implementing large projects and international initiatives. For example:

- The European Commission's SafeLog Project uses intelligent systems to monitor and manage the logistics of dangerous goods, analyse risks, and automatically respond to emergencies [1].
- IBM Watson's Cognitive Logistics optimises routes, predicts incidents, and ensures compliance with safety standards by integrating sensors into vehicles to monitor the condition of cargo [2].
- AUTOPILOT focuses on autonomous driving to reduce human error and improve transport safety by using IoT and AI to control vehicles [3].

Ukraine can adapt these international ideas to improve the safety of dangerous goods transport by integrating intelligent transport systems, cooperating with European projects such as SafeLog, and using AI in line with European ADR standards.

2 Purpose and Objectives of the Study

The study aims to analyse the potential of artificial intelligence (AI) to reduce risks and improve the efficiency of dangerous goods transportation in the context of Ukraine's European integration and its aspirations to join the EU. The study also identifies key areas of research and development to achieve safety standards that meet European requirements.

Research objectives:

1. To assess the current AI use for transporting dangerous goods in Ukraine.
2. Analyse the compliance of Ukrainian standards with European and international requirements in dangerous goods transportation.
3. Identify areas for further research on integrating AI into managing dangerous goods transportation.

3 Using AI to Improve Transport Safety

The transport of dangerous goods is accompanied by several challenges that can seriously affect the environment, people and infrastructure. One of the most critical risks is technical and environmental hazards, such as accidents or toxic leaks, which can lead to significant ecological pollution (Fig. 1). In addition, the human factor also plays an important role, as driver errors, insufficient training or lack of proper information about road conditions increase the risk of accidents. A separate problem is ensuring constant monitoring and control over the condition of transport and cargo, which creates difficulties for timely response to possible threats.

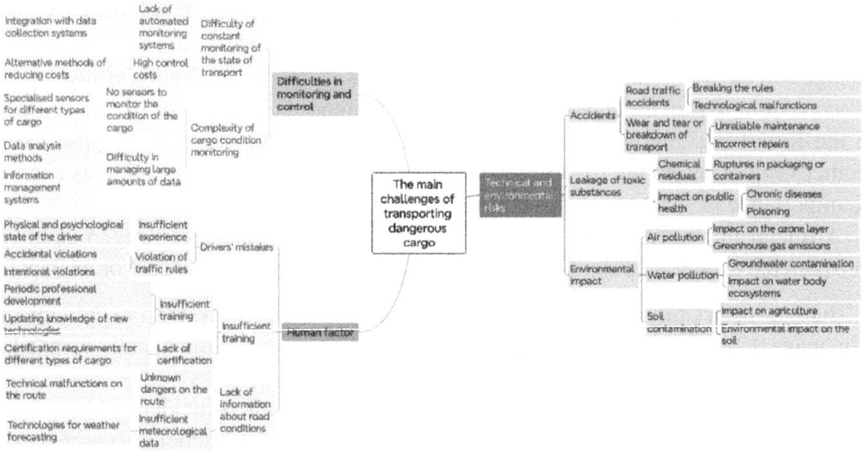

Fig. 1. Assessment of technical, environmental and human factors in the transport of dangerous goods (compiled by the authors)

Transporting dangerous goods is a critical aspect of modern logistics that requires continuous safety improvements. Artificial intelligence (AI) is significantly improving safety in transport by offering innovative solutions (Fig. 2).

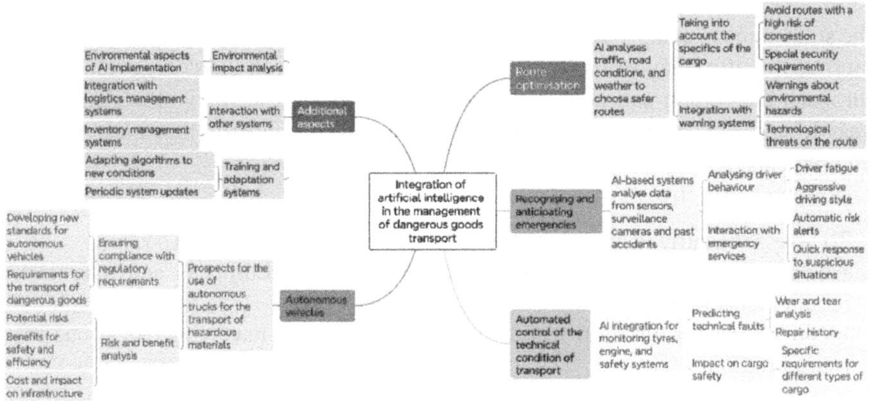

Fig. 2. Integration of Artificial Intelligence into the Safety of Dangerous Goods Transportation (compiled by the authors)

AI optimises routes based on traffic, road conditions and weather by constantly analysing real-time data and adjusting the route accordingly, reducing the risk of accidents. Using AI to recognise and predict accidents is also essential, as it helps detect threats promptly. Automated monitoring of the technical condition of vehicles ensures the reliability of transportation. In the future, the use of autonomous vehicles can significantly improve the safety and efficiency of transportation using hazardous materials.

Here are some examples of how artificial intelligence can be used to improve the safety of dangerous goods transportation (Table 1).

Table 1. Using AI to improve the safety of dangerous goods transport.

Direction	Example	Advantages	Disadvantages
1. Route optimisation	To analyse traffic, road and weather conditions to select safe routes. For example, Waymo and Tesla integrate AI into vehicles to optimise routes [4]	It reduces the likelihood of accidents, avoids dangerous areas, and increases transport efficiency	Dependence on data accuracy, possible technical failures, and the need to constantly update information
2. Recognising and anticipating emergencies	Nauto has developed a system that uses AI to analyse surveillance and sensor data to identify potential risks [5]	Proactive warning of hazards, reducing the number of road accidents	High hardware costs, the need to process large amounts of data, and possible privacy issues
3. Automated control of the technical condition of transport	Condition monitoring systems like Fleet Complete use AI to predict technical faults and prevent breakdowns [6]	Increase transport reliability, reduce downtime, and improve safety	High integration costs, need for regular maintenance and calibration of systems
4. Autonomous vehicles	Trucks from Aurora and Waymo demonstrate the potential for safe transport thanks to precision control [7]	Reducing the human factor, improving the accuracy and reliability of transport	Need for new regulatory standards, high development and implementation costs

Four main application areas are considered: route optimisation, emergency recognition and prediction, automated vehicle maintenance, and autonomous vehicles. Each case study illustrates specific solutions, their advantages and potential disadvantages, which helps better understand the opportunities and challenges of integrating AI into this critical area.

4 Analysis of AI Implementation in the Transport Sector

The introduction of artificial intelligence (AI) in the transport sector opens up new opportunities to improve the safety and efficiency of transportation, particularly in the field of dangerous goods. Let's look at global experience (Table 2) and examples from

Ukraine (Table 3) to analyse successful practices and the potential for further development. Reviewing successful implementations in different countries provides valuable practices for improving Ukrainian systems.

Table 2. Examples of AI implementation in the transport sector: global experience.

Country	An example of AI implementation	Description
USA	Autonomous Trucks for Hazardous Materials	Companies like Waymo and Aurora are testing autonomous trucks to transport dangerous goods. The trucks are equipped with AI systems to detect and respond to hazards and optimise routes [7]
	AI for Real-Time Monitoring	Ryder Systems uses AI to monitor the condition of trucks in real-time, detecting possible leaks or anomalies in the transport of dangerous goods [8]
European Union	Smart Logistics Systems	DHL and Kuehne + Nagel use AI to manage their supply chains, including transporting dangerous goods [9]. AI helps with route planning, risk management, and automatic cargo monitoring
	Intelligent Freight Transport	Volvo and Scania are implementing AI in safety systems for trucks carrying dangerous goods. This includes automatic control of braking systems and cargo monitoring [10]
Japan	Automated Storage and Handling	Toyota develops autonomous systems for warehousing and transporting hazardous materials, using AI to improve the accuracy and safety of operations [11]
	AI in Hazardous Waste Management	Hitachi introduces AI for hazardous materials waste monitoring and management, including systems for risk prediction and management [12]

Ukraine is actively engaged in a diverse range of projects that harness artificial intelligence to enhance the safety and efficiency of transporting hazardous materials. These projects, which encompass pilot initiatives, infrastructure developments, and academic research, demonstrate Ukraine's broad commitment to integrating the latest technologies into the transport sector. The following are the main examples of AI implementation in Ukraine (Table 3).

The introduction of AI in transporting hazardous materials significantly improves worker safety and transportation efficiency by monitoring risks and improving staff training. AI identifies risks for drivers and related personnel by analysing data from sensors, cameras, and monitors. For example, modern systems can warn of hazards on the road or automatically activate auxiliary systems to avoid accidents. This reduces the likelihood of injuries and ensures a safer working environment. AI-based systems are also used to train and monitor drivers by analysing their behaviour, which helps to detect signs of fatigue or deviation in time [13].

For the effective integration of artificial intelligence (AI) in Ukraine, several key measures need to be implemented. These include the development and implementation of relevant legislation and standards, training for professionals and drivers, investments in infrastructure, and the establishment of cooperation with international partners. International partners play a crucial role in sharing best practices, providing technological support, and facilitating cross-border transportation of hazardous materials (Fig. 3).

The following flowchart illustrates the main areas and stages of this process, highlighting their importance for the successful integration of AI in the transport of dangerous goods.

This comprehensive approach will ensure the effective integration of AI into the transport of dangerous goods in Ukraine, increasing the safety and efficiency of transportation by international standards.

5 SWOT Analysis of the Use of Artificial Intelligence to Ensure the Safety of Dangerous Goods Transport

A SWOT analysis of the use of artificial intelligence to ensure the safety of dangerous goods transportation allows us to assess the potential and challenges of technologies in this area (Table 4).

The analysis covers vital aspects: strengths, such as improved safety and route optimisation, and weaknesses, including high cost and dependence on data accuracy. Opportunities relate to development prospects and environmental initiatives, while threats include technical failures and cyber threats. The results will help to develop strategies for the successful implementation of AI in the transport of dangerous goods.

Thus, the SWOT analysis of using artificial intelligence to transport dangerous goods safely provides a comprehensive overview of strengths, weaknesses, opportunities and threats. AI has great potential to improve safety through route monitoring and optimisation and environmental improvements. However, the effectiveness of the systems depends on the quality of the data and is costly. There are also ecological challenges and difficulties in integrating technologies. Opportunities include the development of innovative solutions, but threats such as technical failures and cyber risks can affect security. The analysis is a basis for developing strategies that minimise risks and costs.

The integration of AI to ensure the safety of dangerous goods transportation in Ukraine can be presented as follows (Fig. 4).

The use of AI to enhance transport safety has significant potential to improve the transport of dangerous goods. However, economic, technical, and legal aspects should be considered for effective integration in Ukraine. The key issues are infrastructure development, regulations, and training.

Table 3. AI implementation in the transport sector: projects and research in Ukraine.

Category	An example of AI implementation	Description
Pilot projects in Ukraine	Autonomous Vehicles Testing	Twinning project: This project aims to improve the multimodal transport system of dangerous goods by road, rail, sea and river transport in Ukraine. It includes the improvement of multimodal connectivity in line with European norms and standards [14]
	AI for Safety Monitoring	Monitoring and control systems: The use of artificial intelligence to monitor the condition of vehicles and road conditions. This allows for the detection of potential hazards and prompt response to them, which significantly improves transport safety [15]
Infrastructure initiatives	Development of Smart Infrastructure	Ukraine is developing projects to modernise its infrastructure to support AI technologies. For example, the Ukrainian Railways is introducing new logistics management systems that may include AI elements to monitor and manage the transport of dangerous goods
Academic and scientific research	AI Research in Transport Safety	Ukraine is conducting research in the field of AI to improve transport safety. Kharkiv National Automobile and Highway University is actively exploring the possibilities of introducing AI to manage transport safety, including dangerous goods [16]

6 Legislative Norms and Standards for the Implementation of AI in the Transport of Dangerous Goods: Global Experience and Ukrainian Prospects

When implementing AI in the transport of dangerous goods, it is crucial to consider both global standards and national legislative requirements. Adhering to global standards, as outlined in Table 5, is of utmost importance to ensure the safe and efficient transport of dangerous goods.

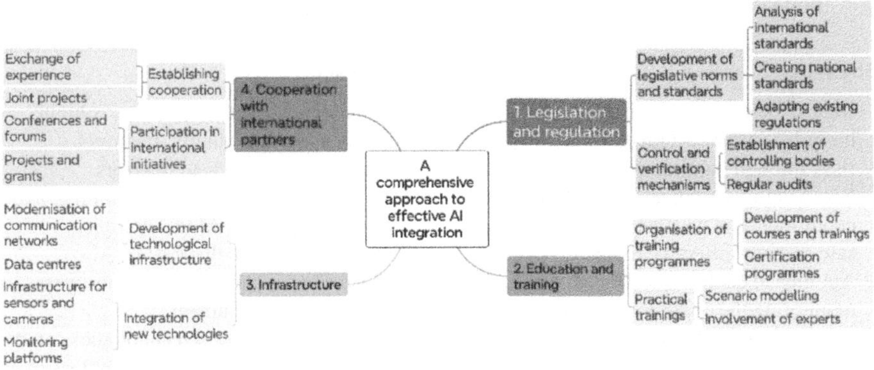

Fig. 3. Integration of AI into the transport of dangerous goods in Ukraine

Global initiatives to introduce intelligent transport systems aim to improve the safety of dangerous goods transport. International documents, such as ADR (European Agreement concerning the International Carriage of Dangerous Goods by Road), set standards for transporting such goods, and artificial intelligence can automate control processes. Ideas for intelligent monitoring and control systems can be implemented in the ADR to automate control processes and improve security. The OECD (Organisation for Economic Co-operation and Development) is working on projects introducing AI to improve transport safety and publishes recommendations for member states.

In Ukraine, there are no specific laws or standards regulating the use of AI in transporting dangerous goods. However, specific general regulations may cover partial aspects of this topic:

- The Law of Ukraine 'On Transport' (No. 232/94-BP) needs to be updated to consider introducing new technologies, such as AI systems.
- The Law of Ukraine 'On Transport Safety' (No. 2806-IV) could be supplemented to regulate the latest technologies in the transport sector.
- CMU Resolution No. 770 'On Approval of the Rules for the Transportation of Dangerous Goods' (2011) urgently requires an update to integrate modern monitoring and control systems, including AI, to ensure the safety and security of transportation.
- Order of the Ministry of Infrastructure No. 515 'On Approval of the Instruction on the Transport of Dangerous Goods by Road' (2012) needs to be amended to accommodate new technologies, including artificial intelligence.
- International ISO standards (ISO 26262 for automotive functional safety, ISO 45001 for occupational health and safety management) may be essential to establish appropriate standards for using AI in transporting dangerous goods.

To effectively integrate artificial intelligence into Ukraine's transport processes, new legislation must be developed covering automation, safety and monitoring while transporting hazardous materials. Key challenges include adapting legislation to new technologies, training specialists to work with AI, and addressing technical and infrastructural issues for practical use.

Table 4. SWOT-analysis of the use of AI for the safe transport of dangerous goods.

Strengths	**Weaknesses**
– Improving safety. AI can detect potential hazards, warn of accidents, and mitigate risks through monitoring and management – Route optimisation: analysing traffic, road conditions, and weather to help you choose safe and cost-effective routes – Automated control of technical conditions, monitoring the condition of vehicles and cargo, which ensures timely detection of technical problems – Reducing environmental impact by optimising routes, reducing fuel consumption and pollutant emissions, and detecting toxic leaks in a timely manner. AI can detect unusual chemical patterns or changes in air quality, which could indicate a toxic leak, allowing for immediate action to be taken.	– The effectiveness of AI systems is highly dependent on data quality. Incorrect, incomplete data can lead to errors – High costs of integrating, maintaining, and constantly updating AI technologies – The production and disposal of AI technologies, including electronics waste, may hurt the environment – Implementation of new technologies can be complex and require significant efforts to integrate with existing systems
Opportunities	**Threats**
– Development of innovative solutions in transport safety and environmental control – Improving global logistics and optimising international transport – Supporting environmental initiatives such as pollution monitoring and natural resource management – The introduction of AI can give companies a competitive advantage and improve the quality and safety of services	– Technical failures in AI systems, such as a self-driving car misinterpreting a stop sign or a chatbot providing incorrect medical advice, can lead to serious problems, including accidents and delays – Cyber threats can affect data security and privacy – International and local regulatory barriers may hinder the adoption of new technologies – Political and economic instability may affect the adoption of AI technologies in international transport

To successfully integrate AI, engaging international experts to develop norms and standards that align with global practices is necessary. An important step is to launch pilot projects to test technologies before implementation. A comprehensive approach that considers international standards and national peculiarities is critical to effectively integrating AI into transporting dangerous goods. Ukraine currently has no specific legislation to regulate AI in this area, which creates challenges for safety and risk management.

7 Conclusions

The role of artificial intelligence in the safety of dangerous goods transportation is essential and consists of ensuring effective monitoring, risk analysis and reducing the impact of the human factor. AI allows for the introduction of autonomous control and

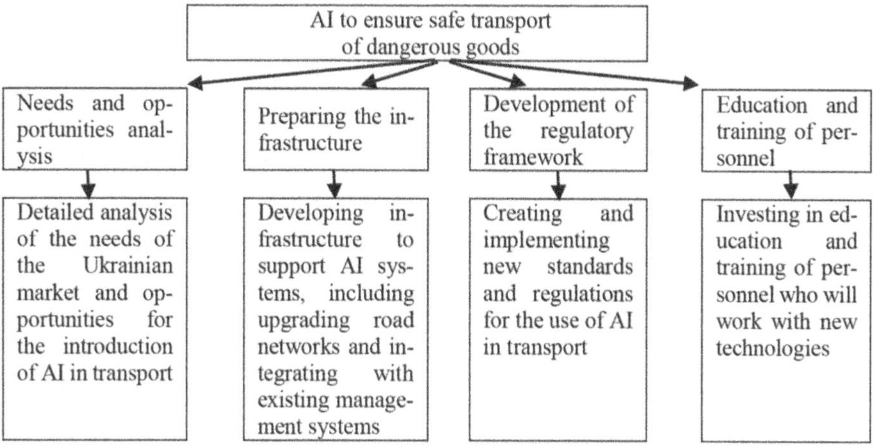

Fig. 4. Integrating AI to ensure the safety of dangerous goods transport

forecasting systems that improve the safety and transparency of transportation. At the same time, introducing AI involves technical, ethical and financial challenges, such as cybersecurity and high technology implementation costs. To fully realise the potential of AI in this area, it is necessary to continue active development and integration of modern technologies into transport processes.

1. The article highlights the importance of artificial intelligence for monitoring the transport of dangerous goods. The article describes how AI technologies help to identify potential risks and reduce the human factor, which increases safety.
2. The need to adapt and develop new legislative norms to regulate the use of AI in transport, particularly in dangerous goods, is substantiated.
3. The importance of international consultations for developing practical standards that meet the best global practices is emphasised.
4. A SWOT analysis of the introduction of artificial intelligence to ensure the safety of dangerous goods transportation was carried out.
5. The need for further development of AI and its integration into transport processes to improve the safety and efficiency of dangerous goods transportation in modern conditions is emphasised.

Table 5. Global regulations for the use of AI in the transport of dangerous goods.

Region	Norm/Standard	Description
European Union	Regulation (EU) 2019/2144	Establishes requirements for the overall safety of road transport and the introduction of new technologies, including safety systems with AI components
	Ethics Guidelines for Trustworthy AI	The European Commission developed the document with recommendations on using artificial intelligence in various industries, including transport
	ISO 26262	International standards for the functional safety of automotive electronic systems may include AI in safety systems
USA	FMCSA (Federal Motor Carrier Safety Administration)	The agency that regulates the safety of freight transport and explores the possibilities of introducing automation and AI technologies in freight transport
	NHTSA (National Highway Traffic Safety Administration)	Defines standards for autonomous vehicles that can be used to transport dangerous goods
International standards	European Transport Conference	The annual conference of the Association of European Transport (Antwerp, Belgium, 2024) brings together policy makers, practitioners and researchers to discuss the use of the latest technologies, including AI, to monitor and secure the transport of dangerous goods
	IMDG Code (International Maritime Dangerous Goods Code)	International Code for the Carriage of Dangerous Goods by Sea, which regulates the requirements for monitoring and control systems, including the possible use of AI

References

1. Transport robots for automotive. SAFELOG project. https://www.safelog.de/en/industries-automotive

2. Hoyt, R.E., Snider, D.H., Thompson, C.J., Mantravadi, S.: IBM Watson analytics: automating visualization, descriptive, and predictive statistics. JMIR Public Health Surveill. **2**(2), e5810 (2016). https://doi.org/10.2196/publichealth.5810
3. Faget, M.-F.: The impact of emerging technologies on logistics: a comprehensive analysis. Proc. Int. Conf. Bus. Excellence **18**(1), 1868–1881 (2024). https://doi.org/10.2478/picbe-2024-0157
4. Jack, W., Butler, J.: Navigating tomorrow: advancements and road ahead in AI for autonomous vehicles. EasyChair, 11955, Manchester, UK (2024)
5. Safer fleets and safer roads for all. Nauto, so much more than an AI dash. NAUTO, https://www.nauto.com
6. One Platform. All your fleet management needs. Connected fleet solutions and data intelligence to drive your business forward. Fleet Complete. https://www.fleetcomplete.com/
7. Lázaro, C.O.: Automates: the future of autonomous cars: MS thesis. Universitat Politècnica de Catalunya (2023). http://hdl.handle.net/2117/393341
8. Ryder System: Terminal industries enhance yard operations with AI. Mexico Business: MBN Staff (2024). https://mexicobusiness.news/mobility/news/ryder-system-terminal-industries-enhance-yard-operations-ai
9. Erdoğdu, M.: Development of logistics management and relationship with industry. Int. J. Eng. Sci. Technol. **3**(2), 91–112 (2021)
10. Safi, A., Khamneh R.A.: Artificial intelligence as a catalyst for supply chain resilience: a qualitative study comparing Scania and Volvo in the construction equipment industry (2023). https://www.diva-portal.org/smash/get/diva2:1761055/FULLTEXT01.pdf
11. Sheffi, Y.: The magic conveyor belt. Supply chains, AI, and the future of work. MIT CTL Media (2023)
12. Uchida, N.: Research and development of AI trust and governance. Hitachi Rev. **71**, 22–29 (2022). https://www.hitachihyoron.com/rev/archive/2022/r2022-sp/01/index.html
13. Buts, Y.V., Krainiuk, O.V.: Topical issues of road safety and analysis of road accidents in Ukraine. Avtoshlyakhovyk Ukrainy. Special Issue **277**, 98–102 (2023)
14. Twinning project 'Support to improve the safety of dangerous goods transport by multimodal transport in Ukraine' Ministry of Communities, Territories and Infrastructure of Ukraine. https://mtu.gov.ua/content/proekt-twinning-pidtrimka-u-pidvishchenni-bezpeki-perevezennya-nebezpechnih-vantazhiv-multimodalnim-transportom-v-ukraini.html
15. Kyrlyk, N.Y.: Artificial Intelligence' and its use in logistics processes. Actual Probl. Econ. **243/244**, 59–66 (2021). https://doi.org/10.32752/1993-6788-2021-1-243-244-59-66
16. Krainiuk, O.V., Kozo, D.S., Barbashyn, V.V.: Evaluation of emergency events during the transport of dangerous goods in the context of the technogenic load of regions. Science and progress of transport. Bull. Dnipro Nat. Univ. of Railway Trans. **3**(75), 27–35 (2018)

Development of a Neural Network Control System with the Function of Predicting the Dynamics of the Parameters of a Complex Ship Object

Vladislav Mykhailenko[1(✉)], Yurii Gunchenko[2], Valerii Leshchenko[1], Larysa Martynovych[2], and Hanna Korenkova[2]

[1] National University "Odesa Maritime Academy", Didrikhson Str. 8, 65029 Odesa, Ukraine
vlamihailenod@gmail.com

[2] Odesa I.I. Mechnikov National University, Vsevoloda Zmiienka St. 2, 65082 Odesa, Ukraine

Abstract. The article addresses the shortcomings of traditional methods used for training neural networks in management systems and emphasizes the importance of exploring the capabilities of the neuro-control method with a predictive function. This method was analyzed using the example of an adaptive air control system within a liquid fuel combustion system of a ship's steam boiler. The neural network training and the modeling of the control system were performed using the Neural Network Toolbox in Matlab, showcasing the advanced features of this platform. The results of the analysis of transition processes demonstrated that the neural network control system successfully meets expected performance criteria. Furthermore, the system showed strong adaptability to changes in the operating modes of the steam boiler, confirming the efficiency of neural control with a predictive function in dynamic conditions. This indicates that such neuro-control methods can significantly improve the stability, precision, and overall performance of complex industrial systems.

Keywords: Neural Network · Automatic Control System · Adaptation · Ship Boiler · Air Excess Coefficient

1 Introduction

The methods of teaching a neural network, which minimize the deviation of the current exit of the object from the input influence, do not always cope well with the task of regulation. Firstly, if the process of transition of the control system lasts longer than one tact, then the neural network (NN), which is minimized by inconsistency, can give a great influence of management, which will lead to re-regulation of the process. Secondly, there is always a delay in the control systems of the neuro network (NNCS) due to feedback, which worsens the quality of regulation [1]. In this regard, an analysis of the method of forecasting neuro-modeling is proposed, which minimizes the functionality of the value of the integrated error predicted by several upcoming cycles [2].

O. Slavinska et al. (Eds.): ITSESQC 2024, LNNS 1335, pp. 248–257, 2025.
https://doi.org/10.1007/978-3-031-87376-8_22

2 Synthesis of a Neural Network Control System that Implements the Predictive Method

Synthesis of the neural network control system was performed using the Neural Network Toolbox Application Package Matlab systems [3]. To solve the task, a neural network regulator is an effective solution - NN Predictive Controller [3, 4]. A neural network regulator uses a model of a non -linear controlled object using the example of the air flow control channel - oxygen content in smoke gases in the fuel combustion system in the firebox of a ship steam boiler (SSB). The model is presented as a neural network to predict the future oxygen content, depending on the consumption of air in the combustion system. At the same time, the air flow entering from the fan to the combustion system depends on the steam load of the SSB. In addition, the regulator calculates the control signal for the fan and, thus, adapts the behavior of the object (adjustable SSB parameter) in this time interval. Consequently, the design of the neurocontroller consists of two stages: the stage of identification of the controlled object and the stage of synthesis of the Law of Management. At the identification stage, a neural network model was developed, which is used at the stage of synthesis for teaching the regulator of the neural network. The scheme of the method used is shown in the Fig. 1 [2].

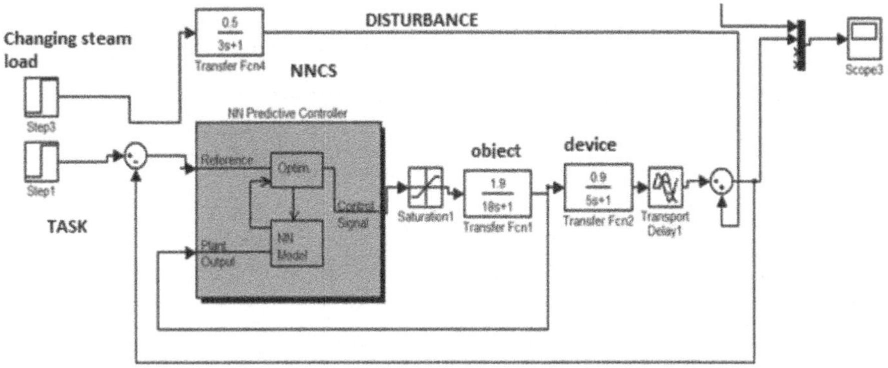

Fig. 1. Modeling the NNCS for the air excess (α')

To predict the future behavior of the regulatory parameter of the SSB and calculating regulation errors, a direct neuroemulator is used, trained using an algorithm for the reverse propagation of the error [3] on the example adapted for a specific SSB, PI regulator. The difference between the proposed method and the typical method of teaching NN [5–7] is the absence of a study neurocontroller. Instead, it uses an optimization module that works in real time. On the k-th stroke, the optimization module receives the desired path y (k + t, j) a few tweed. After that, the module implementing the NN optimization algorithm on a direct neuro -emulator control influences u(k + t, j) to predict the behavior of the object and select the best management strategy.

As a result, the optimal control action is selected that brings the object to the desired state on the target trajectory r(k + 1) (of the given value α). According to a number of researchers [5], the disadvantage of this method is the impossibility of its application in systems with a high sampling rate, because the optimization algorithm will not have time to find the optimal control action in one cycle. It can be noted that since the SSB is an inertial object and the change in the thermal modes of its operation occurs with a significant delay, the method can be successfully tested in ship ACS of steam generating units.

Taking into account the simplicity, convenience and availability of a library of NN elements, specialized software editors built into the electronic library were used to train a neural network controller (NNC) or a regulator in the ACS for burning liquid fuel in the SSB furnace. The NNCS software model for adapting the excess air coefficient (α') responsible for the efficiency of the combustion process by measuring the oxygen content in flue gases, and the interface for setting the neurocontroler are shown in the Fig. 2. Adaptation (α') depends on the thermal load of the SPL, which can vary in a wide range from 15 - to 110% depending on the operating modes of the ship's units and the navigational environment.

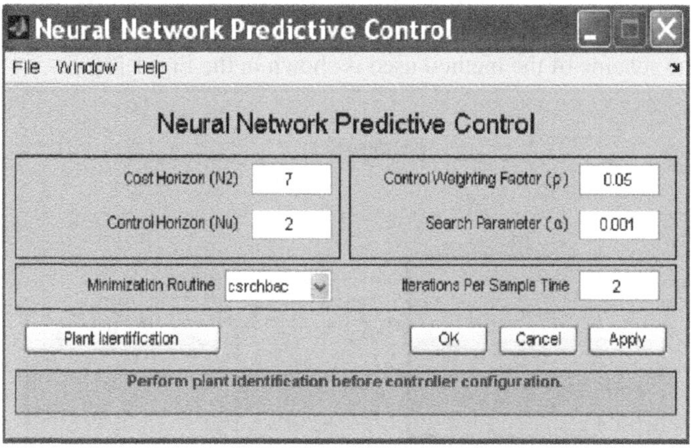

Fig. 2. NNC setup interface

The main parameters of the NNC setting (see Fig. 2), the values of which are selected based on the recommendations [3], are indicated below:

Control Horizon (Nu) – upper summation limit when estimating control power; Control Weighting Factor (ρ) – weight factor for the control power component $\rho = 0.05$; Search parameter (α) is a one-dimensional search parameter that sets the threshold for reducing the quality indicator $\alpha = 0.001$; Minimization Routine - selection of a one-dimensional search procedure; Iteration Per Sample Time - the number of iterations per one cycle of discreteness.

The process of synthesizing the neurocontroller was carried out using the NN Prediction Controller software module. To identify the control object, the Plant Identification software utility was used (see Fig. 3) [3].

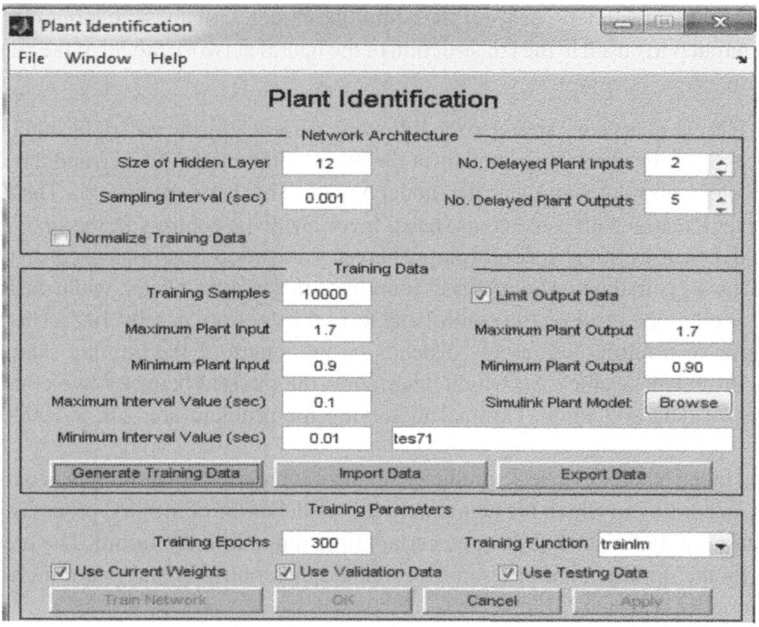

Fig. 3. The interface of the neural network identification program

3 Parameters of the NM Architecture

Size of Hidden Layer (N) – the size of the hidden layer, determined by the number of used neurons; Sampling Interval (Δt) – discreteness clock in seconds, defines the interval between two consecutive moments of data capture; No. Delayed Plant Inputs (Ni) - the number of delay elements at the input of the model; No. Delayed Plant Outputs (N, j) – the number of delay elements at the output of the model; Normalize Training Data. Control windows for normalization of training data to the range [0–1].

Parameters of NN training:

Training Epochs – the number of training cycles; Training function – task of the training function; Use Current Weights – a control window that allows you to confirm the use of current weights of the neural network; Use Validation / Testing Training – selecting these control windows means that 25% of the data from the training sequence will be used for the formation of control and text subsets.

Parameters of the training sequence:

- Training samples (NB) – the length of the training sample (the number of data collection points);
- Maximum Plant Input (φ_{max}) – the maximum value of the input signal;
- Minimum Plant Input (φ_{min}) – the minimum value of the input signal;
- Maximum Interval Value (sec) (t_{max}) – maximum identification interval in seconds;
- Minimum Interval Value (sec) (t_{min}) – minimum identification interval in seconds;

- Simulink Plant Model – tasks of the Simulink model with an indication of the input and output ports used in the construction of the neural network model of the controlled system.

The NN elements correspond to the following parameters: the size of the hidden layer S = 12, the number of delay elements at the input of the model Ni = 2 and the number of delay elements at the output of the model $N_j = 5$. This network is static. The network uses 1 input vector with 6 elements, has 2 layers with 12 neurons in the first (hidden) layer and 1 neuron in the second (output) layer. Activation functions used: hyperbolic tangent (tansig) - in the first layer, linear (purelin) - in the second layer, while the learning error, as well as the error on the control and test set ε does not exceed 10^{-3}. The success of learning or training NN largely depends on the length of the training sample (TS) and the discreteness clock Δt, which determines the interval between two consecutive moments of data collection. Optimal in solving the problem are: TS = 10000, Δt = 0.001 s.

Used in the NNCS α' model, predictive control is based on the principle of a receding horizon, according to which the neural network model of the controlled process assumes the reaction of the control object at a certain time interval in the future. The prediction is used by the optimization program to calculate the control signal that minimizes the control quality criterion [3]:

$$J = \sum_{j=N_1}^{N_2} (y_r(t+j) - y_m(t+j))^2 + \rho \sum_{j=1}^{N_u} (u'(t+j-1) - u'(t+j-2))^2, \quad (1)$$

where the constants N_1, N_2 and N_u set the limits within which the tracking error and the power of the control signal are calculated. The variable u′ describes the test control signal; y_r and y_m are the desired and real responses of the model. The value ρ determines the contribution of power (speed) control.

Depending on the operation mode of the SSB, the proposed neural network automatic control system automatically selects the best value of the coefficient of excess air (α) for the given operation mode, unlike the traditional one, which maintains the value of the coefficient without changes in the modes of operation of the SSB. The efficiency of NNCS operation is achieved by changing the amount of air (theoretical volume of air) supplied by the gas fan to the fuel combustion system in order to achieve the required percentage of oxygen in the output gases or α' [8–11]. Transient processes of the simulated NNCS are shown in the Fig. 4. The analysis of quality indicators of transient processes is shown in the Fig. 5.

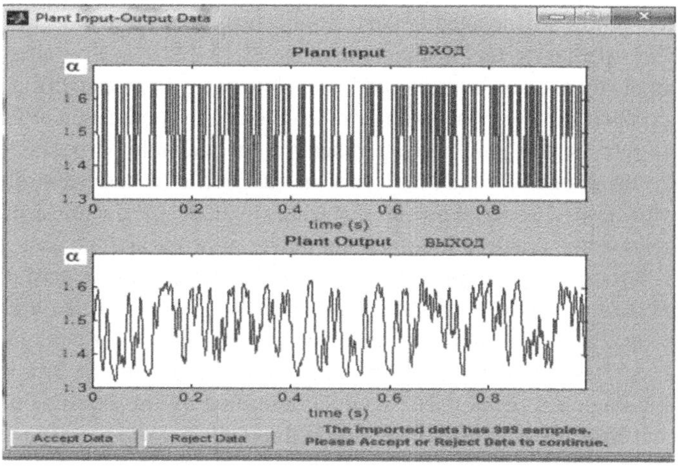

Fig. 4. Modeling of the neural network identification system α'

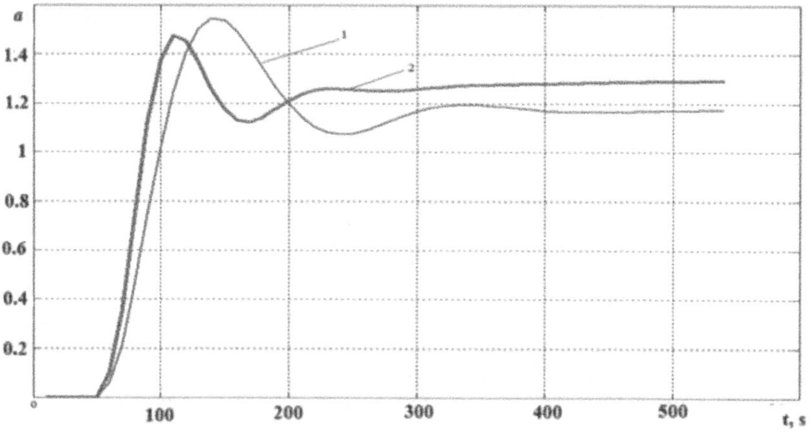

Fig. 5. Processes of the neural network system adaptation of the coefficient of excess air depending on the thermal mode of the SSB: No. 1 – thermal load mode of 55% of the nominally supported NNC; No. 2 - load 75%

4 Development of an Intelligent Automatic Control System for the Combustion of Liquid Fuel in the SSB Furnace

Intelligent automatic control system (ACS), unlike traditional ones, is capable of communicating with operators in a language they understand (fuzzy systems), self-learning, forecasting and working with dynamic objects (neural network and hybrid ACS). According to a number of authors [7–9], intelligent ACS can also be widely used to optimize the process of liquid fuel combustion in steam boilers.

To obtain a mathematical model of the control object (the combustion process through the channel "fuel consumption - O_2 content"), data from the ship's system for monitoring the parameters of the SSB located in the central control room of a modern tanker were used. The SSB operates at fuel loads in the range of 15 - 110% of nominal depending on the modes of steam consumers, in particular ship steam turbines (SST).

For the proposed universal neural network ACS for the combustion of various types of fuel in the furnace of a ship's steam generator, the technology of using two neural networks was used, one of which performs the function of a neurocontroller, and the second - the function of a neuroemulator that learns to model the dynamics of the control object [10]. Since the structure used is a multilayer network and the error backpropagation algorithm can be generalized to any network with signal forward propagation, the proposed neural network controller module (NNC) is trained like a conventional neural network. This will require a training sample in the form of pairs $(\overline{x}, d)$, where $\overline{x} = [x_1, \ldots, x_n]^T$ is the input vector, d^2 is the reference signal. The task is to modify (correct) the parameters of the NNC module described by the expression so that the measure of the error given by the expression is minimal.

$$e = \frac{1}{2}\left[\overline{y}(\overline{x}) - d\right]^2 \tag{2}$$

The learning process of the neural network (NN) will be completed when the error reaches the set minimum value. For the weight $\overline{y}^m$, the expression defining the modification method will look like [11]:

$$\overline{y}^m(t+1) = \overline{y}^m(t) - \eta \frac{\partial e(t)}{\partial \overline{y}^m(t)}, \tag{3}$$

where t = 0, 1, 2, the iteration number, $\overline{y}^m(0)$ denotes the initial value of the weight and the constant (0, 1) is interpreted as a coefficient that determines the learning rate (it is also called the correction step).

For the training of the NNC, a simulation was carried out with a typical PID - regulator, which was tested in work [12]. In the simulation experiment conducted in the MatLab package [11], the PID controller was adapted to changes in the properties of the controlled object. By varying the values of the object along the control channel: T and K, it is assumed to initiate changes in the values of the parameters of the PID - regulator: K_p, T_i, K_d. That is, to adapt them in order to achieve the expected transition process. The object of control is the fuel combustion control system of the auxiliary SSB operating in the nominal mode. The model of the object in the form of a transfer function of an inertial link was obtained in the second section. The model parameter, during the experiment, T changed depending on the steam load of the SSB and the influence of the internal disturbance channel – "change in fuel composition", in the following range (5; 25; 45; 105). Accordingly, the values of the PID settings - the ACS regulator of fuel combustion in the SSB are optimized to obtain the necessary transient process. The values of inconsistency signals (e_i) - inputs to the NNC, and the values of the optimal control influences (u_i) measured at the output of the PID - regulator in the ACS are used as a training sample for the NNC. In the course of the experiment, the following data were obtained, which are a training sample for the neural network regulator (Table 1).

Table 1 data. Are the input and target vectors of the NNC in the self-propelled fuel combustion engine in the SSB furnace.

Table 1. Study sample for NNC

T	5	25	45	105
E	0,000422	0,00069	0,096	0,049
U	3,333	3,335	3,353	1,995

The rules of signal transformation are determined by the mathematical model of the neuron, which is presented in the form of the following analytical expressions [12]:

$$s = \sum_{i=1}^{n} w_i x_i + b;\ \ y = f(s), \tag{4}$$

where w_i is the weight of synapse i is (1, 2,..., n); b – offset value, s – summation result, x_i – input signal vector (i is (1, 2, ..., n); $f(s) = k(s)$ – linear activation function. For example: $w = -2.48$; $b = 3.094$; $k = 1$, the synaptic connection in the neuron is inhibitory (Fig. 6).

A structure with a neuroemulator is used to check the efficiency of the NNC in the SSB fuel combustion control system [12]. The proposed structure of the neuroemulator obtained using the Neural Network program (Fig. 6). The specialized mathematical package MatLab (Simulink and Neural Toolbox) [13] was used to simulate the NN ACS optimization α depending on the steam load of steam units.

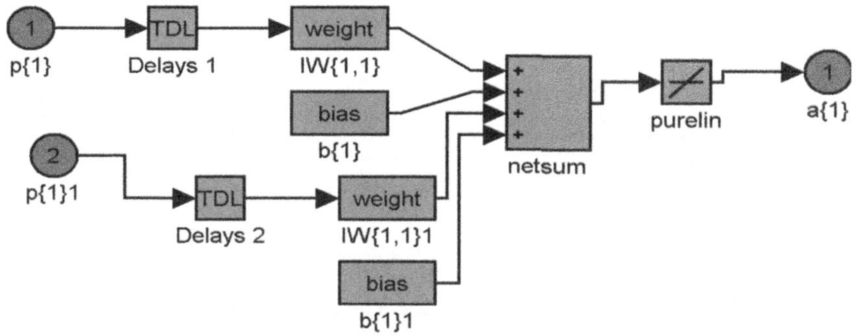

Fig. 6. A fragment of the neuroemulator model implemented in the Neural Network program

The processes of the simulated NN ACS correction α are shown in Figs. 7 and 8.

Approbation of the proposed NNCS adaptation α' that implements the method of forecasting in the combustion process of SSB to thermal load was carried out on two common modes of operation of SSB. The analysis of the type of transient processes demonstrated their compliance with the expected values (adjustment time 300 s, deviation equal to zero) when changing the work modes of the ship's steam units, which was carried out with the help of the disturbance channel. It was also established that the change in the coefficient of excess air during the operation of the SSB allows more

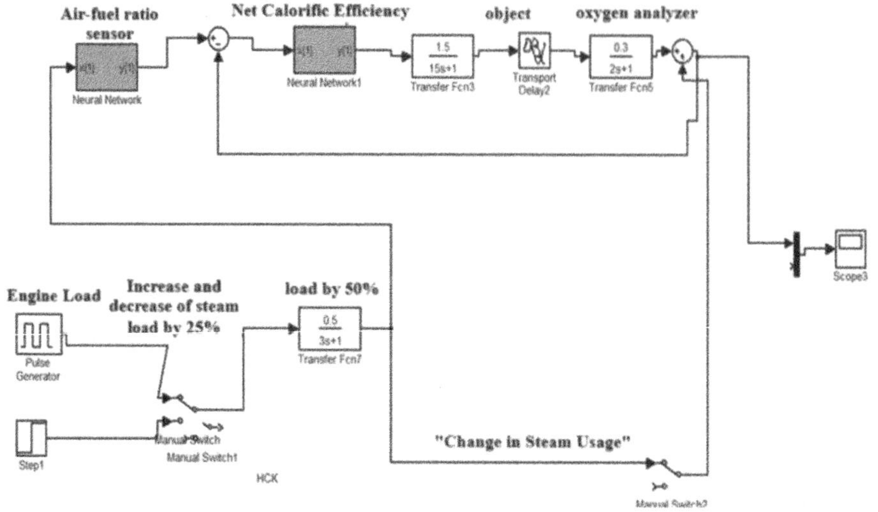

Fig. 7. Modeling of NN ACS optimization α with the aim of increasing the efficiency of SSB

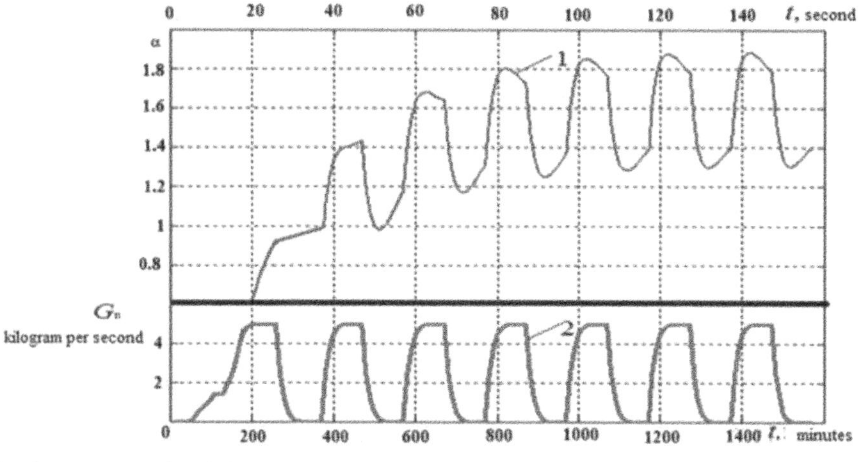

Fig. 8. Processes of the influence of the steam load of the SSB (2) on the optimization of α (1) to achieve the maximum efficiency of the SSB MAS 35

efficient fuel combustion in the boiler furnace and thereby increases the efficiency of the ship power plants.

References

1. Frehse, G., Althoff, M. (eds.): Proceedings of 10th International Workshop on Applied Verification of Continuous and Hybrid Systems (ARCH23). EPiC Series in Computing, vol. 96. EasyChair (2023)
2. Liu, X., Lara-Rosano, F.: Generalized minimum variance control of steam-boiler temperature using neurofuzzy approach. In: Proceedings of the 5th world congress on intelligent control and automation, pp. 134–148. Hangzhou, China, 15–19 June 2018 (2018)
3. Liu, X., Zhang, H.: Modeling of an ultra-supercritical boiler-turbine system with stacked denoising auto-encoder and long short-term memory network. Inf. Sci. **525**, 134–152 (2020)
4. Ekman, M.: Learning deep learning: theory and practice of neural networks, computer vision, natural language processing, and transformers using tensorflow. Addison-Wesley (2022)
5. Bochenek, B., Ustrnul, Z.: Machine learning in weather prediction and climate analyses applications and perspectives. Atmosphere **13**(2), 180 (2022)
6. Åström, K.J., Wittenmark, B.: Computer controlled systems, theory and design. Prentice-Hall, Englewood Cliffs (1984). https://bayanbox.ir/view/8821671619590593184/Adaptive-Control-Astrom-2nd-edition.pdf
7. Marine Boiler and Steam Turbine Generator. https://www.mhimme.com/auxiliary_boilers.html
8. Nakamura-Zimmerer, T., Gong, Q., Kang, W.: QRnet: optimal regulator design with LQR augmented neural networks. IEEE Control Syst. Lett. **5**(4), 1303–1308 (Oct. 2022). https://doi.org/10.1109/LCSYS.2020.3034415
9. Mitsubishi Auxiliary Boiler MAC-B. https://ru.scribd.com/document/334763233/Mitsubishi-Auxiliary-Boiler-MAC-B-pdf
10. Guo, X., Zhou, Z., Zhang, Y., Katz, G., Zhang, M.: Occrob: efficient smt-based occlusion robustness verification of deep neural networks. In: International conference on tools and algorithms for the construction and analysis of systems, pp. 208–226. Springer (2023)
11. Mezhuyev, V., Mykhailenko, V., Martynovych, L., Korenkova, H., Leshchenko, V., Stukalov, S.: Intellectual improvement of the control system for harmful emissions of a ship's utilizing boiler. In: Information Control Systems & Technologies 2023, pp. 153–162. Odesa, Ukraine, 21–23 September 2023, CEUR Vol-3513 (2023)
12. Katz, S.M., Corso, A.L., Strong, C.A., Kochenderfer, M.J.: Verification of imagebased neural network controllers using generative models. J. Aeros. Infor. Syst. **19**(9), 574–584 (2022)
13. Omotayo, T., Bankole, A., Olubunmi Olanipekun, A.: An artificial neural network approach to predicting most applicable post-contract cost controlling techniques in construction projects. Appl. Sci. **10**, 5171 (2020)

Integration of Intelligent Transport and Logistics Systems in the Context of Sustainable Supply Chain Management

Mariia Hryhorak(✉), Maryna Pichugina, and Margarita Chuprina

Igor Sikorsky Kyiv Polytechnic Institute, Beresteyskyi prospect 37, Kyiv 03056, Ukraine
{m_grigorak,pichuginam}@ukr.net

Abstract. The research focuses on the integration of intelligent transport and logistics systems for sustainable supply chain management. The aim is to study the impact of digital technologies on optimizing processes and enhancing sustainability. A robust methodological approach was employed, combining theoretical and empirical research techniques. A comparative analysis of intelligent transport systems (ITS) and intelligent logistics systems (ILS) reveals that their integration can significantly enhance supply chain management. Both systems are designed to optimize processes, increase efficiency, and improve environmental performance. However, the study found that autonomous ITS are often used in isolation, limiting their effectiveness. The research demonstrates that ITS and ILS can effectively interact and integrate to create a comprehensive mechanism that optimizes goods movement, reduces costs, and minimizes environmental impact. Digital technologies like IoT, blockchain, data analytics, AI, automation, and robotization play a crucial role in facilitating this integration. A practical outcome of the research is the analysis of software products related to supply chain management and digital transport and logistics platforms. The study shows that their integration can effectively reduce the carbon footprint. The integration of ITS and ILS provides supply chain managers with valuable insights for making strategic and operational decisions. These decisions can help reduce greenhouse gas emissions throughout the product lifecycle and promote sustainable development strategies for all participants.

Keywords: Intellectualization of Transport and Logistics Activities · Intelligent Transport Systems · Intelligent Logistics Systems · Intersystem Integration · Sustainable Supply Chains · Carbon Footprint Monitoring

1 Introduction

The requirements for the stability and flexibility of commodity movement chains are increasing, which necessitates the search for and implementation of effective management tools and more active use of modern information and communication technologies. At the same time, growing attention to environmental and social aspects of doing business emphasizes the need to integrate sustainable development principles into all aspects of supply chain management. In this context, the intellectualization of sustainable supply chain management becomes particularly relevant.

O. Slavinska et al. (Eds.): ITSESQC 2024, LNNS 1335, pp. 258–269, 2025.
https://doi.org/10.1007/978-3-031-87376-8_23

Most often, the intellectualization of logistics activities and supply chain management is associated with the introduction of modern information and communication technologies, such as artificial intelligence, big data analytics, Internet of Things (IoT) and blockchain. These technologies allow collecting and processing huge amounts of data in real time, which is a key to optimizing processes and reducing negative impact on the environment.

Currently, there are many different information and communication technologies and solutions that contribute to improving supply chain management and can be effectively used to achieve sustainable development goals. However, these solutions are often fragmented and work in isolation from each other, which limits their overall effectiveness. Therefore, it is important to explore the possibilities of integrating existing intelligent systems, especially ITS and ILS. Such integration will allow achieving a comprehensive approach to supply chain management, ensuring more accurate control over resources, route optimization and minimization of carbon footprint, which will directly contribute to sustainable development. The integration of intelligent technologies in sustainable supply chain management not only increases the efficiency and cost-effectiveness of business processes, but also contributes to achieving sustainable development goals. Thus, the intellectualization of supply chain management is an important step towards sustainable development.

The integration of ITS and ILS, especially in the context of sustainable development, is the subject of numerous studies. Scientists and experts are studying ways in which modern technologies such as the Internet of Things, big data and artificial intelligence can be combined to improve supply chain management, reduce environmental impact and achieve sustainable development goals.

Given the growing interest in this topic, the scientific community is actively exploring the possibilities of integrating intelligent systems into supply chains. Numerous studies are being conducted in different countries dedicated to innovative solutions in the field. Below we will consider the most significant publications of recent years that highlight approaches to the implementation of ITS and ILS in sustainable supply chains.

Martin Christopher argued that traditional supply chains focused exclusively on efficiency and cost minimization are no longer sufficient to meet modern requirements. He emphasized that in the era of globalization, supply chains must be flexible and able to adapt to changes in order to ensure long-term sustainability. His research also focused on the importance of integrating modern technologies, including intelligent systems and digitalization, to improve supply chain management and their environmental sustainability [1].

Exploring the development strategies of ILS, Ma Q. et al. (2024) drew attention to the fact that e-commerce became the driver of growing interest in them [2]. The article [3] explores the possibilities of using big data analysis in sustainable supply chains and their transformation in such a way that they are technologically advanced, socially responsible and environmentally friendly. The authors [4–6], proved that artificial intelligence technologies are gaining particular importance in the digital age due to their wide range of applications in logistics and supply chain management. New possibilities of using Internet of Things technology are revealed in publications [7, 8]. Tao Y. (2024) emphasized that intelligent systems use predictive analytics and automated workflows to prevent problems in supply chain management [9]. The authors [10] linked the digitalization of supply chain management with economic, environmental and social criteria and the innovations.

Many studies are devoted to the problems of measuring carbon footprint in supply chains, in particular, the work [11] emphasizes the need to measure greenhouse gas emissions in product life cycle chains, the authors [12] established the relationship between reducing carbon dioxide emissions and business growth, supporting the potential of sustainable business practices to increase competitive advantage and customer loyalty. Research by Chinese scientists proved that every 1% improvement in intelligent infrastructure reduces the intensity of carbon emissions by 0.2810% [13].

The problem of intellectualization of logistics activities attracts the attention of Ukrainian scientists, in particular, in publications [14–16] the directions of digitalization of logistics services and transport-logistics systems are defined. The authors [16] considered the features of using intelligent algorithms and machine learning methods to optimize key aspects of logistics, such as transport routing, inventory management and demand forecasting. The work [17] highlights a logistics digitalization related to the digital documentation and electronic document management.

Thus, the literature review shows the presence of different approaches to defining ILS and their impact on the sustainability of supply chains, digital technologies and intelligent solutions are considered in various fragmentary contexts, the conceptual and categorical apparatus of intellectualization of supply chain management needs to be developed in accordance with the goals and priorities of sustainable development. Often in the scientific literature, the concepts of ITS and ILS are equated. Despite the fairly high level of study of intelligent systems and intelligent solutions in logistics and supply chain management by scientists, the high relevance of the outlined problems requires further interdisciplinary research taking into account industry development trends and sustainable development priorities.

The **aim** of the work is to study the possibilities of integrating ITS and ILS for effective sustainable supply chains management, which involves reviewing the impact of digital technologies on optimizing transport and logistics processes and increasing the supply chains sustainability in the context of sustainable development goals.

2 Methodology

The methodology for studying the integration of ITS and ILS for sustainable supply chain management is based on well-known methods of theoretical and empirical research and includes the following main stages:

1. Thorough analysis of scientific publications related to ITS and ILS, intellectualization of logistics and supply chain management, practices of using modern digital technologies in logistics activities, as well as defining conceptual frameworks and approaches to implementing intelligent systems in supply chain management.
2. Comparative analysis of functional characteristics and digital technologies used in ITS and ILS in terms of their impact on sustainability and efficiency of supply chain management.

3. Analysis of real cases and successful business practices of implementing digital technologies and intelligent solutions in various companies.
4. Development of a conceptual model for integrating existing ITS and ILS for sustainable supply chain management, taking into account the potential for reducing harmful environmental impact and improving environmental standards.
5. Generalization of the obtained data and development of practical recommendations for using the possibilities of integrating intelligent systems, digital technologies and intelligent solutions in supply chain management to reduce carbon footprint.

3 Results and Discussion

Generalization of recent publications made it possible to find out that under the concept of "intelligent system" most often means a structure of technical means and software united by an information process, working in conjunction with the operator or independently of him; capable of synthesizing, generating decisions about action and finding effective ways to solve problems based on information and knowledge in the presence of motivation [18, p. 37]. In our opinion, the view of intelligent systems as only implementing existing technologies is somewhat limited and does not reflect their potential for future developments. Modern intelligent systems not only use existing algorithms and solutions, but are also able to create new knowledge, predict future needs, adapt to changes in the environment and set tasks for new developments. They use artificial intelligence technologies, machine learning algorithms, big data, to solve problems related to automation, process optimization, data-based decision making and efficiency improvement. Intelligent systems of the future may include the ability to self-learn, self-optimize and make autonomous decisions, focusing on tasks that will arise in the future.

In relation to the logistics sphere of activity, the concept of ITS is most often used, which also has different interpretations, but we will rely on the opinion of the authors [19], who believe that it is a combination and application of computer, information and communication technologies for managing vehicles and transport networks in real time. However, unlike the simple integration of existing technologies, modern ITS is also aimed at predicting future challenges and developing new solutions for the future transport environment, taking into account long-term goals such as sustainable development and minimization of carbon emissions. Modern ITS not only adapts to changes, but is also able to initiate new technological trends aimed at forming a responsible approach to the supply chain management process, in order to prevent the destruction of the acquired competitive positions of the enterprise and contributes to the growth of the competitive potential of the enterprise not only in the domestic market, but also in the system of international economic relations.

There are more and more publications that use the concept of ILS. In the work [21] we outlined the contour of intelligent logistics as a synthetic management tool integrated with the end-to-end logistics flow, aimed at optimizing flow processes in logistics systems, supply chains and logistics networks. The authors [22] linked the development of ILSs with Industry 4.0 and smart manufacturing. We really like the opinion of many researchers that the intellectualization of logistics is associated with the formation of a virtual business environment, which involves industry interconnection, shared use and symbiosis of digital technologies through automation, visualization and informatization of logistics services. A powerful driver of growing interest in the patterns of formation of ILSs was the development of e-commerce, which is due to the need for rapid decision-making and network interaction of business partners in real time [2]. Also appropriate is the thesis that ILSs are the cornerstone in modern supply chain management, transforming traditional operations through the integration of advanced technologies such as AI, IoT, big data analytics and automation. At its core, an ILS uses real-time data and automation to address critical issues in SC management [9].

We have generalized the definitions of ITS and ILS and highlighted a system of interconnected structural components of intellectualization of logistics activities (see Fig. 1).

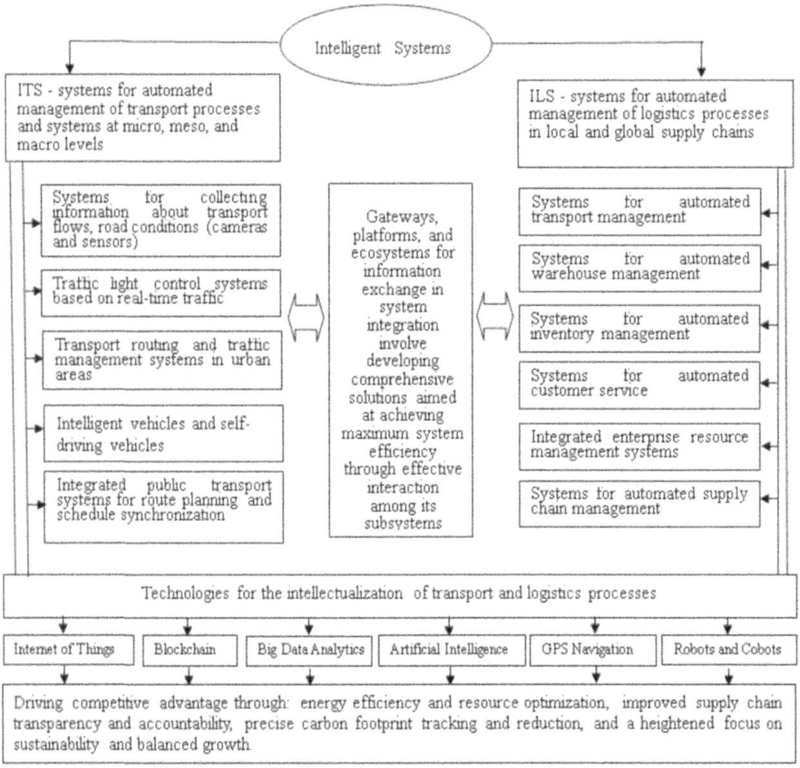

Fig. 1. Conceptual model of ITS and ILS integration in the context of sustainable supply chain management (generalized according to [4, 5, 9, 14, 19, 20])

Thus, the presented conceptual module shows that ITS and ILS have much in common, but differ in their specifics and functionality. Both systems are automated and use similar tools such as artificial intelligence, Internet of Things (IoT) and data analytics to achieve their goals. ITS and ILS aim to optimize processes, increase efficiency and improve environmental performance and sustainable development. These systems can predict future challenges and implement innovations for industry development.

However, significant differences exist. ITS specializes in managing transport processes such as traffic optimization, improving road safety and reducing environmental impact. Its main goal is to ensure efficient functioning of transport infrastructure through automation and advanced technologies. ILS focuses on supply chain management, optimizing the movement of goods, resources and information between different links in the chain, reducing costs and ensuring rapid response to market changes.

ITS and ILS can effectively interact and integrate to improve supply chain management, creating a comprehensive mechanism that ensures optimization of goods movement, cost reduction and environmental impact reduction. ITS provides information about current road situations, allowing ILS to make informed decisions about route planning and delivery times. The integration enables real-time data exchange between all supply chain participants, including transport companies, warehouses and distributors.

This integration creates more sustainable, flexible and environmentally responsible supply chains. ITS can collect and transmit data on emission levels from each vehicle in real time, allowing ILS to track environmental impact and quickly respond to exceeding norms. Combined, these systems can create a comprehensive approach to sustainable management of logistics processes, optimizing each stage to reduce emissions and improve environmental efficiency.

Integration of ITS and ILS significantly improves the accuracy and efficiency of measuring carbon footprint in supply chains through continuous collection, processing and analysis of data on transport operations and logistics processes. This allows for accurate assessment of the impact of each mode of transport, operation and infrastructure on overall CO2 emissions and other greenhouse gases.

Summarizing information from both systems creates a detailed picture of each supply chain link's environmental impact. ILS can automatically calculate the total carbon footprint at each stage, from production and transportation to storage and delivery. This includes both direct emissions from transport use and indirect emissions associated with energy consumption in warehouses or during loading and unloading.

ITS and ILS can track alternative ways to reduce carbon footprint, such as choosing more environmentally friendly routes or modes of transportation. By analyzing data from different modes of transport, the systems can determine the most environmentally efficient solutions for each specific case. ILS can provide transparency of emissions data throughout the product lifecycle using blockchain technology, ensuring authenticity and preventing falsification of information.

These digital technologies and digital solutions are already actively used in intelligent logistics systems for more efficient supply chain management. Table 1 provides data on existing integrated supply chain management systems that use technologies such as data analytics, machine learning, robotics and blockchain to solve specific tasks and improve the overall management process. Some of them have modules that allow calculating carbon dioxide emissions (in CO2 equivalent units) and other greenhouse gases in supply chain management processes.

Table 1. Comparative analysis of the most common intelligent supply chain management systems.

Name of ILS	Brief description	Logistics functions	Emission control functions
SAP Integrated Business Planning (SAP IBP)	Comprehensive SC planning solution that uses big data analytics and machine learning to optimize demand, inventory and production.	Demand forecasting, inventory management, supply planning, production and transport optimization.	Modules for environmental impact assessment, sustainability indicator tracking, data analytics for emission reduction.
Oracle Supply Chain Management Cloud	Integrated supply chain management solutions with a focus on automation and data analytics.	Inventory management, supply planning, production management, transport management, real-time analytics.	Tools for CO2 emissions analysis and reporting, integration with environmental modules for sustainability assessment.
IBM Sterling Supply Chain Suite	Comprehensive supply chain management solution that includes the use of AI and blockchain to improve transparency and risk management.	Product tracking, risk management, data analytics, integration with suppliers and partners.	Emission tracking and reporting, integration with systems for environmental impact management.
JDA Software (now Blue Yonder)	Supply chain management solution that uses machine learning and analytics to optimize inventory, demand and supply.	Demand forecasting, inventory management, supply planning, transport management.	Assessment of CO2 emissions and other environmental indicators as part of demand, inventory and supply management
Kiva Systems (Amazon Robotics)	Robotic warehousing systems for automating the processes of collecting and transporting goods in the warehouse.	Robotic warehousing, order picking automation, warehouse space management.	Improvement of environmental parameters of warehouse complexes

The potential applications and integration of contemporary digital technologies and advanced solutions within intelligent logistics systems include:

1. IoT is a network of physical devices equipped with sensors, software, and other technologies that allow collecting and exchanging data via the Internet. Sensors deployed at various stages of the supply chain measure parameters such as fuel volumes, energy consumption, and CO2 emissions. Data is transmitted to centralized systems or cloud platforms for analysis to determine exact emission volumes. Real-time data collection enables prompt tracking of changes in emission volumes and quick response to deviations from established norms. IoT promotes data integration between different links in the supply chain, allowing for comprehensive carbon footprint analysis across all stages.
2. Data analytics technologies transform large volumes of data into valuable information for informed decision-making. The process begins with collecting data from various sources in the supply chain, including sensors on vehicles, monitoring systems on production lines, and logistics operations information. Data analytics identifies trends and patterns, determining which factors most influence CO_2 emission levels. This allows companies to identify the main sources of emissions and opportunities for improvement.
3. Blockchain technologies ensure transparency, immutability, and distribution of data for accurate accounting of CO_2 emissions. Emission data is collected, recorded in the blockchain, and verified using consensus mechanisms. This guarantees data accuracy and reliability, ensuring distribution and availability of information for all supply chain participants without centralized storage.
4. Artificial Intelligence processes large volumes of data, recognizes complex patterns, and automates decision-making to reduce emissions in supply chains. AI analyzes data from various sources to identify patterns, trends, and anomalies affecting CO2 emission levels. It optimizes logistics routes and schedules, reducing fuel consumption and emissions. AI also improves production processes by analyzing energy consumption and emissions in real-time, identifying inefficiencies, and recommending changes to reduce environmental impact.

These systems and digital technologies demonstrate the diversity of intelligent solutions that can be integrated into supply chain management to increase efficiency, adaptability, and sustainability. Each uses different technologies such as data analytics, machine learning, robotics, and blockchain to solve specific tasks and improve the overall management process.

Also, there are collective use systems and platforms that allow calculating emissions of carbon dioxide (CO_2) and other greenhouse gases for various types of transport, as well as in supply chain management processes. These systems are often used to assess the environmental footprint of businesses, monitor environmental impact, and ensure compliance with environmental standards. In fact, they provide an opportunity for interaction between ITS and ILS to address specific sustainable development issues (Table 1).

These systems help businesses assess and reduce their environmental footprint, as well as improve sustainable development strategies. They provide valuable information for decision-making aimed at reducing CO2 emissions and minimizing negative environmental impact.

Table 2. Description of digital platforms and ecosystems for monitoring greenhouse gas emissions and other environmental indicators.

Digital Platforms	Brief Description	Transport and Logistics Functions
EcoTransIT World	A tool for calculating environmental indicators of freight transportation, including CO2 emissions.	Calculation of CO_2 emissions for various modes of transport (road, rail, sea, air), assessment of the environmental impact of different routes and vehicles.
Sustainable Freight Transportation (SFT) Tool	Calculation of CO2 emissions in freight transportation and logistics.	Assessment of the environmental impact of goods transportation, calculation of greenhouse gas emissions based on transport and route data.
Carbon Trust Footprint Calculator	A calculator for assessing CO2 emissions for businesses, including logistical processes.	Calculation of CO2 emissions taking into account energy consumption, transport costs, and other factors, assistance in implementing emission reduction strategies.
Greenhouse Gas Protocol (GHG Protocol) Tools	Tools and methodologies for accounting and reporting greenhouse gas emissions.	Includes tools for calculating emissions in various areas such as energy, transport, production, and logistics.
Enablon Sustainability Management	A platform for sustainability management, including calculation of CO2 emissions and other environmental indicators.	Monitoring and reporting of CO2 emissions, environmental data management, support in achieving sustainable development goals.
SAP Environment, Health, and Safety (EHS)	Calculation of CO2 emissions in the context of environmental and safety management.	Emissions monitoring, reporting on environmental impact, support in compliance with regulatory standards.
EcoEfficiency Calculator by Siemens	Calculator for assessing environmental efficiency, including CO2 emissions calculation.	Calculation of environmental indicators based on resource use and energy consumption.
Caleb – Carbon Footprint Calculator	A solution for calculating CO2 emissions that covers various aspects of business, including logistics.	Assessment of emissions based on data on energy consumption and transport costs.

Thus, the integration of ITS and ILS allows for a systematic approach to measuring the carbon footprint in supply chains, providing detailed data on emissions at each stage, from transport to logistics, and creating opportunities for their reduction through optimization of transport and logistics processes in real-time (Table 2).

4 Conclusion

1.A comparative analysis of the essence and functional features of ITS and ILS revealed their common and distinctive characteristics, as well as determined their impact on the management of key business processes in supply chains. It is shown that the integration of these systems allows for increasing the efficiency of goods and services flow management, reducing operational costs, improving process transparency, and minimizing negative environmental impact.
2.Considering modern challenges of sustainable development, particularly the need to reduce greenhouse gas emissions, which can be successfully addressed through the application of innovative technologies, an analysis of available digital technologies and intelligent solutions in the activities of various companies was conducted. It is proven that the use of the Internet of Things, blockchain, artificial intelligence, and big data analytics contributes to accurate measurement, monitoring, and reduction of harmful emissions at all stages of the supply chain.
3.The integration of ITS and ILS allows for creating new approaches and new business models for supply chain management, ensuring their transparency, efficiency, and sustainability. As a result, supply chains become more flexible, allowing for quick responses to external challenges and requirements for carbon footprint reduction.
4.To achieve sustainable development goals, it is necessary to optimize the interaction between different management systems in supply chains. This involves not only technical integration but also adaptation of organizational processes, which will contribute to increasing the environmental and economic efficiency of supply chains. A comparative analysis of existing intelligent systems and software products that promote balanced management of supply chains in terms of economic, environmental, and social requirements and allow measuring and monitoring the carbon footprint and developing measures to reduce it through inter-system integration and collaboration of various supply network links was conducted.
5.The proposed recommendations for implementing intelligent solutions can serve as a basis for further improvement of sustainable supply chain management. The integration of intelligent systems creates additional opportunities for reducing environmental impact and promotes more sustainable development of enterprises on a global scale.

Further research may focus on a deeper analysis of the integration of ITS and ILS to improve the efficiency and environmental friendliness of supply chains. In particular, the development and implementation of digital solutions and tools based on artificial intelligence that can predict carbon emissions and other environmental parameters in real-time. This will increase the efficiency of decisions to reduce the carbon footprint. On the other hand, it is worth investigating the impact of intelligent systems integration on the use of the economic potential of sustainable supply chains and economic incentives for those participants in the movement of goods who will have the greatest progress in reducing greenhouse gas emissions and achieving sustainable development goals.

References

1. Christopher, M.: Logistics and Supply Chain Management (6th ed.). Pearson, UK (2022)
2. Ma, Q., Luo, S., Li, J.: Research on the development strategy of intelligent logistics system of live streaming e-commerce supply chain. Front. Bus. Econ. Manag. **13**(3), 123–127 (2024)
3. Jabbour, C.J.C., Fiorini, P.D.C., Ndubisi, N.O., Queiroz, M.M., Piato, É.L.: Digitally-enabled sustainable supply chains in the 21st century: a review and a research agenda. Sci. Total. Environ. **725**, 138177 (2020)
4. Jonathan, H., Magd, H., Khan, S.A.: Artificial intelligence and augmented reality: a business fortune to sustainability in the digital age. Navigating the Digital Land-scape: Understanding Customer Behaviour in the Online World, 85–105. Emerald Publishing Limited, Leeds (2024)
5. Richey, R.G., Jr., Chowdhury, S., Davis-Sramek, B., Giannakis, M., Dwivedi, Y.K.: Artificial intelligence in logistics and supply chain management: a primer and roadmap for research. J. Bus. Logist. **44**(4), 532–549 (2023)
6. Adesoga, T.O., Ajibaye, T.O., Nwafor, K.C., Imam-Lawal, U.T., Ikekwere, E.A., Ekwunife, D.I.: The rise of the "smart" supply chain: how AI and automation are revolutionizing logistics. Int. J. Sci. Res. Arch. **12**(2), 790–798 (2024)
7. Yang, X., Shen, T., Ting, L.: Research on the application of Internet of Things technology in intelligent logistics. Logist. Sci-Tech. **47**(08), 39–42 (2024)
8. Zrelli, I., Rejeb, A.: A bibliometric analysis of IoT applications in logistics and supply chain management. Heliyon **10** (2024)
9. Tao, Y.: The application of intelligent logistics systems in supply chain manage-ment and its challenges: case studies of automated warehousing and unmanned delivery. Front. Bus. Econ. Manag. **16**(2), 100–107 (2024)
10. Zhu, Y., Zhang, Z.: Supply chain digitalization and corporate ESG performance: evidence from supply chain innovation and application pilot policy. Financ. Res. Lett. **67**, 105818 (2024)
11. Truant, E., Crocco, E., Corazza, L., Borlatto, E.: Life cycle thinking and carbon accounting in sustainable supply chains: a structured literature review and research agenda. Sustain. Account., Manag. Policy J. (2024)
12. Hasan, M.R., Islam, M.Z., Sumon, M.F.I., Osiujjaman, M., Debnath, P., Pant, L.: Integrating artificial intelligence and predictive analytics in supply chain manage-ment to minimize carbon footprint and enhance business growth in the USA. J. Bus. Manag. Stud. **6**(4), 195–212 (2024)
13. Yi, M., Chen, D., Wu, T., Tao, M., Sheng, M.S., Zhang, Y.: Intelligence and carbon emissions: The impact of smart infrastructure on carbon emission intensity in cities of China. Sustain. Cities Soc. **112**, 105602 (2024)
14. Hryhorak, M., Harmash, O., Popkowski, T.: Artificial intelligence in supply chain management: opportunities and threats for professional competence. Intellect. Logist. Supply Chain. Manag. **19**, 24–44 (2024)
15. Grytsenko, S., Glushchenko, S., Shulga, A.: Innovative and technological support for optimization of logistics values. Econ. Her. Donbass **3**(73), 26–31 (2023)
16. Pichuhina, Y., Maksymov, O.: Use of artificial intelligence and modern information technologies in the city logistics systems. Econ. Soc. **62** (2024)
17. Medvedev, Ye., Popova, Yu., Petrenko, O.: Planning of logistics systems using artificial intelligence. Econ. Soc. **59** (2024)
18. Smerichevska, S., Prodanova, L., Yakushev, O.: Digitization of logistics and supply chain management. Intellect Logist. Supply Chain Manag. **26**, 113–123 (2024)
19. Avrunin, O., Vladov, S., Petchenko, M., Shushlyapina, N.: Intelligent Automation Systems. NovaBuk, Kremenchuk (2021)

20. Kostyuk, Y., Romanenko, O.: Theoretical foundations of intellectualization of urban transport and logistics systems. Econ. Innov. **24**(4), 60–69 (2022)
21. Stepanov, O., Wenger, A.: Intelligent transport systems in the transport process. Munic. Econ. Cities. **161**, 212–217 (2023)
22. Grygorak, M.: Intellectualization of the Logistics Services Market: Concepts, Methodology, Competence. Sik Group Ukraine, Kyiv (2017)

Integration Development of Logistics Activities of Corporate Enterprises Based on Intellectualization and Management Technologies

Oleksii Hutsaliuk[1(✉)], Iuliia Bondar[2], Oleksandr Kalinin[3], Viktor Sokolovskiy[4], and Alla Navolokina[5]

[1] Private Higher Education Institution «Rauf Ablyazov East European University», 16, Nechuya-Levytskogo Street, Cherkasy 18036, Ukraine
alex-g.88@ukr.net

[2] Volodymyr Vynnychenko Central Ukrainian State University, 1, Shevchenko Street, Kropyvnytskyi 25006, Ukraine

[3] Mykolas Romeris University, 20, Ateities Street, 08303 Vilnius, Lithuania

[4] Flight Academy of the National Aviation University, 1, S. Choban Street, Kropyvnytskyi 25005, Ukraine

[5] International European University, 42 V., Hlyshkova Street, Kyiv 03680, Ukraine

Abstract. The research is dedicated to the current conditions of globalization and technological progress in the development of logistics activities of corporate enterprises, which is becoming increasingly important. The main purpose of the research paper is to determine areas of integration development of logistics based on the introduction of intelligent systems and management technologies. It should be noted that the use of modern logistics management to manage logistics processes can save logistics costs equivalent to 1% and increase sales by 10%. Logistics is one of the most dynamic and significant sectors in the EU economy, accounting for approximately 10% of GDP. The article analyzes current trends in logistics, in particular, artificial intelligence and process automation, as well as their impact on the efficiency of corporate enterprises. It is determined that advanced technologies, such as artificial intelligence, machine learning, the Internet of Things (IoT) and big data, can reduce logistics costs by up to 30%, while AI can increase the efficiency of logistics processes by up to 50%. The article examines the increase in the competitiveness of corporate enterprises by optimizing logistics processes, reducing costs, improving the quality of customer service and making efficient managerial decisions.

Keywords: Logistics Activity · Corporate Enterprises · Management Technologies · Innovations · Blockchain · Integration Development · Intelligent Transport Technologies

O. Slavinska et al. (Eds.): ITSESQC 2024, LNNS 1335, pp. 270–286, 2025.
https://doi.org/10.1007/978-3-031-87376-8_24

1 Introduction

In today's world, logistics plays a critical role in corporate enterprise operations, ensuring efficient management of the flow of goods, information and resources. In the context of globalization, increased competition and rapid development of digital technologies, enterprises are forced to look for new approaches to improving their logistics processes. One of these approaches is the integration development of logistics activities based on intellectualization and introduction of cutting-edge management technologies.

The integration development implies coordinated interaction of all links in the logistics chain, from the supply of raw materials to the delivery of final products to the consumer. It requires efficient coordination of enterprise's internal processes and cooperation with external partners. At the same time, the intellectualization of logistics processes based on artificial intelligence, big data, automation and analytics can substantially enhance management efficiency, reduce costs and increase forecasting accuracy.

The intellectualization of logistics activities, including artificial intelligence, big data analytics, automation and the Internet of Things (IoT), allows increasing forecasting accuracy, reducing costs, as well as ensuring the high speed and quality of service. It is critical to secure a competitive advantage on the market where even minor delays or errors can lead to loss of customers and revenue.

Modern business conditions affected by globalization changes have designed a new paradigm of the development of global technology. The paradigm is based on assumptions of the new technology industry: the Internet of Things, big data processing, artificial intelligence, digitalization and robotics, etc. In 2019–2023, the number of enterprises that used social media to receive customer feedback or respond to customer orders increased by almost 30%; to engage customers in development or innovative activities increased by 29.6%; and to cooperate with various stakeholder groups by 26.9%. During this period, the number of enterprises that purchased customer relationship management software increased by 44%. And their share in the total number of this group of enterprises that purchased cloud computing services increased by 3.5 percentage points [1, 2].

Innovations and inventions in information technology can significantly increase computer capabilities; increase the speed of information processing and data transmission; reduce the cost per unit of processing, which leads to the active and efficient use of digital information.

The proof of this hypothesis is confirmed by the objective inability of an individual business entity to maintain a proper competitive position and ensure the market share growth on its own. Accordingly, the foundation for innovation adoption is the emergence of various types of corporate enterprises.

At the same time, such corporate enterprises can have different legal forms (holding companies, group of companies or associations) and can be limited to certain institutional rules for conducting innovative activities by individual enterprises or business entities. When analyzing the share of joint-stock companies in the corporate sector of Ukraine in 2019–2023, we can say that the dynamics slightly decreased from 1.09% in 2019 to 0.92% in 2023, but the share of limited liability companies was 97.8% in 2019 and was steadily increasing to 98.13% in 2023 [1, 2]. In the latter case, the development and adoption of innovations will be referred to strategic alliances or clusters of enterprises.

In any case, corporate enterprises are considered to be a highly complicated system with a heterogeneous composition and probabilistic nature of the behavior of participants.

Besides, such global challenges as changes in the global economy, the impact of pandemics, natural disasters and other factors are increasing the importance of integrating and intellectualizing logistics processes. Enterprises that promptly introduce such approaches have a better chance of success and sustainable development despite external threats and instability.

Thus, the integration development of logistics activities based on intellectualization and management technologies is an extremely relevant topic, as it reflects current trends in business development and responds to challenges faced by corporate enterprises in the context of globalization and digital transformation.

2 Purpose and Methods of Research

The purpose of the study is to develop theoretical and methodical principles and practical recommendations for improving the efficiency of logistics activities of corporate enterprises through the integration of modern management technologies and intellectualization of business processes.

To achieve this goal, the authors used the following methods in the study: dialectical, abstract and logical one to summarize theoretical foundations of functioning and development of logistics activities of corporate enterprises; analysis and synthesis when studying logistics activities of corporate enterprises as a system and determining its structural components; system analysis to substantiate the methodical approach to assessing the success of the corporate integration process; graphical method to visualize the results of the study in corporate enterprise activities; logical method to build the structure of the study.

3 Literature Review

The integration development of logistics activities based on intellectualization and management technologies is an interdisciplinary topic that combines economics, management, logistics, information technology and other knowledge areas. Well-known foreign and Ukrainian scientists have made a significant contribution to the study of this topic.

A corporation as an integrated business structure is considered by [3, 4] as an economic space of economic interaction and a set of inextricably related and intertwined segments that use alternative modes of resource allocation – market and administrative. At the same time, the institutional function of the corporate organization revealed above obtains a more significant economic content and acquires a new meaning consisting in increasing the flexibility of the economic structure of the economic mechanism of society, which results in two ways of economic relations: a market one that ensures the best use of resources but can be accompanied by substantial unproductive transaction costs and limited opportunities for social regulation in the distribution of wealth; a corporate one, which is a set of localized economic entities whose internal economic structure is based on administratively controlled redistribution of resources but whose economic interaction is carried out using the market mechanism. Thus, corporate organizations (corporations) in economic and structural terms represent a certain extremely mobile link between market economic and administrative organizational areas of the economic structure of society. This approach to a certain extent embodies the specific corporate ideology of economic convergence of various economic mechanisms (market and administrative) originated in classical works by [5, 6].

With each of these approaches and tools focusing only on certain components of the integration dynamics, we suggest that decisions on setting parameters of the integration project should be based on such a criterion as success of the corporate integration development process. The criterion of success is generally not used in the economic literature primarily due to the lack and difficulty of identifying the content of the concept of success. However, it should be noted that there are some attempts to use this concept of success in relation to the integration development made by [7] and co-authors. These authors used a fairly broad list of criteria for assessing interregional integration projects but did not introduce a definition of the success of integration transformations.

Such a consolidated representation of the subject of innovative activities in the form of an association of economic agents based on deepening relations between them requires determining the place of integration processes in the innovation activity of a corporate enterprise and their development processes. The most illustrative solution here is the work by [8], which highlights the organization of integration development management of the enterprise. A positive aspect of this work is the definition of constituent elements of the process of developing a mechanism for managing the so-called integrated business structure.

The combination of integration and innovation processes in the context of single management of the consolidated development of individual business entities requires additional consideration of solutions by [9] regarding the independence of decision-making by individual participants in corporate enterprises and the possibility of changing their chosen development areas directly during the implementation of innovative transformations.

Most authors consider logistics as an efficient tool and mechanism for managing, organizing and generating material flows at the enterprise (micro-level). The new type of logistics is considered by [10] as market logistics (macro-level) that takes into account the interests of all business entities, the state and consumers, which is one of the conditions for the development of efficient commodity market logistics.

Martin Christopher is one of the world's leading experts in logistics and supply chain management. Martin Christopher actively investigated and developed the concept of value chains, stressing the importance of integrating all stages of supply, from production to the end consumer. He notes that it is necessary to use state-of-the-art management technologies, such as enterprise resource planning (ERP) systems, supply chain management (SCM) systems and intelligent demand forecasting tools, in order to achieve efficient integration [11].

Klaus Schwab is the Founder and Executive Chairman of the World Economic Forum. He comprehensively covers the Fourth Industrial Revolution including the intellectualization and digitalization of logistics and management processes [12].

Erik Brynjolfsson and Andrew McAfee are well-known researchers in digital economy, particularly in the impact of digital technologies on business and the economy. Their research is dedicated to topics that directly affect the integration development of logistics activities of corporate enterprises. Researchers also focus on how companies use intelligent systems to make decisions, including managing logistics processes. They emphasize the importance of combining human abilities with artificial intelligence capabilities, allowing businesses to respond more efficiently to market changes and optimize their logistics operations [13].

It should be mentioned that the corporate integration association is primarily a merger of capitals of different economic entities carried out to achieve goals set by these entities. In relation to such goals, it should be noted that researchers mainly focus on achieving common goals set for the corporate enterprise as a whole.

These scientists have made an important contribution to the development of theoretical foundations and practical approaches to the integration of logistics processes and the introduction of the latest management technologies at enterprises, which allows efficiently using intelligent systems to increase business productivity and competitiveness. However, there are still questions regarding the creation of a holistic concept of the integration development of logistics activities of corporate enterprises that would improve their logistics activities of corporate enterprises due to the efficient use of intellectual and management technologies in the current dynamic business environment.

4 Results and Discussion

The logistics development of corporate enterprises in their integration into the digital economy is a continuous process creating a new efficient supply chain that is better than the previous one. These changes are caused by the impact of the external environment, structural technical and technological characteristics, development and adoption of cutting-edge information technologies, use of intellectual resources, etc.

The development, transformation and regulation of logistics activities of corporate enterprises is a cyclical process that includes six components: areas and mechanisms of regulation, assessment of logistics activities, market chain links and logistics gaps, efficiency of regulatory processes and forecasting. These components can be affected by the set goals and objectives of regulation, state control and adjustment of mechanisms and regulatory measures ensuring the efficiency and a positive market response to management influence. Flexibility and adaptability of regulatory mechanisms ensure the dynamic development of market processes.

The World Bank offers a fairly thorough analytical tool for studying the development of logistics activities in the world, namely the Logistics Performance Index (LPI). The Logistics Performance Index is an interactive tool for the comparative analysis of countries' logistics profiles, allowing countries to compare challenges and opportunities they face in developing corporate logistics activities and determine the ways to improve the efficiency in logistics activities. The LPI is based on a global survey of local operators (global freight forwarders and express carriers) and quantitative data on the efficiency of key components of the supply chain [14].

in 2007–2023. Source: Designed by the authors according to the World Bank [14].

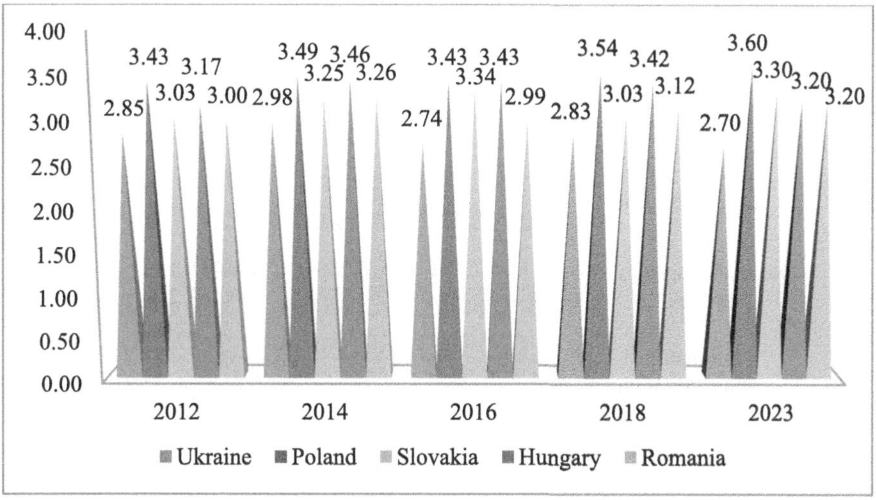

Fig. 1. Dynamics of Ukraine's Logistics Performance Index and its neighboring countries. Source: Designed by the authors according to the World Bank [14].

Comparing Ukraine's Logistics Performance Index with its neighboring countries, it should be noted that this country is significantly inferior to Poland, Slovakia, Hungary and Romania in this indicator, as Ukraine has the lowest index of 2.70 as of 2023 [14] (Fig. 1).

Figure 1 shows that the Logistics Performance Index of Poland, Slovakia, Hungary and Romania has steady growth trends, as each of the analyzed countries showed a higher indicator in 2023 according to the World Bank ranking. Such trends mean the sustainable policy on developing logistics in these countries. Changes in the dynamics of Ukraine's Logistics Performance Index correspond to the dynamics of the political and economic situation in the country.

Figure 2 shows that the lowest level of efficiency in Ukraine's logistics network is demonstrated by the customs and infrastructure.

It should be noted that the situation at the customs is quite objective, as the biggest problem of customs operations is smuggling of goods across the customs border of Ukraine. In particular, based on the mirror statistics method to analyze export and import operations, the volume of smuggling of goods across the customs border of Ukraine amounted to about 4 billion USD [15]. Therefore, it is not only the inefficiency of customs activities but also the inefficiency of public policy in fiscal administration and customs logistics.

Intelligent transport technologies and smart logistics are crucial components of the modern transport infrastructure at corporate enterprises. They are aimed at improving the efficiency, safety and environmental sustainability of transportation systems (Fig. 3).

Intelligent transport technologies (ITT) include various information and communication technologies to manage transportation systems. Some of the key aspects of ITT include [16, 17]:

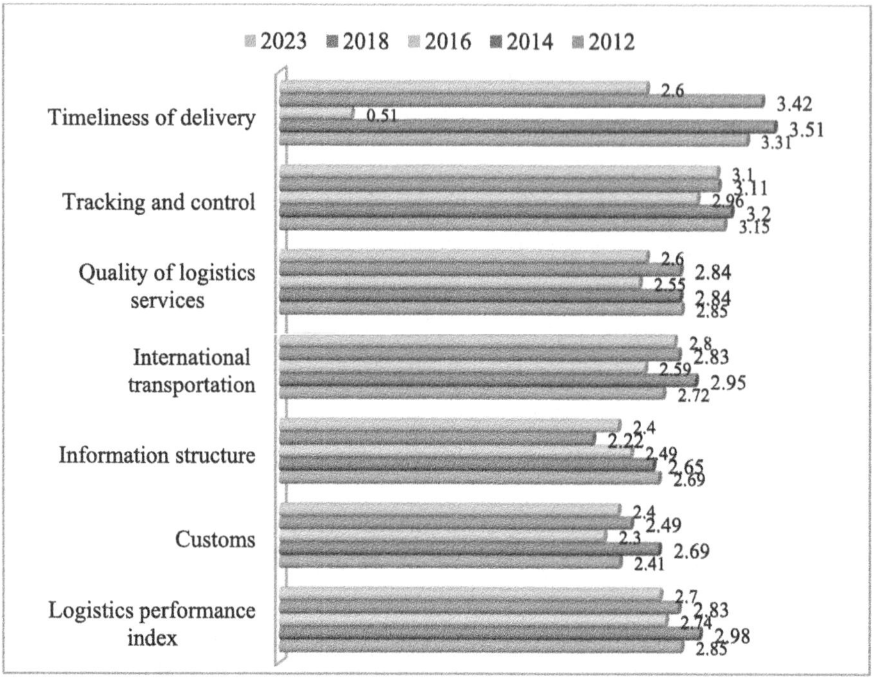

Fig. 2. Structure of Ukraine's Logistics Performance Index in 2012–2023. Source: Designed by the authors according to the data [14, 15].

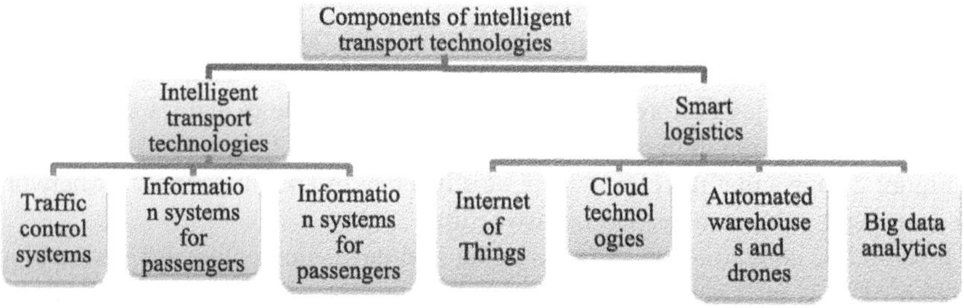

Fig. 3. Intelligent transport technologies. Source: Designed by the authors

1. traffic control systems, i.e. the use of sensors, cameras and other devices to monitor and regulate road traffic in order to reduce congestion and improve safety.
2. information systems for passengers to provide passengers with real-time information about traffic, delays, the most appropriate routes, etc.
3. automated fare collection systems are contactless payment systems that simplify the fare payment process and reduce delays.
4. information systems for passengers means the automated driving technologies, advanced driver assistance systems (ADAS) and other innovations to improve safety and comfort.

Smart logistics implies the use of state-of-the-art technologies to optimize the transportation and delivery of goods. The main components of smart logistics include:

1. the Internet of Things (IoT), i.e. the use of sensors and other devices to track the location and condition of cargo in real time.
2. big data analytics involves analyzing large amounts of data to predict demand, optimize routes and reduce costs.
3. automated warehouses and drones include the use of robots and drones to automate warehousing and delivery processes in order to increase the efficiency and reduce the risk of errors.
4. cloud technologies, namely the use of data storage and processing cloud platforms to ensure flexibility and speed of access to information.

Innovative development of logistics activities of corporate enterprises is also driven by blockchain technology that benefits freight forwarders, carriers and logistics service providers. Blockchain platforms allow updating all internal corporate systems via the unchanging data chain. The blockchain market in supply chains was estimated at 0.7 billion USD in 2022, and it was estimated to grow to 1.0 billion USD in 2023. In 2023, the blockchain market was expected to grow to 20.5 billion USD [18] (Fig. 4).

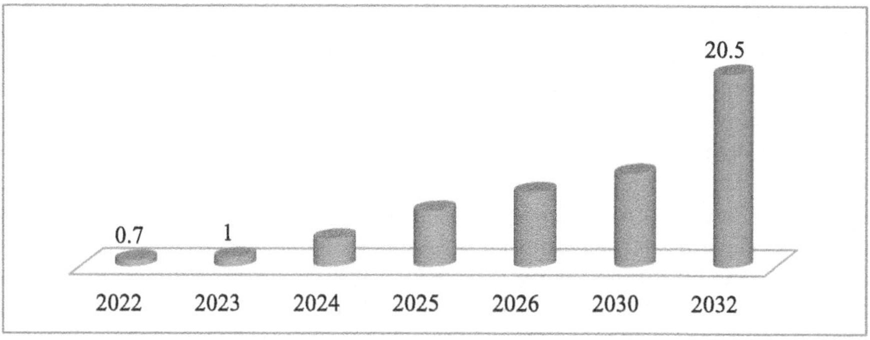

Fig. 4. Dynamics of the blockchain market in supply chains. Source: analyzed by the authors based on [18].

The Internet of Things (IoT) is developing quite rapidly. IoT devices allow corporate enterprises to control and increase visibility of their supply chains. Equipping warehouses, containers and vehicles with IoT technology, business leaders and supply chain managers can control all logistics processes and improve inventory management and delivery procedures. IoT sensors allow enterprises to create reliable digital networks that enable more efficient asset management, reduce costs and increase return on investment (ROI) while increasing resilience to negative disruptive events. The global IoT market in the logistics sector was estimated at 37.41 billion USD in 2021 and is expected to grow to 93.85 billion USD in 2028 [19] (Fig. 5).

Cloud-based transportation control software has shown a great impact on supply chain management, with almost 54% of companies planning to increase their investments in cloud computing to improve the sustainability of logistics activities [20].

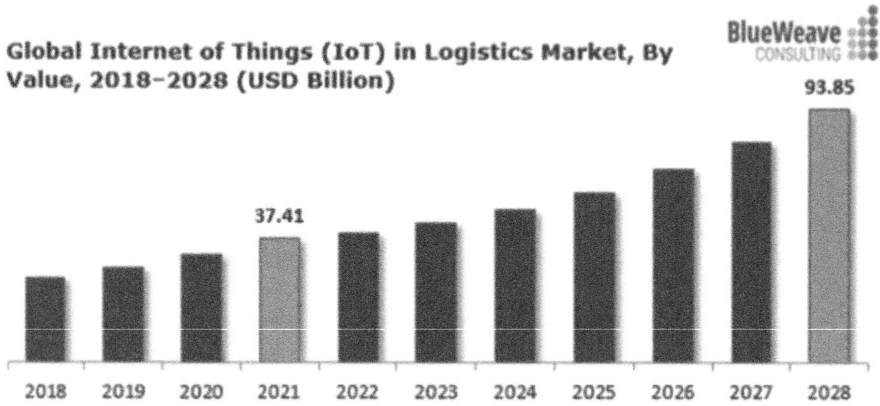

Fig. 5. Global IoT market in logistics. Source: designed by the authors based on [19].

Information logistics is of particular importance because of connecting the production, distribution, exchange and consumption of information, i.e. information flows. Currently, information is the foundation for the development and transformation of the market logistics chain, its transparency and ability to adapt to the current conditions of corporate enterprises.

Nowadays, information is becoming the dominant factor of production, while various social and economic phenomena, including economic cycles, are information-based and have versatile information content. The transition to the digital economy has increased the scale of digital traffic and the volume of information transfer, while information technologies are transmitted through the global network covering most countries worldwide. Digital strategies, models, operations, marketing methods, missions and goals are aimed at accelerating and increasing the results using digital technologies. Working with intangible assets, information and relationships creates opportunities for their use in case of involving highly skilled personnel. The interaction of all aspects of the new economy establishes a digital platform uniting the interests of all market participants.

Both the introduction of innovations and the regulation of cooperation and integration relations will involve certain transformational actions. Their direction, scale, radicality and duration will be determined by local goals of corporate enterprise participants (the degree of opportunistic behavior and asymmetry of the established association of enterprises). The internal competition for innovative knowledge and relevant resources will also have a significant impact on parameters of the organization of the management of corporate enterprise innovation development. Here, the degree of integration of corporate enterprise participants, which should be divided into internal and external, is of great importance. Thus, integration processes can lead to the establishment of a common information environment and a clear alignment of previously separated functions, the allocation of end-to-end business processes and the creation of unified management procedures for corporate enterprises as a whole. All of this will ultimately lead to improved performance of the innovation development process.

Given the above mentioned, two fundamentally different approaches to the implementation of processes of innovative development of corporate enterprises can arise.

The first approach is limited to the development of an innovative product or justification of potential innovative changes only within the existing corporate enterprise. In this case, the organization of innovation development management will hardly differ from the management of the separate enterprise. The approaches, principles and rules for organizing such management are quite well presented in the economic literature [21]. That is why further attention of this study will be focused on the second approach to the organization of management of corporate enterprise integration development.

The second approach is based on the creation of new cooperation networks or strategic alliances. The methodological basis of this approach is the idea of promoting knowledge required to develop and adopt innovations outside the corporate enterprise. Consequently, the organization of management of innovative development of a corporate enterprise primarily requires the exchange of knowledge resources between corporate enterprise participants (both existing and potential), development of interaction rules for carrying out joint innovative activities by corporate enterprise participants, determination of areas of innovative potential growth of the newly established corporate enterprise. It should be noted here that the adoption of innovations is possible only if the subject of innovative activities has sufficient potential for its implementation.

The organization of management of corporate enterprise innovative development requires taking into account parameters of combinatorial advantages for carrying out innovative activities and the synergistic effect of mutual complementation of resources and technologies arising from the interaction between participants in integration processes. At the same time, the system of information support of management processes is undergoing a certain transformation, which should focus on the creation, distribution and use of consolidated innovative knowledge. In this case, the subsystem of information support of corporate enterprise management will focus not only on collecting information used to monitor activities of corporate enterprise participants or to calculate plan-fact deviations.

It is necessary to create information support for processes of creating, disseminating, processing and using knowledge within the established corporate enterprise. It is also important to develop a system of criteria for selecting members of corporate enterprises and controlling their participation in the implementation of innovative transformations. In this case, it would be appropriate to expand the existing approaches to assessing the efficiency of innovation projects in terms of determining the level of use of consolidated innovation resources. Any of the available methodical approaches [22, 23] should take into account parameters of diffusion of innovations between corporate enterprise participants, opportunities for corporate synergy or areas of using consolidated financial resources.

Summarizing the above-mentioned prerequisites for implementing processes of innovative development of the corporate enterprise, we present a scheme for organizing the management of such processes. It should be noted that innovative development should be carried out on an ongoing basis.

The relevant scheme requires a cyclical presentation. In this context, it is appropriate to use the apparatus of Petri nets [22] that represent the organization of innovation development management as a discrete event process.

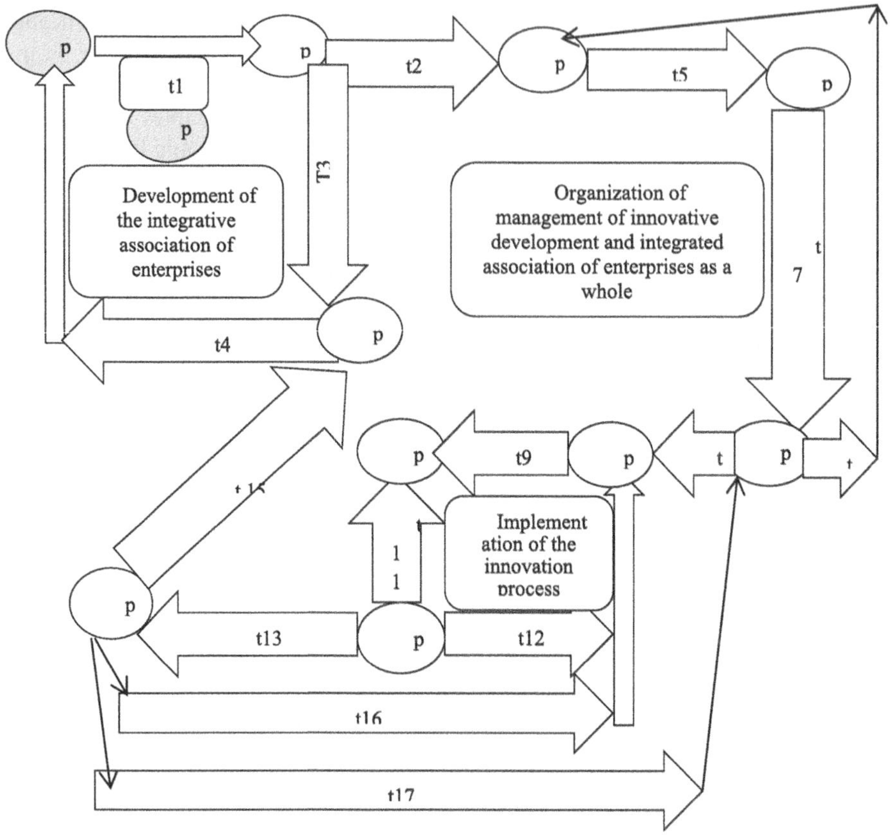

Fig. 6. Model of the process of organizing the management of corporate enterprise integration development. Source: designed by the authors based on [23, 24].

In general, the Petri net is a set of events (defined by the set *{p}*) and transitions (defined by the set *{t}*) that initiate the occurrence of these events. The suggested scheme for organizing the management of innovative development of the corporate enterprise is shown in Fig. 6.

The scheme shown in Fig. 6 is based on four sets of events and transitions. The first set is responsible for the selection of participants in the integration and innovation process and the organization of their management. The start of this process is modeled by position p_1. The limitation for this model is that there is a certain enterprise providing the start of the integration process. At the same time, the position p_0 can correspond to an organizationally dedicated corporate center organized according to one of the types developed by N. Fut [21]. Directly the selection of participants in innovative development is modeled by the transition t_1, which receives suggestions from potential corporate enterprise participants (the desire to join integration processes is confirmed). The entire list of potential corporate enterprise participants is contained in the component p_1. The generated set of corporate enterprise participants is provided in the position p_2.

It is clear that corporate enterprise operations are possible only if the values of individual participants are consistent. The verification of such compliance is implemented by transitions t_2 (in case of compliance) and t_3 (in case of non-compliance). If the values are not consistent, parameters of the discrepancy are analyzed (position p_3) and transferred to the start of the integration and innovation process (transition t_4).

The second set of events in the scheme shown in Fig. 6 describes the process of organizing the management of innovative development of the already established corporate enterprise. The main elements of the organizational process (represented by the position p_4) are reduced to formalization of the subject and object of management, development of proper management procedures (determination of levers and tools for implementing management influences), development of organizational regulations and principles of interaction, determination of rights and responsibilities for the implementation of development processes, etc.

The fulfillment of the transition t_5 indicates the consolidation of components of the organizational process in the form of regulations or relational (institutional) norms accepted for implementation. The chip in the position p_5 reflects project documents or the first version of the principles of interaction between corporate enterprise participants. If it is impossible to fix such norms for a particular configuration of the corporate enterprise, the transition t_5 is triggered and a new stage of the integration process is initialized. Accordingly, the chip in the position p_7 indicates the formation of these norms. If it is necessary to revise the rules of interaction between corporate enterprise participants or change parameters of the organization of the management process, the transition t_8 is triggered. Otherwise, the transition t_9 triggers the start of the innovation development process.

The third set of events is directly responsible for preparing the introduction and implementation of innovative transformations. The guidelines for such transformations are modeled by the position p_7. The fulfillment of the transition t_{10} models the coordination of parameters of innovative transformations. The formed parameters are reflected in the position p_8, and the transition t_{11} models the implementation of innovative changes. The fulfillment of this transition transmits the resulting characteristics of changes to the position p_9, which are compared with the desired vector of corporate enterprise's goals within this position. If there are no significant deviations, the existing mechanism for managing innovative development keeps operating. The transition t_{12} is responsible for this. In case of deviations, the procedure for assessing the compliance of the corporate enterprise's composition with the desired requirements for the results of innovative development is initialized (the start is determined by the fulfillment of the transition t_{13}).

This is where the fourth set of events and transitions comes into play. Thus, the position p_{10} contains the results of positioning of all corporate enterprise participants on the plane: innovation potential – resource potential – approach to protection of innovations. After the fulfillment of the transition t_{14}, the results of the pairwise correlation of characteristics of innovative activities of corporate enterprise participants are transferred to the position p_{11}. There are several possible options of further actions: the complete revision of the composition of corporate enterprise participants (initialized by the fulfillment of the transition t_{15}), changing parameters of the organization of innovation development management (initialized by the fulfillment of the transition t_{16}) and adjusting the requirements for the results of corporate enterprise's innovative activities as a whole.

Table 1. Key areas of integration development of logistics activities

Intellectualization of logistics processes	Automation	The use of automated systems for inventory management, warehousing, transportation and order processing
	Internet of Things	Introduction of the Internet of Things (IoT) to track the movement of goods, the condition of vehicles and warehouses in real time
	Machine learning and artificial intelligence	The use of analytical tools for demand forecasting, route optimization and decision automation
Management technologies	ERP systems	ERP (Enterprise Resource Planning) integration to ensure efficient resource management, including finance, human resources, production and logistics
	Supply chain management (SCM) systems	The use of SCM to coordinate all supply chain participants, from suppliers to end users
	TMS (Transportation Management Systems)	Transportation management systems for planning, executing and optimizing the physical movement of goods
Integration of information flows	Digital platforms	Development of unified digital platforms for the exchange of information between all participants in the logistics process
	Cloud technologies	The use of cloud solutions for data storage and processing to provide quick access and flexibility in managing information flows

(*continued*)

Table 1. *(continued)*

Partnership and cooperation	Outsourcing of logistics services	Cooperation with logistics operators to optimize logistics processes and reduce costs
	Integration with suppliers and customers	Close interaction with suppliers and customers to ensure transparency and efficiency of the supply chain

Source: [17, 25, 26]

Integration development of logistics activities of corporate enterprises based on intellectualization and management technologies is the key element of successful operations of modern companies. It implies the use of innovative technologies to optimize logistics processes, increase efficiency and competitiveness of enterprises (Table 1).

Advantages of integration development based on intellectualization and management technologies include the following [27, 28]:

- increased efficiency, namely optimizing resources, reducing costs and increasing productivity;
- improving the quality of customer service, i.e., faster order fulfillment, reduced errors and increased customer satisfaction;
- flexibility and adaptability: the ability to respond quickly to market changes and customer needs;
- competitive advantages: increased competitiveness due to innovative solutions and efficient management of logistics processes.

The integration of logistics activities of corporate enterprises requires the introduction of state-of-the-art technologies and management practices to ensure efficient coordination of all aspects of logistics processes. The main areas of this development include digitalization, flexibility, partnership, intellectualization and sustainable development.

The digitalization of logistics processes using integrated information systems, such as ERP and SCM, allows businesses to improve the accuracy and speed of operations, as well as ensure the transparency of the entire supply chain. The flexibility and adaptability of logistics systems are becoming essential in the face of rapid market changes, enabling businesses to respond promptly to new challenges and customer needs.

The partnership between supply chain participants results in strong and long-term cooperation that increases the overall efficiency of logistics processes. The intellectualization of logistics by introducing artificial intelligence and big data analytics allows automating routine tasks, optimizing routes and improving forecasting accuracy, which leads to lower costs and higher customer service.

Taking into account the conducted research, the following recommendations should be highlighted for improving the efficiency of the logistics activities of corporate enterprises, namely:

increasing competitiveness, i.e. effective logistics allows you to reduce costs, improve the quality of customer service and ensure faster delivery of goods;

increasing customer satisfaction, in particular, timely delivery and high quality of service directly affect the level of customer satisfaction. Improvement of logistics allows better response to consumer requests;
well-developed logistics strategies help reduce risks associated with delivery delays, product shortages or unforeseen situations;
efficient logistics can contribute to reducing the impact on the environment by optimizing transport routes and reducing resource consumption;
innovation and the adoption of new technologies such as automation, artificial intelligence and analytics that increase overall efficiency and productivity.

Therefore, the development of practical recommendations for increasing the efficiency of logistics activities is an important step to ensure the stable development and successful operation of corporate enterprises.

Thus, the integration development of logistics activities of corporate enterprises is a prerequisite for achieving high competitiveness, efficiency and sustainable development on the modern market. The adoption of these areas allows corporate enterprises to successfully adapt to new conditions, ensuring their stable development and growth.

5 Conclusions

The research shows that the integration development of logistics activities of corporate enterprises based on intellectualization and management technologies is extremely relevant in the modern world where globalization, digitalization and the rapid pace of technological development are radically changing conditions for business operations. In this environment, logistics is no longer just a process of moving goods, it is transforming into a complex system that includes supply chain management, resource optimization, process efficiency and, ultimately, the creation of added value for companies. It should be noted that the use of modern logistics management to manage logistics processes can save logistics costs equivalent to 1% and increase sales by 10%. Logistics is one of the most dynamic and significant sectors in the EU economy, accounting for approximately 10% of GDP [29].

It is determined that the integration development of logistics activities involves the coordination and interaction of various functional units of the enterprise, as well as the establishment of close relations with external partners and contractors. It requires the introduction of cutting-edge management technologies, such as ERP systems, analytical tools for data management, as well as integrated platforms ensuring the coordinated work of all participants in logistics processes.

The intellectualization of logistics implies the use of advanced technologies, such as artificial intelligence, machine learning, the Internet of Things (IoT) and big data, which can reduce logistics costs by up to 30%, while AI can increase the efficiency of logistics processes by up to 50% [30]. State-of-the-art technologies reduce supply risks at all stages of the logistics process, ensuring the quick response to any changes. These technologies can significantly improve planning processes, demand forecasting, inventory management, as well as optimization of routes and transportation costs. Besides, the intellectualization increases the flexibility and adaptability of logistics systems, which is critical in the face of volatile markets and changing consumer needs.

The described model of the process of organizing the management of integration development of the corporate enterprise reflects logistics activities of the innovation and integration process of interaction between economic entities during their consolidated development. It is determined that such logistics activities require further research in terms of revealing characteristics of the internal content of transitions for the success of the corporate integration development process.

Thus, the integration development of logistics activities based on intellectualization and modern management technologies not only increases the competitiveness of corporate enterprises but also creates new opportunities for their sustainable growth and development. This approach ensures the efficient use of resources, increased labor productivity, cost reduction and, most importantly, increased customer satisfaction. The success of enterprises in the current business environment largely depends on how efficiently they can integrate these advanced solutions into their operations and use them to create additional value.

References

1. Use of information and communication technologies at enterprises of Ukraine for 2021: statistical bulletin. State Statistics Service of Ukraine, https://www.ukrstat.gov.ua/menu/index_%20ukr_internat/iukr_inter_ish
2. Use of information and communication technologies at enterprises of Ukraine for 2022: statistical bulletin. State Statistics Service of Ukraine, https://www.ukrstat.gov.ua/menu/index_%20ukr_internat/iukr_inter_ish.htm
3. Furubotn, E., Richter, R.: Institutions and Economic Theory: Achievements of the New Institutional Economic Theory. The University of Michigan Press Manufactured (2005)
4. Aliekperov, A.: Creating Business and Corporate Strategy An Integrated Strategic System. Routledge Focus (2021)
5. Berle, A., Means, G.: The Modern Corporation and Private Property. N. Y.: Transaction Publishers (1932)
6. Galbraith, J.: The New Industrial State. Princeton University Press (2007)
7. Kurushina, E., Petrov, M.: Criteria for the success of spatial development projects based on interregional integration. Reg. Econ. **14**(1), 176–189 (2018)
8. Ivanov, Y., Pylypenko, A.: Integrative development of business entities: theoretical justification and management organization: monograph.VD "INZHEK", Kharkiv (2012)
9. Arefieva, O., Vasyutkina, N.: Corporate management: evolution, formation, development. Lira-K, Kyiv (2013)
10. Lysyuk, V., Topalova I.: Anitsa Hunet. Reproducible principle of market logistics theory formation. Economic innovations: coll. of science works Odesa: IPREED of the National Academy of Sciences of Ukraine, 67, 119–129 (2018)
11. Martin, Ch.: Logistics and Supply Chain Management: Creating Value-adding Networks. Pearson Education Limited (2005)
12. Schwab, K.: The Four Industrial Revolution. Cologny (2016)
13. Brynjolfsson, E., McAfee, A.: The Second Machine Age: Work, Progress, and Prosperity in a Time of Brilliant Technologies Reprint Edition. W. W. Norton & Company; Reprint edition (2016)
14. Logistics Performance Index. The World Bank, https://lpi.worldbank.org/international/global
15. Ughrin, V., Bondarenko, E., Shevchuk, A.: Effectiveness of the work of the customs authorities of Ukraine in the conditions of martial law. Econ. Anal. **32**(3), 171–182 (2022). https://doi.org/10.35774/econa2022.03.171

16. Hutsaliuk, O., Bondar, I., Kotsiurba, O., Ostapenko, O., Skoptsov, K., Boiko, O.: Factor-criteria assessment of greening prerequisites for transport infrastructure development in Ukraine. IOP Conf. Ser. Earth Environ. Sci. **1126**(1), 012009 (2023). https://doi.org/10.1088/1755-1315/1126/1/012009
17. Kwilinski, A., Hnatyshyn, L., Prokopyshyn, O., Trushkina, N.: Managing the Logistic Activities of Agricultural Enterprises under Conditions of Digital Economy. Virtual Econ. **5**(2), 43–70 (2022). https://doi.org/10.34021/ve.2022.05.02(3)
18. Gupta, A.: Blockcahin in Supply Chain Market Research Report. Market Research Future (2023). https://www.marketresearchfuture.com/reports/blockchain-supply-chain-market-6702
19. Internet of Things (IoT) in Logistics Market to Grow at a CAGR of 14.0%, during Forecast Period. BlueWeave Consulting, https://www.globenewswire.com/en/news-release/2022/10/06/2529734/0/en/Internetof-Things-IoT-in-Logistics-Market-to-Grow-at-a-CAGR-of-14-0-during-ForecastPeriod-BlueWeave-Consulting.html
20. The 2022 MHI Annual Industry Report. MHI (2022). https://www.mhi.org/publications/report
21. Fut, N., Hensley, D., Landsberg, M., Morrison. R.: For more on the role of the corporate center, see: The Role of the Corporate Center. McKinsey Bull. **1**(3) (2003). https://www.management.com.ua/strategy/str067-3.html
22. Barad, M.: Petri nets - a universal modeling model. Appl. Math. **7**(9), 829–839 (2016). https://doi.org/10.4236/am.2016.79074
23. James, L. Peterson: Petri Net Theory and the Modeling of Systems (2019)
24. Ya-Fen, Kuo, Lin, Yi-M., Chien, H.: *Corporate* social responsibility, *enterprise* risk *management*, and real earnings *management*: Evidence from managerial confidence. Finance Res. Lett. 101805 (2020). https://doi.org/10.1016/j.frl.2020.101805
25. Zaloznova, Y., Trushkina, N.: Management of logistic activities as a mechanism for providing sustainable development of enterprises in the digital economy. Virtual Econ. **2**(1), 63–80 (2019). https://doi.org/10.34021/ve.2019.02.01(4)
26. Trushkina, N.: Improvement of the organizational-and-economic mechanism of management logistic activity of enterprise. Agricul. Resour. Econ. **5**(4), 156–172 (2019). https://doi.org/10.51599/are
27. Hutsaliuk, O., Levchenko, A., Storozhuk, O., Zalevskyi, A., Doroshenko, T., Hryhorash, S.: Directions for increasing the level of environmental friendliness of innovative and investment attractiveness of transport and logistics companies. IOP Conf. Ser. Earth Environ. Sci. **1126**(1), 012028 (2023). https://doi.org/10.1088/1755-1315/1126/1/012028
28. Hutsaliuk, O., Bondar, I., Doroshenko, T., Zarubina, A., Onoyko, Y., Semeniuk, L.: Ecological Air Tourism in the System of Environmental Engineering. IOP Conf. Ser. Earth Environ. Sci. **1269**, 012031 (2023). https://doi.org/10.1088/1755-1315/1269/1/012031
29. Statistical Office of the European Union (Eurostat), http://eurostatica.com/services/product/ua-eu-prod.php.
30. The Challenges of Logistics Digitalization. DFreight, https://dfreight.org/blog/the-challenges-of-logisticsdigitalization (2022)

Intelligent Air Transport Technologies Resources Network Planning in Regards with the Individuals' Preferences Functions Entropy Conditional Optimization

Andriy Goncharenko(✉) and Viktor Iliushyn

National Aviation University, Liubomyra Huzara Avenue 1, Kyiv 03058, Ukraine
andrii.honcharenko@npp.nau.edu.ua

Abstract. The paper is dedicated to the problem of the network planning in the aspect of an optimal time ordering. The traditional Johnson's algorithm is modified with the individuals' subjective preferences functions of the available alternatives. It is made an attempt to apply the theory of the subjective entropy conditional optimization to the intelligent development. The simplest case with a two-by-two air transport technologies time resources matrix is considered. In order to evaluate the impact of the subjective preferences upon the optimal time ordering it is proposed to apply a nonlinear model. The time variation occurs depending upon the preference function value belonging to an acceptably wide diapason of the alternatives' preferences. The conducted computer simulation in the view of the numerical experimentations demonstrates the plausible scenarios of the technological decompositions, rearrangements, and recombination. The necessary diagrams are plotted. The illustrated approach allows choosing the optimal resources recombination (decomposition) for minimal time depending upon the preferences in the explicit view.

Keywords: Intelligent Transport · Transport Technologies · Air Transport · Resources Recombination · Network Planning · Subjective Entropy · Individual Preferences · Conditional Optimization · Multimodal Transportation · Routing · Transnational Routes · Decomposition · Uncertainty · Time Ordering · Scenario

1 Introduction

1.1 Literature Survey

Intelligent transport systems are indispensable in the contemporary spur developing societies.

In fact, subjective entropy of individuals' preferences [1] plays a serious role in air transport technologies implementation. Theoretically, subjective analysis [1] follows mathematical concepts of the famous Jaynes' principle [2–4] however directed into the sphere of psychology. As for the social component it is also combined with economical issues mentioned in books [5, 6].

O. Slavinska et al. (Eds.): ITSESQC 2024, LNNS 1335, pp. 287–295, 2025.
https://doi.org/10.1007/978-3-031-87376-8_25

It is without any doubt such approaches are applicable to air transport technologies [7] in particular, as well as to their general ideas [8–11] and transport systems [12–17] and their elements with the adjacent problematic [18–27].

1.2 Problem Setting

In the problems of the network planning with the determined values the solution to the time ordering problem could be found with the use of the Johnson's algorithm in the case of two operational chains of technological links (machines) according to the provisions by book [6].

Here it is proposed to take into consideration the element of the uncertainty in the view of the individuals' subjective preferences functions of the available alternatives introduced in monograph [1].

2 Theoretical Developments

In the general concept of subjective analysis [1], the objective functional is constructed in the following form of

$$\Phi_\pi = \alpha H_\pi + \beta\varepsilon + \gamma \mathrm{N}, \tag{1}$$

where Φ_π – the objective functional which takes into account the cognitive features of the subjective individuals' preferences functions π; α – the cognitive parameter of the individuals' psych (cognitive social attitude to the air transport management system, for example) in regards with the entropy H_π (subjective entropy) of the subjective preferences functions π of the achievable air transport managerial alternatives; β – the other cognitive parameter of the social mind-set with the correspondingly subjectively estimated value of ε; ε – subjective function of efficiency; γ – one more cognitive parameter of the subjective psych concerning the normalizing condition of N for the subjective preferences functions of π.

Having reduced the objective functional of (1) for one of the cognitive parameters of α, β or γ, it is possible to get a more compact view of the objective functional (1).

For instance, reducing the objective functional of (1) for the cognitive parameter of α, one can get

$$\Phi_\pi = H_\pi + \beta\varepsilon + \gamma \mathrm{N}. \tag{2}$$

Surely, in notation of (2), the rest of the cognitive parameters of β and γ have a relative meaning with respect to α, compared to the notation of the objective functional of (1).

Subjective entropy can regularly be interpreted in the traditional Shannon's type form:

$$H_\pi = -\sum_{i=1}^{N} \pi_i \ln \pi_i. \tag{3}$$

The functional of (3) of the subjective entropy H_π of the individuals' preferences functions π_i is constructed with respect to the set of the attainable alternatives.

The number of the alternatives is N.

Then, the problem is formulated in the view of the following statement:

Taking into account the conditional optimum of the distribution of the individuals' preferences functions π_i in the framework of the conditional extremization of the subjective entropy functional H_π subject to the constraints of the objective functionals of (1) or (2), find the rational air transportation technologies resources elements recombination (decomposition).

In order to evaluate the impact of the subjective preferences upon the optimal time ordering it is proposed to apply the following model:

$$T(\pi) = h[(\pi - \pi_{\min})(\pi - \pi_{\max}) - Exx], \tag{4}$$

where $T(\pi)$ – the individuals' subjectively dependable time intervals for the air transport technological elements (resources) realization; h – the augmentation coefficient; $\pi_{\min}$ and $\pi_{\max}$ – the border values of the individuals' subjective preferences functions in case of their own minimal and maximal magnitudes taken into account for a specified problem setting conditions or circumstances; *Exx* – the allowance parameter calibrating the deviations of the operational air transport technologies resources time diapasons of $T(\pi)$.

As for the element of *Exx* figuring in formula (4) it is expedient to implement an explicit view expression:

$$Exx = -ex + \left(\frac{\pi_{\max} + \pi_{\min}}{2} - \pi_{\min}\right)\left(\frac{\pi_{\max} + \pi_{\min}}{2} - \pi_{\max}\right), \tag{5}$$

where ex – the parameter related to the *Exx* values.

One more complicated case will be when some extra coefficients are introduced into the dependencies of (4) and (5).

3 Computer Simulations

In the simplest form of

$$UC = \begin{pmatrix} 40 & 30 \\ 20 & 15 \end{pmatrix}, \tag{6}$$

where UC – the operational air transport technologies resources time matrix.

The vector-columns in matrix (6) are:

$$B_1 = UC^{\langle 0 \rangle} = \begin{pmatrix} 40 \\ 20 \end{pmatrix}, \tag{7}$$

where B_1 – the vector-column pertaining to the fist operational air transport technologies resource (aircraft, board etc.) times of the matrix column of $UC^{\langle 0 \rangle}$ of (6).

$$B_2 = UC^{\langle 1 \rangle} = \begin{pmatrix} 30 \\ 15 \end{pmatrix}, \tag{8}$$

where B_2 – the vector-column pertaining to the second operational air transport technologies resource (aircraft, board etc.) times of the matrix column of $UC^{\langle 1 \rangle}$ of (6).

Correspondingly,

$$U_1 = B_{1_0} = 40, \tag{9}$$

where U_1 – the time spent for the first element (aircraft, board etc.) in operation of U (unloading the air cargo, disembarkation of passengers etc.) i.e. the first element of the first vector-column (7) of the matrix column of $UC^{\langle 0 \rangle}$ of (6).

$$C_1 = B_{1_1} = 20, \tag{10}$$

where C_1 – the time spent for the first element (aircraft, board etc.) in operation of C (customs office operations for the unloaded air cargo, passport control procedures for the disembarked passengers etc.) i.e. the second element of the fist vector-column (7) of the matrix column of $UC^{\langle 0 \rangle}$ of (6).

Analogously,

$$U_2 = B_{2_0} = 30, \tag{11}$$

where U_2 – the time spent for the second element (aircraft, board etc.) in operation of U (unloading the air cargo, disembarkation of passengers etc.) i.e. the first element of the second vector-column (8) of the matrix column of $UC^{\langle 1 \rangle}$ of (6).

$$C_2 = B_{2_1} = 15, \tag{12}$$

where C_2 – the time spent for the second element (aircraft, board etc.) in operation of C (customs office operations for the unloaded air cargo, passport control procedures for the disembarked passengers etc.) i.e. the second element of the second vector-column (8) of the matrix column of $UC^{\langle 0 \rangle}$ of (6).

Total times in two scenarios of (6)–(12) will count up to

$$T_1 = U_1 + C_1 + (U_2 - C_1) + C_2 = 85. \tag{13}$$

And optionally,

$$T_2 = U_2 + C_2 + (U_1 - C_2) + C_1 = 90. \tag{14}$$

The variant of (13) on conditions of (6)–(12) is optimal (shorter, better than the combination of (14) duration).

The idle times will be

$$DT_1 = U_1 + (U_2 - C_1) = 50 \tag{15}$$

in the first scenario. And

$$DT_2 = U_2 + (U_1 - C_2) = 55 \tag{16}$$

in the second scenario.

The situation can be drastically changed if taking into consideration the preference dependable times, likewise in (1)–(5).

That means that instead of (6)

$$UC = \begin{pmatrix} U_1 & U_2 \\ C_1 & C_2 \end{pmatrix} = \begin{pmatrix} 40 & 30 \\ 20 & 15 \end{pmatrix}, \tag{17}$$

one can take into account

$$UC(\pi) = \begin{pmatrix} U_1 & U_2 \\ C_1 & C_2 = T(\pi) \end{pmatrix} = \begin{pmatrix} 40 & 30 \\ 20 & T(\pi) \end{pmatrix}. \tag{18}$$

Accordingly to (4) and (5), and using

$$ex = 0.05 \text{ and } h = 200 \tag{19}$$

it is possible to get the result illustrated in the diagram (see Fig. 1).

In Fig. 1 it is depicted Prf for the individuals' subjective preferences function: π_i obtained in the conditional optimization procedure of (1)–(3); T(Prf) stands for the individuals' subjectively dependable time intervals for the air transport technological elements (resources) realization: $T(\pi)$ as the function of preferences π by (4) with (5) and (19).

In such circumstances (see Fig. 1) the total times of the entire air technological resources operations need a recalculation and after that it may require the resources recombination.

Indeed,

$$T_1(\pi) = U_1 + C_1 + (U_2 - C_1) + T(\pi). \tag{20}$$

The time becomes dependable upon preferences.

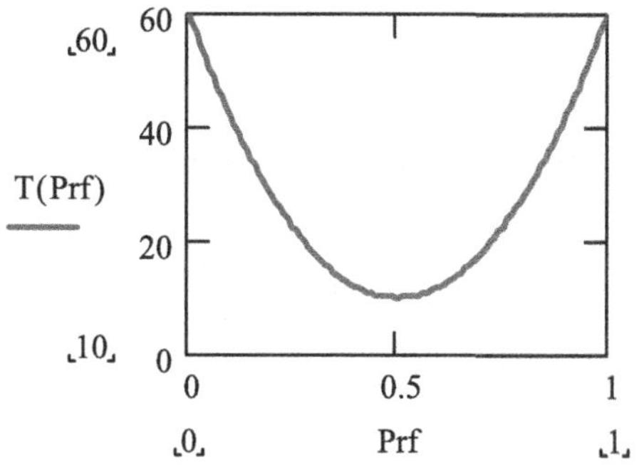

Fig. 1. A preference dependable time.

The results are shown in Fig. 2.

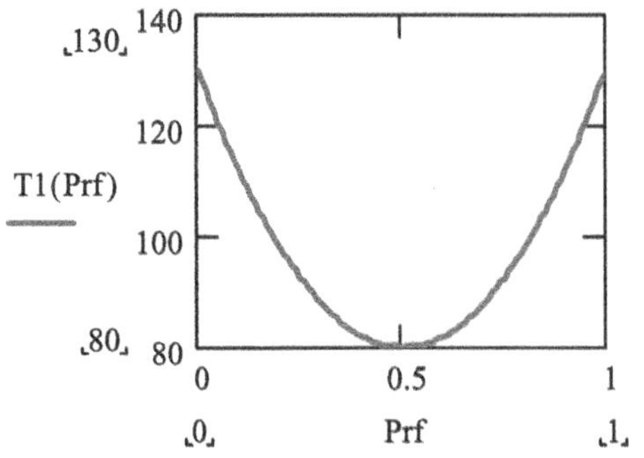

Fig. 2. The entire operation time depending upon preference in the first scenario.

In Fig. 2 it is depicted T1(Prf) for the individuals' subjectively dependable time intervals for the air transport technological elements (resources) realization by the first scenario: $T_1(\pi)$ as the function of preferences π by (20).

As for the second scenario the picture will be as follows:

$$T_2(\pi) = \begin{vmatrix} U_2 + T(\pi) + (U_1 - T(\pi)) + C_1 & \text{if} \quad T(\pi) \le U_1; \\ U_2 + T(\pi) + 0 \cdot (U_1 - T(\pi)) + C_1 & \text{otherwise.} \end{vmatrix} \tag{21}$$

Therefore, it results in the diagram shown in Fig. 3.

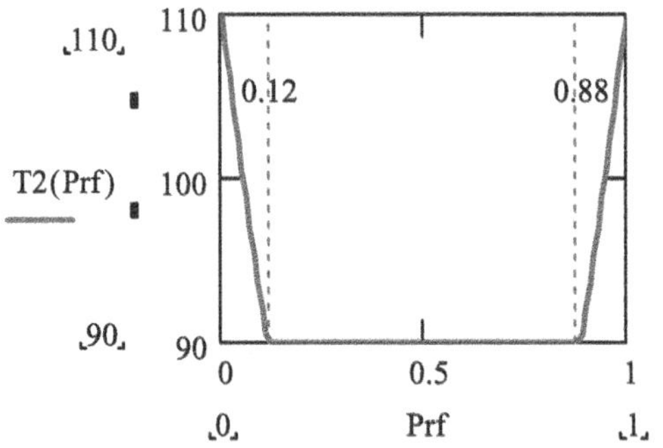

Fig. 3. The entire operation time depending upon preference in the second scenario.

In Fig. 3 it is depicted T2(Prf) for the individuals' subjectively dependable time intervals for the air transport technological elements (resources) realization by the second scenario: $T_2(\pi)$ as the function of preferences π by (21).

4 Analysis and Discussion

Comparing two scenarios' results, it is possible to analyze the conditions for a rational recombination (decomposition).

Plotting both diagrams of the first and second scenarios: $T_1(\pi)$ and $T_2(\pi)$ as the function of preferences π by (20) and (21), one gets the following picture.

The results are visible in Fig. 4.

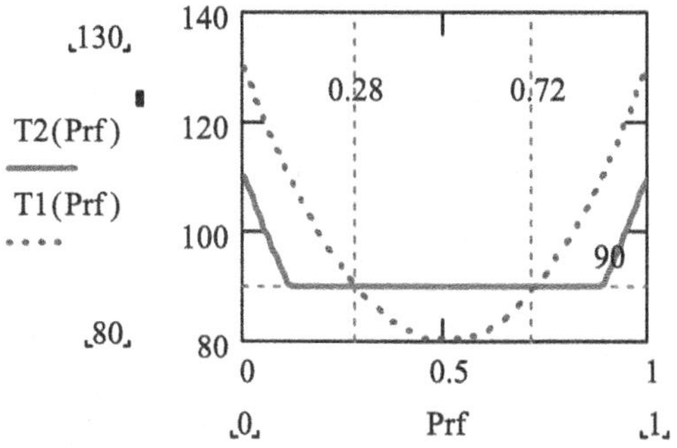

Fig. 4. The combination of the entire operation time depending upon preference.

In Fig. 4 it is demonstrated that in the intervals of preferences $\pi = 0 \ldots 0.28$ and $\pi = 0.72 \ldots 1$ it is better to realize the second scenario; whereas $\pi = 0.28 \ldots 0.72$ is optimal for first scenario.

5 Conclusion

It is discovered how the individuals' subjective preferences influence the time for the air transport technological elements (resources) realization. Optimal resources recombination (decomposition) for minimal time depends upon the preferences in the explicit view. It is a significant component for the air transport technologies improvements. The implications of such findings will benefit to the airline operation.

In further research it is promising to take into account various factors and the dynamics of the operational air transport technologies processes.

References

1. Kasianov, V.: Subjective Entropy of Preferences Subjective Analysis. Institute of Aviation Scientific Publications, Warsaw, Poland (2013)
2. Jaynes, E.T.: Information theory and statistical mechanics. Phys. Rev. **106**(4), 620–630 (1957)
3. Jaynes, E.T.: Information theory and statistical mechanics. II. Physical review. **108**(2), 171–190 (1957)
4. Jaynes, E.T.: On the rationale of maximum-entropy methods. Proc. IEEE **70**, 939–952 (1982)
5. Silberberg, E., Suen, W.: The Structure of Economics. A Mathematical Analysis. McGraw-Hill Higher Education, New York (2001)

6. Taha, H.: Operations Research: An Introduction, 8th edn. Prentice Hall, Upper Saddle River, NJ (2007)
7. Ivannikova, V., Shenchuk, D., Konovalyuk, V., Borets, I., Vysotska, I.: Estimation of the innovative technologies influence on passengers processing procedures at the airport. Transp. Res. Procedia **59**, 127–136 (2022)
8. Goncharenko, A., Iliushyn, V.: Rational air transportation technologies resources recombination on condition of the generalized values. 7th International Workshop on Computer Modeling and Intelligent Systems. CMIS-2024, 86–96 (2024). https://ceur-ws.org/Vol-3702/paper8.pdf
9. Goncharenko, A., Teterin, S.: Computer modeling of discrete systems in the case of linear and non-linear restrictions on the optimal speed of the aircraft based on the Lagrange method. 7th International Workshop on Computer Modeling and Intelligent Systems. CMIS-2024, 157–168 (2024). https://ceur-ws.org/Vol-3702/paper13.pdf
10. Goncharenko, A.: Multi-optional hybridization for UAV maintenance purposes. 5th International IEEE Conference on Actual Problems of Unmanned Aerial Vehicles Developments. APUAVD-2019, 48–51 (2019). https://doi.org/10.1109/APUAVD47061.2019.8943902
11. Goncharenko, A.: An entropy model of the aircraft gas turbine engine blades restoration method choice. In: International Conference on Advanced Computer Information Technologies. ACIT-2018, 2–5 (2018)
12. Kunytska, O., Persia, L., Gruenwald, N., Datsenko, D., Zakrzewska, M.: The Sustainable and Smart Mobility Strategy: Country Comparative Overview. Smart Technologies in Urban Engineering. STUE 2022. Lecture Notes in Networks and Systems, 536. Springer, Cham (2023). https://doi.org/10.1007/978-3-031-20141-7_59
13. Danchuk, V., Danchuk, M.: Synergetic model for sustainability of enterprise personnel management in conditions of increased risk of entrepreneurial activity. Stud. Syst. Decis. Control. **545**, 957–967 (2024)
14. Taran, I., Bikhimova, G., Danchuk, V., Tursymbekova, Z., Oliskevych, M.: Improving the methodology for optimizing multimodal transportation delivery routes and cyclic schedules in a transnational direction. Transp. Probl. **19**(1), 157–170 (2024)
15. Olivari, E., Gurrì, S., Caballini, C., Carotta, T., Chiara, B.D.: Ports go green: a cost-energy analysis applied to a case study on evaluating the electrification of yard tractors. Open Transp. J. **18**(1) (2024)
16. Iwan, S., Wagner, N., Kijewska, K., Jensen, S.A.: Concept of the knowledge-based city logistics: problems and solutions. PLoS ONE **19**(6), e0305563 (2024). https://doi.org/10.1371/journal.pone.0305563
17. Comi, A., Idone, I.: The use of electric vehicles to support the needs of the electricity grid: a systematic literature review. Appl. Sci. **14**(18), 8197 (2024). https://doi.org/10.3390/app14188197
18. Comi, A., Gruenwald, N., Danchuk, V., Kunytska, O., Vakulenko, K., Zakrzewska, M.: How Cities and Universities Approach the Sustainable Development Goals. Handbook of Sustainability Science in the Future. Springer, Cham (2023). https://doi.org/10.1007/978-3-030-68074-9_155-1
19. Pavăl, F.F.; Bricicaru, I.: Improving road safety on the national road network in Romania through pilot solutions. J. Eng. Sci. **30**(2), 59–71 (2023). https://doi.org/10.52326/jes.utm.2023.30(2).05
20. Ionaşcu, I., Ionaşcu, M., Nechita, E., Săcărin, M., Minu, M.: Digital transformation, financial performance and sustainability: evidence for European Union listed companies. Amfiteatru Econ. **24**(59), 94–109 (2022). https://doi.org/10.24818/EA/2022/59/94
21. Kiermayer, M.: Weiß, C,: Grouping of contracts in insurance using neural networks. Scand. Actuar. J., Taylor Fr. J. **2021**(4), 295–322 (2021)

22. Steffens, K.G.: The History of Approximation Theory: From Euler to Bernstein. Birkhäuser Boston, MA (2007)
23. Alonso, F., Faus, M., Tormo, M.T., Useche, S.A.: Could technology and intelligent transport systems help improve mobility in an emerging country? Challenges, opportunities, gaps and other evidence from the caribbean. Appl. Sci. **12**(9), 4759 (2022). https://doi.org/10.3390/app12094759
24. Alonso, F., Faus, M., Esteban, C., Useche, S.A.: Who wants to change their transport habits to help reduce air pollution? A nationwide study in the caribbean. J. Transp. Health **33**, 101703 (2023)
25. Ballarin, P., Airoldi, A., Aceti, P., et al.: Experimental identification of frictional effects on interlaminar toughness of composite laminates in 4ENF test. Exp. Mech. **62**, 1135–1145 (2022). https://doi.org/10.1007/s11340-022-00860-8
26. Gunaydin, K., Türkmen, H.S., Airoldi, A., Grasso, M., Sala, G., Grande, A.M.: Compression behavior of EBM printed auxetic chiral structures. Materials **15**(4), 1520 (2022). https://doi.org/10.3390/ma15041520
27. Gorbachev, S., et al.: Distributed secure consensus control of nonlinear multi-agent systems under sensor and actuator attacks. J. Franklin Inst. **360**(11), 7501–7534 (2023)

Unmanned Aerial Vehicles in Ukraine: Modernity, Challenges, and Development Prospects

Vitaliy Zaytsev(✉)

Alfred Nobel University, Sicheslavska Naberezhna Street, 18, Dnipro 49000, Ukraine
vitaliy.zaytsev@duan.edu.ua

Abstract. The development of unmanned aerial vehicles (UAVs) and systems in Ukraine was one of the key industries with a large historical legacy and significant potential for further growth. Analyzing the historical aspects, it can be noted that Ukraine has deep roots in the aviation industry, which serves as a basis for modern UAV developments. Aspects of the UAV development in Ukraine and their role in the military operations of the war between Ukraine and the Russian federation are considered. Modern UAVs are an important element in military operations and play a significant role in ensuring safety and comfort in civilian applications. The use of UAVs for infrastructure monitoring, environmental control, goods delivery, and services increases the quality of life and contributes to urban infrastructure development. An important component of success is the role of education and science, which ensure highly qualified personnel training and contribute to the development of the technologies. Scientific research and international cooperation open new horizons for Ukraine's integration into the world scientific community and the UAV market. The issue of certification is critical for ensuring safety, compliance with international standards, and entering global markets. The UAV development in Ukraine has all the prerequisites for successful growth – to realize this potential, active support from the state, development of educational and scientific bases, regulatory policy improvement, and integration into the international market are necessary. This will allow Ukraine to take a worthy place among the countries in UAVs and contribute to innovation in global processes.

Keywords: Unmanned Aerial Vehicles · Drones · Technological Development

1 Introduction

Unmanned aerial vehicles (UAVs) and other unmanned systems play an important role in various industries, including defense, civil infrastructure, environmental monitoring, agriculture, etc. In everyday life, a UAV is often called a «drone» [1]. The variety of types of UAVs allows for wide opportunities for their use in various human activity spheres [2]. In recent years, Ukraine has become one of the leaders in developing and implementing these technologies due to internal needs and external challenges. The development of

O. Slavinska et al. (Eds.): ITSESQC 2024, LNNS 1335, pp. 296–307, 2025.
https://doi.org/10.1007/978-3-031-87376-8_26

unmanned systems in Ukraine is an important aspect of modern technological achievements [3]. Below we will consider aspects of the development of UAVs in Ukraine, their role in military operations, technological prospects, as well as challenges and prospects that the country faces on this path.

2 Historical Aspects of the Ukrainian UAVs Development

The history of the development of unmanned aerial vehicles in Ukraine is multi-layered and closely related to the general history of the country's aviation and space industry. Ukraine was one of the aviation and space industry-leading centers in the USSR.

After Ukraine gained independence in 1991, the development of unmanned technologies in the country faced significant difficulties. The economic crisis, the Soviet industrial infrastructure collapse, and the reduction of government funding for scientific research significantly slowed down progress in this area.

However Ukrainian engineers and scientists continued to work on the preservation and development of acquired knowledge and technologies. One of the first significant independent Ukraine projects in the UAVs was the creation of unmanned aerial vehicles for civilian use, in particular for aerial photography and agro monitoring.

2014 was a turning point in the UAV development. In response to new challenges, companies, and scientific institutes began to actively develop combat drones that could effectively perform tasks in war conditions. In particular, much attention was paid to the creation of UAVs various types for reconnaissance and adjusting artillery fire.

Against the background of military operations, numerous volunteer initiatives appeared that contributed to the UAV's development. Volunteers, engineers, and programmers joined forces to develop and improve UAVs of various models and types, which are used at the front to ensure the Ukrainian force's tactical advantage. This, in turn, contributed to the further development of the Ukrainian defense industry and the involvement of foreign partners in cooperation.

To date, the UAV development in Ukraine has reached a significant level. Ukrainian enterprises and scientific institutions are actively working on new UAV projects. Ukraine participates in international projects and programs that contribute to the integration of its technologies into global markets (the world UAV market is more than 130 billion US dollars).

The history of the development of UAVs in Ukraine is evidence of a constant desire for innovation and technological progress. Despite numerous challenges, Ukrainian engineers and scientists demonstrate a high capacity for adaptation and development, which opens up new prospects for the further growth of this important industry.

3 UAVs Role in Military Operations in Ukraine

The role of UAVs in military operations has grown significantly in recent years, and their importance cannot be overestimated [4]. UAVs have become an integral part of modern warfare, providing the ability to effectively perform a variety of military tasks, including reconnaissance, surveillance, adjusting artillery fire, as well as directly participating in combat operations as strike platforms. The war unleashed by the russian federation in Ukraine became a catalyst for the widespread use and development of unmanned technologies in the Armed Forces of Ukraine (AFU).

One of the key tasks performed by UAVs is reconnaissance and surveillance. Thanks to the ability to stay in the air for a long time, drones can quickly provide the command with up-to-date information about the situation on the battlefield, the movement of enemy forces, and the location of military facilities. This allows for a timely response to changes in the situation and making strategic decisions. In particular, the Ukrainian military actively uses drones to monitor the front line, identify enemy positions, and evaluate the results of shelling.

In addition, UAVs play an important role in adjusting artillery fire. With the help of cameras and other sensors mounted on drones, the military can more accurately target artillery strikes, which significantly increases the effectiveness of fire and reduces casualties among the civilian population. The use of drones to adjust fire allows to reduce the consumption of ammunition and provides higher accuracy in hitting targets, which is especially important in the conditions of modern wars.

The use of attack UAVs has become another important aspect of modern military operations. FPV drones are capable of directly attacking enemy objects, under the control of the operator and at high speed attacking the enemy equipment and manpower.

Unmanned aerial vehicles are also used to deliver ammunition, medicine, food, and other necessary resources to hard-to-reach areas or directly to the front line. This significantly increases the mobility of the troops and allows them to be supplied with the necessary materials even in difficult combat conditions.

The war stimulated the development of counter-drone technologies. The increase in the use of UAVs forced the Ukrainian military and scientists to work on creating means to counter enemy drones. This includes both the physical destruction of drones and the development of electronic warfare systems capable of jamming control signals or intercepting control of enemy UAVs. These technologies are an important element of modern defense, as they allow neutralizing threats originating from enemy drones.

Another important military operations aspect is their use of UAVs for psychological pressure on the enemy. Drones' continuous patrols over the war zone create a persistent sense of surveillance, which can lead to the demoralization of enemy forces. In addition, strike drones, capable of launching surprise attacks, increase the stress level in enemy forces, which can lead to disorganization and erroneous decisions on their part.

The Ukrainian defense industry actively cooperates with foreign partners to exchange technologies and expand the capabilities of Ukrainian UAVs. This cooperation makes it possible to use advanced achievements of the world market and integrate them into domestic developments. Ukrainian companies are creating artificial intelligence systems that enable drones to autonomously perform complex tasks on the battlefield.

Also, the quick response of the industry to the requirements of the military launched a system of research work in manufacturing companies and constant updating and improvement of samples of unmanned military equipment. The experience of Ukraine is adopted by the world's leading companies – manufacturers of unmanned equipment. Ukraine first created the AFU's Forces of Unmanned Systems in 2024.

4 Technological Prospects of UAV Development in Ukraine

The development of UAVs in Ukraine shows significant potential and opens up wide prospects for both military and civil applications. Ukraine, having rich experience in the aerospace industry, continues to actively develop its technological capabilities, in particular in the direction of creating innovative solutions for UAVs.

One of the main directions of technological development of UAVs in Ukraine is increasing the autonomy and intellectual drone capabilities. The use of artificial intelligence (AI) and machine learning (ML) allows drones to perform complex tasks without operator intervention, which is especially important in combat situations or when performing critical civilian tasks [5]. For example, the development of algorithms for the automatic recognition of objects on the battlefield or for autonomous decision-making during missions is a priority area of research.

In addition, significant efforts are aimed at improving UAV navigation systems. Ukraine is actively developing technologies that allow drones to navigate in space with high accuracy even in difficult conditions, such as urban areas or areas with a weak GPS signal. This includes the use of combined navigation methods combining data from various sensors, such as inertial systems, cameras, and LIDARs, which allows drones to operate autonomously even in difficult conditions [6].

Another important direction is the development of communication technologies for UAVs. Ukrainian developers are working on the creation of reliable data transmission systems, and CRPA antennas that provide stable communication between the drone and the operator over long distances or in conditions of radio-electronic warfare. The advanced communication protocols and data encryption use allow to reduce the drone control interception risks, which is critically important in military operations.

Improving the aerodynamics and structures of UAVs is another important area of development. Engineers are working on design solutions for the devices, which will allow them to reduce their weight and, accordingly, increase the duration and height of the flight. New designs of wings and propellers are also being developed, which provide greater maneuverability and stability in conditions of turbulence or strong winds.

An equally important aspect is the expansion of UAV functionality due to the integration of various sensors and modules [7]. Modern drones can be equipped with high-resolution cameras, thermal imagers, radar stations, as well as systems for dropping ammunition or delivering cargo. This allows them to be used for a wide range of tasks, from military operations to scientific research, for example, in ecology or geology. UAVs must be able to connect to other devices and systems, such as sensors, mobile devices, or network infrastructures [8]. This will ensure the collection, transfer, and analysis of data in real-time. However, it is also necessary to ensure the protection of data transmitted through IoT networks from cyber threats [9]. This includes data encryption and access control [10, 11]. The use of UAVs to collect large amounts of data, such as images, videos, and indicators of the state of infrastructure, and further process them for decision-making. At the same time, the use of analytical tools to predict possible infrastructure malfunctions or predict maintenance needs based on data collected by UAVs will allow increasing the efficiency ones in the new infrastructure.

Another promising field is the creation of swarms of drones [12], that is, the coordination of the work of a large number of drones that can act as a single organism.

This allows complex operations, such as large-scale reconnaissance missions, attacks on enemy targets, or search and rescue operations, with increased efficiency and minimal losses. In Ukraine, research is being conducted in this area, in particular, with the use of artificial intelligence algorithms to coordinate the actions of a swarm of drones.

5 Use of UAVs to Ensure the Safety and Comfort of Civil

UAVs are about dual-use technologies. They play a significant role not only in ensuring safety but also in increasing the comfort of the population. Their use makes it possible to significantly increase the efficiency of task performance, reduce risks to human life, and provide quick access to critical information in real-time [13]. In Ukraine, these technologies have already found wide application in various fields, from national security and law enforcement to ensuring comfortable living in urban conditions.

One of the main areas of UAV use is national security and border protection [14]. The use of UAVs in wartime was mentioned above. Drones are used to patrol the state border, which allows the detection of violations, illegal movements, and smuggling. Drones, thanks to the possibility of staying in the air for a long time and being equipped with high-precision cameras and thermal imagers, can record the activity of violators even in hard-to-reach areas, where the location of stationary means of surveillance is difficult or impossible [15].

UAVs are used by law enforcement agencies to maintain public order counter crime, and document destruction during military operations or natural disasters [16]. They can be used during mass and sporting events, demonstrations, and other large gatherings to monitor the situation and identify potential threats. Thanks to the ability to quickly receive video information from the scene of events, law enforcement officers can quickly respond to emerging situations and prevent violations of order. This is especially important in the conditions of large cities, where control of the situation is difficult due to high population density and limited access to certain areas.

In rescue operations, UAVs have also found their important application [17]. They are used to search and rescue people in hard-to-reach areas, such as mountain ranges, forests, or areas affected by natural disasters. Thanks to thermal imaging equipment, UAVs can detect people even in the dark or under the debris of buildings, which increases the successful completion chances of the rescue operation.

UAVs are also actively used to control environmental safety, in particular, to monitor the state of the environment and identify environmental threats [18]. Thanks to the use of specialized sensors, drones can detect air, water, or soil pollution, which allows timely measures to eliminate negative effects on the environment.

In addition to ensuring safety, UAVs are also used to increase the comfort of the population, especially in the conditions of modern cities. One of the promising areas is the use of drones for the delivery of packages, mail [19], food, medicine, and other necessary goods. This is especially true in large cities with traffic jams or in emergencies where access to certain areas may be difficult.

UAVs are also used in urban planning and infrastructure management [20, 21]. Drones can monitor the condition of road infrastructure, bridges, tunnels, and other objects, detecting potential threats such as cracks in structures or damage to the road surface. This makes it possible to carry out repairs promptly and prevent accidents, which contributes to the improvement of the quality of life in the city.

GIS technologies today cannot be imagined without the support of UAVs [22].

In agriculture, UAVs are used to monitor the condition of crops, apply fertilizers, control pests, and protect plants. This makes it possible to significantly increase the efficiency of agricultural production, reduce costs, and increase productivity, which contributes to ensuring the country's food security and improving life quality.

Using UAVs to automate delivery processes in smart logistics or for warehouse inventory management. Integration of UAVs with other automated and robotic systems to perform complex tasks, such as control and management of industrial processes [23].

The use of drones in health care is also gaining importance. UAVs can be used to deliver medicines and medical equipment to hard-to-reach areas, which is especially important during pandemics or natural disasters.

In addition, drones can be used to organize entertainment events [24], such as light shows or aerial photography during festivals and other cultural events. This not only creates new entertainment opportunities but also increases the level of satisfaction of the population with life in cities.

Below are some examples of the integration of UAVs into transport technologies:

1. Unmanned air taxis [25].
2. Inspection of transport infrastructure.
3. Traffic monitoring [26, 27].
4. Environmental monitoring of transportation routes.
5. Logistics and cargo delivery. Intelligence missions for smart logistics
6. Emergency and rescue operations on the roads and railways.
7. Supporting infrastructure recovery.
8. Intelligent parking management system etc.

6 Challenges and Prospects for the UAVs Development

The development of unmanned systems in Ukraine is one of the key areas that requires significant efforts and resources, especially in the conditions of modern challenges and the geopolitical situation. Despite the significant progress achieved in recent years, this industry faces numerous challenges that need to be resolved to ensure the further development and strengthening of Ukraine's position on the world market.

One of the main challenges facing the development of unmanned systems in Ukraine is insufficient funding. The economic difficulties the country is experiencing limit opportunities for investment in R&D. The war and the restrictions associated with it complicate the opportunities for international cooperation and access to global markets.

Another challenge is the shortage of specialists. Although Ukraine has a high level of education and training of engineers and scientists, the unpopularity among applicants of technical specialties has led to the fact that few specialists are trained in Ukraine in aerospace engineering. By the way, this trend is also observed in the world – on average, the aircraft production market lacks up to 30% of design engineers and up to 20% of production engineers [28].

Many talented specialists go abroad in search of better working conditions. This also makes it difficult to provide a sufficient number of qualified personnel to implement innovative projects in UAVs.

Technological competition from other countries is also an important challenge. Ukraine competes with major international players who have significant financial resources, a technological base, and access to global markets. This puts Ukrainian companies in a less favorable position, especially on the world stage, where the requirements for products are becoming increasingly strict.

Ukraine faces cyber security issues that are relevant in the context of UAV development. Threats from hacker attacks, attempts to intercept control of drones or data theft are serious risks that can negatively affect the UAV's security and operation. These challenges require constant improvement of protection systems and the introduction of the latest cybersecurity technologies.

The development of international cooperation is promising. Ukraine already cooperates with many countries and international organizations in UAVs. Attracting foreign investments, and participation in international programs and initiatives contribute to Ukraine's integration into the UAV world market and allow us to compete on equal terms with leading players.

Another perspective is the AI and ML technologies development, which can significantly expand the UAV functionality. The integration of AI makes it possible to increase the autonomy of drones, improve their ability to make decisions in real time, and ensure the performance of complex tasks without operator intervention. This opens up new opportunities for the use of drones from military to civil.

No less important is the development of the domestic market of unmanned systems. Support from the government, in particular in the form of favorable regulatory policies, tax incentives, and subsidies for startups and enterprises engaged in UAV development, can stimulate industry growth.

Despite numerous challenges, the UAV development in Ukraine has enormous potential both in expanding the UAV market in various industries (such as agro, energy, logistics, transport, construction, environmental monitoring, etc.), and for the development of the market for services, such as related to UAVs (aerial photography, infrastructure monitoring, delivery of goods, entertainment and many other areas). The introduction of new business models and services based on unmanned technologies can become an additional source of income and contribute to the growth of the economy.

7 The Role of Education and Science in the UAV Development

Education and science play a fundamental role in the development of UAVs in Ukraine. They provide not only the training of highly qualified personnel but also contribute to the creation of new technologies that form the basis for innovation and the country's competitiveness in the world market [29]. Given the rapid pace of development in this field, the importance of education and science becomes even more obvious.

1. Teach engineers.

The teaching of qualified specialists is a key factor in the development of the industry of unmanned systems. Many universities in Ukraine teach engineers, and specialists in aviation technologies, robotics, information technologies, and other fields necessary for the creation and maintenance of UAVs.

2. Scientific research and innovation.

Scientific research is the driving force behind technological progress in UAVs. Ukrainian researchers are actively engaged in the development of new materials creation, control systems, energy-saving technologies, AI, autonomous systems, and swarms.

3. Cooperation with international scientific institutions.

Ukrainian universities and research institutes actively cooperate with foreign partners, participating in joint projects, and exchange of knowledge and experience. Such cooperation allows our scientists to have access to the latest technologies and research, which contributes to the integration of Ukraine into the world scientific community.

4. The influence of education and science on the commercialization of developments.

Universities and research institutes are often incubators of innovative ideas that can then be implemented in industry. Cooperation between research institutions, universities, and private companies allows for the rapid introduction of the latest technologies into production. This partnership is a key factor in ensuring Ukraine's technological leadership in UAVs.

5. Challenges and prospects for the development of scientific education.

Despite the significant potential, the development of education and science in UAVs faces many challenges. One of them is insufficient funding of scientific research and educational programs. This limits opportunities for introducing new technologies, modernizing the educational base, and conducting experimental research. An important problem is also the «brain drain» when young talented specialists leave abroad in search of better conditions for development.

Education and science are fundamental components of the successful UAV development in Ukraine. They provide training of highly qualified personnel, contribute to the creation of innovation, and also play a key role in the commercialization of scientific developments. Given the rapid industry development, supporting education and science is critical to ensuring Ukraine's long-term success in the UAVs.

8 Issues of Certification of UAVs and Systems

Separately, the issue of certification of UAVs and their systems, because this is a critically important aspect of the safe and effective integration of these technologies into the airspace, as well as for ensuring compliance with international standards. The certification process includes assessment of compliance with technical requirements, verification of reliability and safety, as well as obtaining official permits for operation. In Ukraine, the issues of certification of UAVs are becoming more and more relevant, as this industry continues to develop rapidly [30, 31].

1. The importance of certification for the market.

UAV certification is an important prerequisite for their commercial use. It provides confidence that the device meets all safety requirements and can be safely used in civil and military airspace. The lack of certification limits the opportunities for commercialization of products in foreign markets, which restrains the development of the industry.

2. Certification challenges and issues.

In Ukraine, the process of certification of unmanned systems is at the stage of development. In times of war, there is a simplified certification procedure for UAVs for military use. The lack of developed and approved national standards for unmanned systems that would meet international requirements is still a difficulty for UA's manufacturers.

3. The need to develop national standards and certification centers network.

Given these challenges, there is an urgent need in Ukraine to develop and implement national standards for UAV certification and develop a certification centers network. The development of standards requires the active participation of both government bodies and representatives of the industry, including manufacturers, scientific institutions, and experts. Solving this problem will create a basis for the sustainable development of the industry and allow Ukraine to take a worthy place among the UAV world leaders.

9 Conclusion

The development of unmanned aerial vehicles and systems in Ukraine is one of the key industries with significant potential for further growth and innovation. Analyzing the historical aspects, it can be noted that Ukraine has deep roots in the aviation industry, which serves as a reliable basis for modern UAV developments. Industry participants face numerous challenges such as funding, regulatory barriers, and staff shortages, but at the same time have great prospects due to the high scientific and technical potential of the country and the growing demand for these technologies in both domestic and international markets.

Modern UAVs are becoming not only an important element of military operations, but also play an increasingly important role in ensuring the safety and comfort of the population in civilian areas. The use of drones for infrastructure monitoring, environmental control, and delivery of goods and services improves the quality of life and contributes to the development of urban infrastructure. In Ukraine, this technology continues to develop, and every year there are new opportunities for its application in various fields, which contributes to increasing the level of safety and comfort of citizens' lives.

An important component of success is the role of education and science, which ensure the training of highly qualified personnel and contribute to the development of the latest technologies. Scientific research and international cooperation open new horizons for the integration of Ukraine into the world scientific community and the UAV market.

UAV certification is a key factor in ensuring the safety and competitiveness of Ukrainian developments in the domestic and foreign markets.

In conclusion, the UAV development in Ukraine has all the prerequisites for successful growth, but the realization of this potential requires active support from the state, development of the educational and scientific base, improvement of regulatory policy, and integration into the international market.

References

1. Knysh, B., Brovko, P., Popil D.: Classification of known types of unmanned aerial vehicles. Modern engineering and innovative technologies, **1**(2), 34–39 (2017). https://www.moderntechno.de/index.php/meit/article/view/g117-004
2. Tymochko, O. et al.: Classification of unmanned aerial vehicles. Weapon systems and military equipment **1**(9), 61–66 (2007). https://openarchive.nure.ua/server/api/core/bitstreams/878899d8-b7a7-4481-af22-9835c0748ba0/content
3. Kutovy, O.P.: Development trends of unmanned aerial vehicles. Sci. Weapons **4**, 39–47 (2014)
4. Bursala, O., Horoshko, D., Kulba, P., Chupryna, V.: Unmanned aerial vehicles of air combat. Collection of scientific works of the State Research Institute of Testing and Certification of Armaments and Military Equipment **1**(1), 50–57 (2019). https://dndivsovt.com/index.php/journal/article/view/184
5. Soori, M., Arezoo, B., Dastres, R.: Artificial intelligence, machine learning and deep learning in advanced robotics, a review. Cognitive Robot. **3**, 54–70 (2023). https://doi.org/10.1016/j.cogr.2023.04.001
6. Fesenko, H., et al.: Flying sensor and edge network-based advanced air mobility systems: reliability analysis and applications for urban monitoring. Drones **7**(7), 409 (2023)
7. Glotov, V., Tserklevych, A.: Analysis and perspectives of aerial photography from an unmanned aerial vehicle. Bulletin of the National University «Lviv Polytechnic». Ser.: Modern achievements of geodetic science and production **1**(27), 131–136 (2014)
8. Pasha, J., et al.: The drone scheduling problem: a systematic state-of-the-art review. IEEE Trans. Intel. Transp. Syst. **23**(9), 14224–14247 (2022). https://doi.org/10.1109/TITS.2022.3155072
9. Yazid, Y., Ez-Zazi, I., Guerrero-González, A., El Oualkadi, A., Arioua, M.: UAV-enabled mobile edge-computing for IoT-based on AI: a comprehensive review. Drones (2021). https://doi.org/10.3390/drones5040148
10. Yahuza, M., et al.: Internet of drone's security and privacy issues: taxonomy and open challenges. IEEE Access **9**, 57243–57270 (2021). https://doi.org/10.1109/ACCESS.20
11. Nguyen, T., Katila, R., Gia, T.N.: An advanced internet-of-drone system with blockchain for improving quality of service of search and rescue: a feasibility study. Future Generate. Comput. Syst. **140**, 36–52 (2023). https://doi.org/10.1016/j.future.2022.10.002
12. Abdelkader, M., Güler, S., Jaleel, H.: Shamma, JS: Aerial swarms: recent applications and challenges. Curr. Robot. Rep. **2**(3), 309–320 (2021). https://doi.org/10.1007/s43154-021-00063-4
13. Voronko, I., Voronko, V.: Fields of application of unmanned aerial vehicles. Current state of scientific research in IT technologies, electronics, engineering, nanotechnologies, and transport spheres: col. of science monograph. Vinnitsa **160** (2020). https://publishing.logos-science.com/index.php/books/issue/view/14/9
14. Zaytsev V.Ye.: The territory monitoring complex based on unmanned aerial vehicles // Aviation, industry, society: materials IV International science and practice conf. (Kremenchuk,

May 18, 2023). Ministry of Internal Affairs of Ukraine, Kharkiv national university of internal affairs, Kremenchuk's Flight College, Science Park «Science, and Safety». Kharkiv: KhNUVS. 87–90 (2023). https://dspace.univd.edu.ua/items/e288d376-8bcc-40ec-a46f-2cd2fb5706ba

15. Maladyka, I. et al.: Payload of the drone as a basis for substantiating the flight range. Theory and practice of fire extinguishing and liquidation of emergencies: Materials of the 12th International. Scientific practice conference, Cherkasy, January 8, 2021, 91–92 (2021)
16. Bilous, V.: Classification of unmanned aerial vehicles and its significance for forensic practice. Theory and practice of forensic examination and forensics, **16**, 47–57 (2016). https://doi.org/10.32353/khrife.2016.06
17. Mosov, S. et al.: Classification, functions, and tasks of unmanned aviation in the sphere of civil defense of Ukraine. Scientific bulletin: Civil defense and fire safety, **2**(12), 54–68 (2021). https://doi.org/10.33269/nvcz.2021.2(12).54-68
18. Yarovyi, O.: Control systems for unmanned aerial vehicles for monitoring ground objects. Control, navigation and communication systems. Collection of scientific papers, **3**(49), 33–38 (2018). https://doi.org/10.26906/sunz.2018.3.033
19. Zaytsev, V.Ye., Voronko, V.V., Voronko I.O.: Analysis of the issue of mail delivery by unmanned aerial vehicles. Integration de las ciencias fundamentale y aplicadas en email paradigm de la society post-industrial: Colección de documents cientificos «ΛΌΓΟΣ» con acts de la Conference Internacional Científica y Práctica (Vol. 2), 24 de April de 2020. Barcelona, España: Plataforma Europea de la Ciencia. 57–60 (2020). https://doi.org/10.36074/24.04.2020.v2.14
20. Ning, Z., et al.: Mobile edge computing and machine learning in the internet of unmanned aerial vehicles: a survey. ACM Comput. Surv. **56**(1), 1–31 (2023). https://doi.org/10.1145/3604933
21. Lin, CC, Chianca, B., Bereholschi, LD, Chen, JJ, Silvestre G.: Cost-Effective Offloading Strategies for UAV Contingency Planning in Smart Cities. 32nd International Conference on Computer Communications and Networks (ICCCN), 24–27, 1–7 (2023). https://doi.org/10.1109/ICCCN58024.2023.10230212
22. Messous, MA, Hellwagner, H., Senouci, SM, Emini, D., Schnieders, D.: Edge Computing for Visual Navigation and Mapping in a UAV Network. IEEE International Conference on Communications (ICC), 7–11 June 2020, 1–6 (2020). https://doi.org/10.1109/ICC40277.2020.9149087
23. Awasthi, S., et al.: Micro UAV swarm for industrial applications in indoor environment: a systematic literature review. Logistics Res. **16**(1), 1–43 (2023)
24. Mykytyuk, M.: The role and place of UAVs in ensuring the safety of people during mass events. Scientific notes of the Lviv University of Business and Law. Legal series, **18**, 41–47 (2017). https://nzlubp.org.ua/index.php/journal/article/view/33
25. Hemmati, A., Zarei, M., Souri, A.: UAV-based Internet of Vehicles: a systematic literature review. Intel. Syst. Appl. **18**, 200–226 (2023). https://doi.org/10.1016/j.iswa.2023.200226
26. Agbaje, P., Anjum, A., Mitra, A., Oseghale, E., Bloom, G., Olufowobi, H.: Survey of interoperability challenges in the internet of vehicles. IEEE Trans. Intel. Transp. Syst. **23**(12), 22838–22861 (2022). https://doi.org/10.1109/TITS.2022.3194413
27. Butilă, E.V., Boboc, R.G.: Urban traffic monitoring and analysis using unmanned aerial vehicles (UAVs): a systematic literature review. Remote Sens. **14**(3), 620 (2022)
28. McKinsey Technology Trends Outlook 2024, https://www.mckinsey.com/capabilities/mckinsey-digital/our-insights/the-top-trends-in-tech/
29. Zaytsev, V.Ye. Maksymovich, O.V.: Prospects for the creation of unmanned aerial vehicles in Ukraine. Aviation, industry, society: materials of the III International. Science and practice conf. (Kremenchuk, May 12, 2022): Ministry of Internal Affairs of Ukraine, Kharkiv national

university of internal affairs, Kremenchuk flight college, Science Park «Science and Security». Kharkiv: KhNUVS, 72–74 (2022). http://dspace.univd.edu.ua/xmlui/handle/123456789/12148
30. On the approval of the Rules for the technical operation of unmanned aircraft complexes of the I class of the state aviation of Ukraine dated August 10, 2018, № 401: Order of the Ministry of Defense of Ukraine, https://zakon.rada.gov.ua/laws/show/z1062-18#Text
31. On the approval of the Rules for the execution of flights by unmanned aircraft complexes of the state aviation of Ukraine dated 08.12.2016 № 661: Order of the Ministry of Defense of Ukraine, https://zakon.rada.gov.ua/laws/show/z0031-17#Text

Empirical Analysis of Urban Flooding's Effects on Traffic Flow Parameters: Modifying the Volume-Delay Function for Enhanced Transportation Modeling

Lance Kenneth D. Mamuyac(✉), Jon Robin D. Delos Reyes, Louisse Shaola L. Lumanglas, Eian Lanz C. Rebotiaco, and Alexis M. Fillone

De La Salle University, 2401 Taft Ave, Manila, Philippines
lance_mamuyac@dlsu.edu.ph

Abstract. Flooding has been widely regarded as the most destructive natural hazard, causing significant disruptions in transportation. While previous empirical observations focused on the effects of rainfall on traffic, they often overlooked the direct impact of flooding on urban transportation systems. This study addressed this gap by investigating the effects of flooding on traffic flow in Metro Manila. By analyzing 15 roads under various flood conditions, the study developed empirical relationships between flood depth and key traffic parameters. The findings revealed significant reductions in road capacity and vehicle volume as flood depth increased. New polynomial functions were developed to model the relationship between flood depth and average vehicle speed for different road types. Moreover, the effects of flood-induced lane closures on road capacity were also highlighted. These insights led to the modification of Volume-Delay Functions (VDFs) to incorporate flood-related factors, enhancing the accuracy of traffic predictions during flood events. The research provided a foundation for improved flood-based transportation within Metro Manila, with potential applications in other flood-prone urban areas globally.

Keywords: Flooding · Modelling · Volume-Delay Functions · Metro Manila

1 Introduction

Flooding is widely regarded as the most destructive natural hazard due to its immediate impact on human life and property [1]. In urban areas, intense precipitation commonly leads to flooding, submerging up to 40% of a city's road network and causing heavy traffic disruptions [2]. The Philippines, with its tropical climate, experiences frequent heavy rainfall and flooding [3]. Urban flooding in Metro Manila often impacts key roads in major economic hubs, leading to substantial economic losses due to traffic congestion, estimated at 2.4 billion pesos per day [4].

A previous study by [5] examined the effects of various rainfall intensities on traffic metrics, establishing quantitative links between rainfall and key traffic parameters.

O. Slavinska et al. (Eds.): ITSESQC 2024, LNNS 1335, pp. 308–319, 2025.
https://doi.org/10.1007/978-3-031-87376-8_27

However, rainfall intensity alone is not a reliable predictor of traffic impacts, as flooding depends not only on rainfall but also on local factors like drainage infrastructure, road conditions, and landscape [6, 7]. Flooding is highly localized and varies even within the same region. While rainfall generally affects traffic by prompting drivers to reduce speed as a safety measure—due to decreased traction, longer braking distances, and impaired visibility—flooding poses more direct physical barriers [8]. Floodwaters can cause vehicles to lose control, hydroplane, or become stranded due to hydrostatic forces, with depth and movement of water significantly obstructing vehicle speed [1]. Thus, there is a gap in the current understanding of the effects of flooding on vehicle movement within Metro Manila.

Previous flood-based traffic models by [9] and [10] heavily relied on assumptions for their traffic parameters. Using empirical observations to study the impact of different flood conditions on vehicle movement would greatly improve the accuracy of Volume-Delay Functions (VDF). This, in turn, leads to more precise flood-based traffic models that better reflect real-world conditions, enhancing their reliability. The main objective of this study is to assess the impact of flooding on traffic, specifically examining changes in number of lanes, vehicle speed, light vehicle percentage, and capacity to evaluate the effects of varying flood depths in Metro Manila. This approach aims to provide a clearer understanding of how flooding disrupts transportation and to inform future flood-based traffic models.

2 Methodology

2.1 Study Area

Metro Manila, the capital of the Philippines, experiences a tropical maritime climate with significant rainfall, averaging 2,450 mm annually, leading to frequent urban flooding [3, 11]. Despite ongoing flood control initiatives, including over 5,000 projects such as drainage expansions and trash clean-ups [12, 13], recent typhoons and the southwest monsoon have caused widespread flooding in Metro Manila and nearby provinces [14, 15].

[4] identified flood-prone streets in Metro Manila, which helped pinpoint key areas for study by cross-referencing with the Metro Manila Development Authority's (MMDA) CCTV locations. The study focused on 15 Metro Manila roads, covering both primary and secondary hierarchies and including both free-flowing sections and intersections (see Fig. 1). Thirty-minute CCTV videos recorded between September and October 2023 were used, capturing both flooded and non-flooded conditions. Due to the limited MMDA video archive, all available footage of flooded roads from this period was gathered. To ensure quality and consistency, further filtering was applied: night-time videos were excluded for better visibility, only flood events from regular rain conditions (not from typhoons or storms) were included, and only footage from regulated intersections connected to primary roads was analyzed. This refined footage served as the primary data for the study.

Road information was collected using Google Maps, with the number of lanes determined through Street View images. Exclusive lanes, such as the bus lanes on Epifanio de

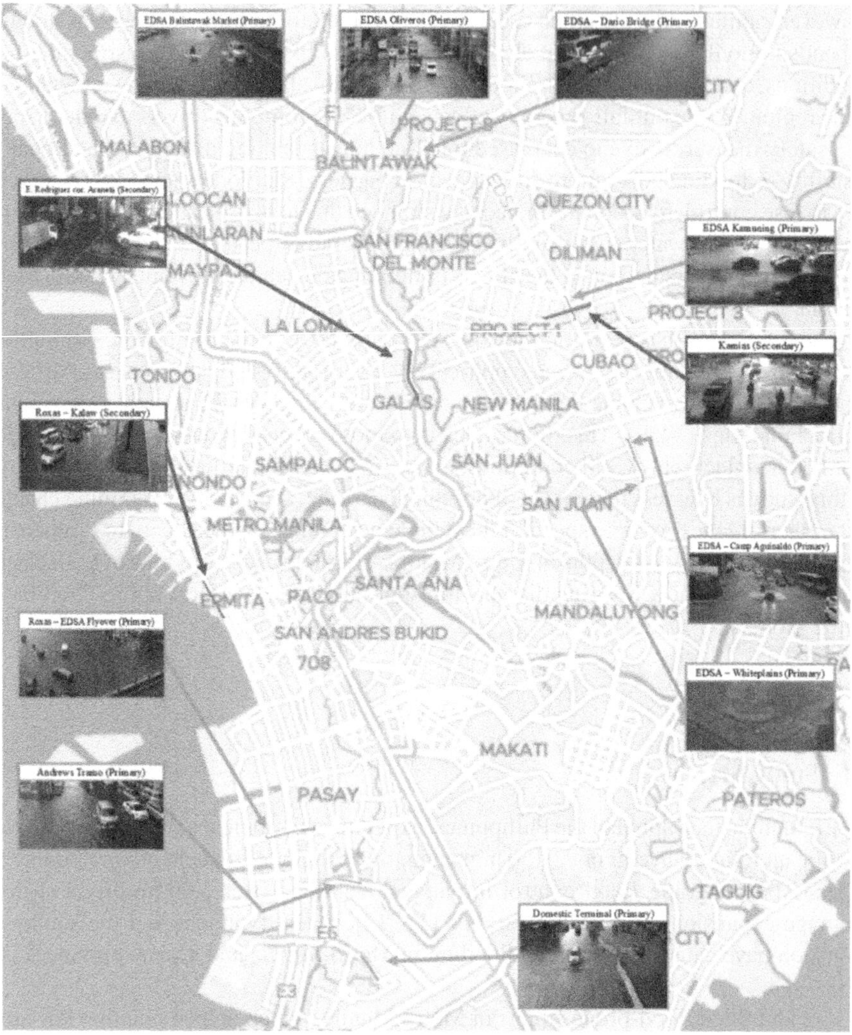

Fig. 1. Characteristics of the observed roads and their location within Metro Manila

los Santos Avenue (EDSA), were excluded from the lane count. As noted by [16], dedicated lanes serve specific vehicles and follow different rules, enabling them to avoid congestion. This separation causes these lanes to operate independently from adjacent traffic. To maintain consistency, only general-purpose lanes were considered in the dataset. Exclusive lanes were defined as those physically or regulatorily separated from other lanes, while lanes without physical barriers, such as bike lanes, were classified as general-purpose because they are still affected by surrounding traffic. Table 1 shows the number of lanes and the presence of exclusive lanes for each road studied.

Traffic conditions on each road were assessed by measuring road segments using Google Maps' Measure Distance feature. Two distinct landmarks, such as streetlights,

Table 1. Road details

Road name	Number of lane per direction	Bus lane	Bike lane
EDSA - Balintawak Market	5	✓	x
EDSA - Camp Aguinaldo	5	✓	✓
EDSA - Dario Bridge	5	✓	x
EDSA - Oliveros NB	5	✓	x
EDSA - White Plains	4	✓	✓
MIA Domestic Terminal 2 (NB)	3	x	x
MIA Domestic Terminal 2 (SB)	3	x	x
Roxas - EDSA Flyover	5	✓	✓
Andrews – Tramo	3	x	x
EDSA - Kamuning (NB)	3	x	x
E. Rodriguez (EB)	4	x	x
E. Rodriguez (WB)	4	x	x
Kamias (EB)	2	x	x
Kamias (WB)	2	x	x
Roxas – Kalaw	5	x	x

pedestrian overpasses, or elevated transit line columns, were used for accurate measurements. The distance between these landmarks represented the segment covered by a vehicle and formed the basis for traffic condition analysis.

2.2 Flood Data Collection

To assess the effects of varying flood depths on road traffic characteristics, both maximum and sampling flood depths were measured. Maximum flood depth was estimated using the MMDA Flood Gauge, as shown by Fig. 2 [17]. This involved approximating the highest water level on the road, converted from inches to centimeters, based on the MMDA's classification system. Flood depth at the most submerged section was recorded at five-minute intervals, with the highest value considered the maximum depth for that segment.

In contrast, the sampling flood depth measured only the water level directly traversed by sampled vehicles. Following [8], flood depths were estimated using known object dimensions, specifically measuring water levels (in cm) relative to the tire diameter of passing vehicles (see Fig. 3). To maintain accuracy, vehicle dimensions were obtained from Zigwheels Philippines, which provides model-specific data. This process was repeated for each sampled vehicle.

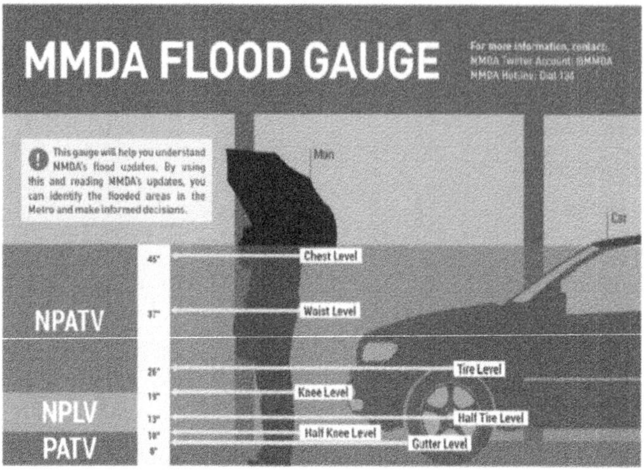

Fig. 2. MMDA flood gauge (Abana et al., 2019)

Fig. 3. Flood depth estimated using vehicle wheel diameter

2.3 Video Analysis

Road Demand. Vehicle volumes were manually counted under both flooded and non-flooded conditions. All vehicles using the observed road were counted, regardless of travel direction (i.e., left, right, or straight). The peak 15-minute volume was recorded and multiplied by four to estimate the road's peak hourly volume, denoted as V_{hr} (See Eq. 1).

$$V_{hr} = 4 \times V_{15} \tag{1}$$

wherein,

V_{15}= Peak 15-min volume

This counting method has inherent measurement uncertainties. Larger vehicles may obstruct smaller ones due to camera positioning, creating minor discrepancies. To improve reliability, multiple observers recorded vehicle counts independently and averaged their results to account for human variability, which remained minimal. Additionally, rain and flooding can occur unpredictably, not limited to peak hours, complicating precise traffic flow capture. To address this, flood and non-flood data were consistently collected on the same weekday and time, maintaining comparability between conditions.

Vehicle Speed. [8] outlined a method for manually calculating vehicle speed by measuring the distance traveled over a set period. A random sample of 30 vehicles per mode per segment was collected. These samples were uniformly distributed throughout the footage. Speed was calculated by analyzing the distance traveled by each vehicle over a fixed time. Velocity (in km/h), denoted as v, was calculated using the following formula:

$$v = \frac{3.6 \times \mathrm{d}}{t} \tag{2}$$

wherein,

d = Distance covered (m)

t = Elapsed time (s)

3.6 = Used for Conversion (M/s to km/h)

Flow Rate. The flow rate was determined by manually counting the number of vehicles passing a specific point on the road within a given time frame. The flow, denoted as F and measured in passenger car units per hour per lane (PCU/hr/lane), was calculated using the following formula:

$$F = \frac{\frac{f}{t} \times 3600}{n} \tag{3}$$

wherein,

f = Vehicle volume of road segment (PCU)

t = Elapsed time (s)

n = Road segment's number of lanes (lane)

3600 = Used for Conversion (s to hr)

It should be noted that the vehicle counts were converted to their Passenger Car Unit (PCU) using the Passenger Car Equivalent Factors (PCEFs) supplied by [18].

3 Results and Discussion

3.1 Traffic Flow Parameters

The impact of maximum flood depth on traffic flow is summarized in Table 2. The data is categorized into three flood depth ranges: less than 25 cm, 25–40 cm, and 40 cm or greater. As flood depth increased, traffic flow was increasingly disrupted. Lane closures rose from 0 to 3 as flooding worsened, while average road capacity decreased significantly, by 43.40% in mild flooding and up to 72.54% in severe conditions. In terms of road density, it was observed that values ranged from 159.34% to 210.23%. Traffic volume also declined, with reductions ranging from 16.66% to 57.56%. The proportion of light vehicles was particularly affected, decreasing by 29.37% in mild flooding and by 78.70% in severe flooding. These results clearly show that flooding significantly disrupts traffic, with the greatest impact occurring at depths of 40 cm or more.

To maintain uniformity in the data collection process, comparisons between flood and non-flood scenarios were made at the exact same time frame on the same day of the week. Given that flooding and rainfall events can occur at any time of day—not solely during peak hours—this approach ensured that traffic flow was compared consistently, rather than at potentially variable peak and non-peak hours. Additionally, the analysis utilized empirical data obtained from video observations rather than relying solely on theoretical maximum capacity values from local studies, allowing for a more accurate reflection of real-world conditions.

Regarding the traffic flow structure, the traffic was generally free-flowing under non-flood conditions but transitioned to a congested state as flood waters began to accumulate. This progression illustrates how flooding can initially slow traffic gradually before intensifying congestion as flood depths increase.

Table 2. Key traffic flow parameters under various maximum flood depths

	Maximum flood depth		
	$x < 25$	$25 \leq x < 40$	$x \geq 40$
Lane closures (n)	0	1–2	3
Average density (PCU/km/lane)	159.34%	179.28%	210.23%
Average capacity (PCU/hr/lane)	−46.94%	−53.34%	−72.54%
Average change in light vehicles (veh/hr)	−29.37%	−44.21%	−78.70%

A polynomial regression analysis was conducted to simulate the impact of flooding on vehicle speed, with average speed (y) modeled as a function of flood depth (x). This approach aligns with the study by Pregnolato et al. (2017), who also applied a quadratic model in similar flood-depth studies. Using a quadratic model reflects the non-linear relationship between flood depth and vehicle speed, which a linear model would fail to capture. In real conditions, as water depth increases, vehicle speed doesn't decline at a constant rate. Shallow flooding may initially slow vehicles, but as depth rises, increased water resistance further reduces speed at a faster rate, eventually leading to risks like hydroplaning or stalling. Moreover, higher-order polynomials were avoided to prevent overfitting and unnecessary complexity, which can introduce oscillations that do not align with the steady decrease in speed as depth rises. Therefore, the quadratic model not only mirrors the real-world impact of flooding on vehicle speeds but does so with physical realism.

Table 3 presents the equations describing the relationship between flood depth and vehicle speeds. The R-squared values highlight a distinct variation in the correlation based on road type, indicating that the effect of flood depth on vehicle speed differs across various road conditions.

Table 3. Equation of flood depth-vehicle speed function

Type	Equation	R-squared
Primary (free-flow)	$y = 0.0001x^2 - 0.1101x + 27.0786$	0.9165
Primary (intersection)	$y = 0.0002x^2 - 0.1409x + 22.768$	0.9347
Secondary (intersection)	$y = 0.00013x^2 - 0.1791x + 27.257$	0.9356

It is important to note that the sample flood depth refers to the specific depth a vehicle traversed, rather than the maximum flood depth on the road. As roads are typically sloped, the crown of the road experiences lower flood depths while the curb faces the highest. Consequently, vehicles often encounter flood depths smaller than the maximum recorded depth. This sample flood depth directly influences vehicle speed, as it pertains to the portion of the road the vehicle passes over.

Table 4 presents the changes in average vehicle speed compared to dry conditions, measured at 5 cm flood depth intervals (5, 10, 15, 20, 25, and 30 cm) using the equations. The data reveals a consistent decrease in average speed as flood depth increases across different road types. On primary roads, speed declines by 5.28% at 5 cm of flooding, rising to 14.80% at 30 cm. For primary roads with intersections, speed reductions range from 8.32% at 5 cm to 22.22% at 30 cm. On secondary roads, the decrease is 6.21% at 5 cm, reaching 21.73% at 30 cm. These results demonstrate the significant impact of flooding on vehicle speeds, with all road types showing substantial reductions as flood depth increases.

Table 4. Key traffic flow parameters under various sample flood depths

Type	Flood depth (cm)					
	5	10	15	20	25	30
Primary (free-flow)	−5.28%	−7.22%	−9.14%	−11.04%	−12.93%	−14.80%
Primary (intersection)	−8.32%	−11.18%	−14.00%	−16.78%	−19.52%	−22.22%
Secondary (intersection)	−6.21%	−9.36%	−12.49%	−15.59%	−18.67%	−21.73%

3.2 VDF Factors

VDFs are equations used to quantify the delay or increase in travel time caused by traffic congestion. These functions describe the relationship between traffic volume and travel time across varying congestion levels, typically exhibiting a nonlinear pattern where delays rise sharply as road capacity nears or exceeds its limit. In EMME 4 modeling, the VDF for each link type is expressed by the equation:

$$Tx = length\left(\frac{60}{el1}\right)\left(1 + \alpha\left(\frac{volau + volad}{lanes*el2}\right)\right)^{\beta} \tag{4}$$

wherein,

Tx = Travel Time at a Volume/Capacity Ratio
α = Value for each link type
β = Value for each link type
el1 = Design speed
el2 = Lane capacity
volau = Auto volume
volad = Additional volume

However, the findings indicate that several parameters of the original VDF equation, such as capacity, design speed, vehicle volume, and number of lanes, are affected under flooded conditions. To adjust for these variations, modification factors were incorporated into a revised VDF equation, as shown in Eq. #. Tables 5 and 6 present the data for these factors, categorized by different flood depths. These factors were derived by calculating the percentage differences between conditions with varying flood depths and those under dry conditions.

$$Tx = length\left(\frac{60}{el1(S)}\right)\left(1 + \alpha\left(\frac{volau(V) + volad}{lanes{*}el2(C)}\right)\right)^{\beta} \tag{5}$$

wherein,

C = Capacity factor
V = Volau factor
S = Design speed factor

Table 5. Summary of factors based on maximum flood depth

	Maximum flood depth		
	x < 25	25 ≤ x < 40	x ≥ 40
Capacity factor (C)	0.5306	0.4666	0.2746
volau factor (V)	0.7063	0.5579	0.2130

Table 6. Summary of factor based on sampling flood depth

	Type	Flood depth (cm)					
		5	10	15	20	25	30
Design speed factor (S)	Primary (free-flow)	0.9472	0.9278	0.9086	0.8896	0.8707	0.8520
	Primary (intersection)	0.9168	0.8882	0.8600	0.8322	0.8048	0.7778
	Secondary (intersection)	0.9379	0.9064	0.8751	0.8441	0.8133	0.7827

From the table, it is evident that as the maximum flood depth increases, the adjustment factors consistently decrease across all road types and conditions. The capacity factor (C) and vehicle volume factor (V) were presented for three flood depth ranges. For flood depths below 25 cm, C was 0.566 and V was 0.7063. These values dropped to 0.4986 and 0.5579 for depths between 25 and 40 cm, and further to 0.2746 and 0.213 for depths of 40 cm or more.

For the design speed factor (S), the values varied depending on road classification and flood depth. For primary roads under free-flow conditions, the speed factor ranged from 0.9472 at 5 cm flooding to 0.8520 at 30 cm. Primary roads at intersections showed factors between 0.9168 and 0.7778, while secondary roads at intersections had factors ranging from 0.9379 to 0.7827 for the same flood depth range. Regarding the number of lanes, there is no general factor to apply; instead, lane closures must be manually edited in the network model based on the maximum flood depth. For roads with fewer lanes than indicated closures, it can be assumed the road is fully closed. For instance, if 40 cm of flooding results in 3 lane closures but the road only has 2 lanes, the road can be considered impassable for most vehicles.

4 Conclusion

Flooding is a major issue in Metro Manila, where heavy rainfall frequently submerges roads and disrupts traffic [2, 3]. While past studies have linked rainfall intensity to traffic metrics, they often neglect localized factors like drainage and road conditions that influence flooding and traffic [6, 7]. Building on previous flood-based traffic models by [9] and [10], this study empirically assessed flooding's impact on traffic at various depths. It aimed to improve Volume-Delay Functions (VDF) and provide insights into how flooding affects Metro Manila's transportation infrastructure.

The analysis of 15 roads revealed a modified Volume-Delay Function (VDF) equation is necessary to account for the impacts of flooding on traffic flow. Key parameters, such as capacity (C) and volume (V), showed a significant decline as flood depths increased. For example, at flood depths below 25 cm, C and V were 0.566 and 0.7063, respectively. These values dropped to 0.4986 and 0.5579 for depths between 25 and 40 cm, and further decreased to 0.2746 and 0.213 for depths above 40 cm. Similarly, new mathematical functions to represent the relationship between flood depth and average vehicle speed for different road types was developed for the design speed factor (S). These functions exhibited a polynomial downward trend. Lastly, the number of lanes became a critical factor during floods, as higher flood depths led to lane closures, rendering roads impassable in severe cases. These findings emphasize the need to integrate flood-related factors into traffic models for more accurate representation of traffic conditions during flooding. These findings have crucial implications for the development of flood-based transportation models and policies in flood-prone urban areas of the Philippines. By establishing empirical relationships between flood depth and vehicle movement, the study offers a foundation for more accurately predicting traffic disruptions during flood events. Integrating these relationships into transportation models can lead to better estimates of congestion levels, travel times, and overall network performance under flooded conditions. This would significantly improve emergency planning and response strategies, aiding in more effective resource allocation and traffic management during floods. Future studies could expand this research by broadening the geographic scope to include the entire Greater Manila area, offering a comprehensive view of flooding's impact across different cities and provinces for more effective region-wide solutions. Additionally, exploring psychological and behavioral factors through surveys and interviews could reveal how Filipino drivers assess risks and decide on detours or waiting during floods, leading to improved public safety campaigns and traffic management strategies.

Ultimately, this study marks a significant initial step in assessing traffic dynamics during floods, setting a precedent for similar research in other urban areas globally that face the growing threat of extreme weather events.

Acknowledgement. The authors wish to extend their heartfelt appreciation to the Transportation Engineering (TRE) department of De La Salle University for their indispensable support, which was instrumental in conducting this study. They also express sincere gratitude to the Metropolitan Manila Development Authority (MMDA), particularly Mr. Jonnel Cervantes and Director Milagros Silvestre, for their generous assistance and cooperation in providing the essential footage used in this research. Their contributions were invaluable in making this study possible.

References

1. Shah, S.M., Mustaffa, Z., Martínez-Gomariz, E., Yusof, K.W., Al-Qadami, E.H.: A review of Safety Guidelines for vehicles in floodwaters. Int. J. River Basin Manag. **19**(1), 25–41 (2019). https://doi.org/10.1080/15715124.2019.1687487
2. Diakakis, M., Boufidis, N., Salanova Grau, J.M., Andreadakis, E., Stamos, I.: A systematic assessment of the effects of extreme flash floods on transportation infrastructure and circulation the example of the 2017 Mandra Flood. Int. J. Disaster Risk Reduct. **47**, 101542 (2020). https://doi.org/10.1016/j.ijdrr.2020.101542
3. PAGASA: Climate of the Philippines, https://www.pagasa.dost.gov.ph/information/climate-philippines#:~:text=The%20Climate%20of%20the%20Philippines
4. Lagmay, A.M., et al.: Street floods in Metro Manila and possible solutions. J. Environ. Sci. **59**, 39–47 (2017). https://doi.org/10.1016/j.jes.2017.03.004
5. Mejia, H.N., Sigua, R.G.: Impacts of different rainfall intensities on key traffic flow parameters at North Luzon expressway using underwood's exponential model. Philipp. Transp. J. **1**(2) (2018). https://ncts.upd.edu.ph/tssp/wp-content/uploads/2018/08/Mejia18.pdf
6. Hooper, E., Chapman, L., Quinn, A.: Investigating the impact of precipitation on vehicle speeds on UK motorways. Meteorol. Appl. **21**, 194–201 (2012). https://doi.org/10.1002/met.1348
7. Sabir, M., van Ommeren, J., Koetse, M.J., Rietveld, P.: Welfare effects of adverse weather through speed changes in car commuting trips. SSRN Electron. J. (2008). https://doi.org/10.2139/ssrn.1269333
8. Pregnolato, M., Ford, A., Wilkinson, S.M., Dawson, R.J.: The impact of flooding on road transport A depth-disruption function. Transp. Res. Part D: Transp. Environ. **55**, 67–81 (2017). https://doi.org/10.1016/j.trd.2017.06.020
9. Abad, R.P.B., Fillone, A.M.: Vulnerability assessment of alternate traffic routes to flooding in Metro Manila, Philippines. J. East. Asia Soc. Transp. Stud. **12**, 167–181 (2017). https://doi.org/10.11175/easts.12.167

10. Bacero, R.S., Fillone, A.M.: Impact of flooding on truck movement in Metro Manila Philippines. Asian Transp. Stud. **9**, 100113 (2023). https://doi.org/10.1016/j.eastsj.2023.100113
11. Afuang, B.: Floods and the city, https://www.philstar.com/business/motoring/2001/05/16/83330/floods-and-city
12. Clapano, J.R., Villeza, M.E.: Drainage capacity overpopulation blamed for metro manila flooding, https://www.philstar.com/headlines/2024/07/28/2373528/drainage-capacity-overpopulation-blamed-metro-manila-flooding
13. Patinio, F.: DPWH over 5k new flood control projects ongoing, https://www.pna.gov.ph/articles/1229922
14. Agcaoili, N., Rojas, A.: How super typhoon carina enhanced habagat according to an atmospheric scientist, https://news.abs-cbn.com/news/2024/7/28/how-super-typhoon-carina-enhanced-habagat-according-to-an-atmospheric-scientist-1440
15. Disaster Response Operations Monitoring and Information Center: DSWD DROMIC Report #14 on the Effects of Southwest Monsoon and Super Typhoon Carina as of 29 July 2024 6AM, https://dromic.dswd.gov.ph/wp-content/uploads/2024/07/DSWD-DROMIC-Report-14-on-the-Effects-of-Southwest-Monsoon-and-Super-Typhoon-Carina-as-of-29-July-2024-6AM.pdf
16. Arasan, V.T., Vedagiri, P.: Planning for dedicated bus lanes on roads carrying Highly heterogeneous traffic. 50th Annual Transportation Research Forum (2009). https://doi.org/10.22004/ag.econ.207621
17. Abana, E.C., Dayag, C.V., Valencia, V.M., Talosig, P.H., Ratilla, J.P., Galat, G.: Road flood warning system with information dissemination via social media. Int. J. Electr. Comput. Eng. (IJECE) **9**(6), 4979 (2019). https://doi.org/10.11591/ijece.v9i6.pp4979-4987
18. DPWH: Department Order No 22, https://www.dpwh.gov.ph/dpwh/sites/default/files/issuances/DO_022_S2013.pdf

Determination of the Environmental Component Life Cycle of a Vehicle

Volodymyr Volkov[1], Volodymyr Kuzhel[2](✉), and Tetiana Volkova[1]

[1] Kharkiv National Automobile and Highway University, Yaroslava Mudrogo Street, 25, Kharkiv 61002, Ukraine

[2] Vinnytsia National Technical University, Khmelnytske Highway, 95, Vinnytsia 21000, Ukraine

kuzhel_v@vntu.edu.ua

Abstract. The authors explored a complete life cycle model and examined the potential for developing information systems to support car operation processes' organizational and functional aspects. They integrated information related to the stages of a car's life cycle and its technical control systems, which are essential for monitoring and diagnosing the condition of vehicles. Their study is based on a comprehensive approach considering the interaction between the "car - driver - operating conditions - infrastructure for car operation (including transport and highways)." They focused on implementing "systems with full responsibility," such as FADEC (Full Authority Digital Electronic Control) systems, aiming to create integrated onboard systems for managing the operational processes of various vehicle components. The research also looked into control and diagnostic systems, as well as systems for organizational and functional support, such as IPV (Impact Protection Vehicle), CALS (Continuous Acquisition and Life Cycle Support), and PLM (Product Lifecycle Management). The authors proposed and developed various information software complexes (ISC) and algorithms for their operation. They presented the results of their studies on fuel consumption under controlled experimental conditions.

Keywords: Life Cycle Model · Information Systems · Diagnostic System · Car Operation Process · Full Authority Digital Electronic Control System

1 Introduction and Literature Review

Motor transport is a crucial sector of the Ukrainian economy, supporting nearly all sectors and various segments of the population. It plays a significant role in enhancing mobility and improving the overall quality of life. Currently, Ukraine's car fleet consists of over 14 million vehicles, which are categorized as follows: trucks comprise 15.5%, buses account for 2.6%, and passenger cars make up 81.9% [1].

Moreover, motor transport was one of the first industries to undergo extensive privatization in Ukraine, with a particular emphasis on road transport enterprises that focus on cargo transportation [2].

O. Slavinska et al. (Eds.): ITSESQC 2024, LNNS 1335, pp. 320–331, 2025.
https://doi.org/10.1007/978-3-031-87376-8_28

A significant number of carriers with various forms of ownership have emerged in the market for motor transport services. However, many of these carriers lack experience in organizing transportation, the necessary professional knowledge, and an adequate technical infrastructure. Additionally, their vehicles often do not meet transportation safety requirements. The state transport policy in this industry is enforced through the laws of Ukraine, a regulatory and technical framework, and improvements to the system of state management and regulation governing the relations and activities of business entities. State regulation of the transport services market is implemented through the licensing of motor transport activities. Currently, approximately 140,000 licenses have been issued to legal entities and individuals in Ukraine, with around 400,000 vehicles in operation. According to data from the Main State Inspection of Road Transport, 61% of carriers operate only one vehicle, 22.4% operate up to three vehicles, 7% operate up to five vehicles, 5.4% operate up to ten vehicles, and 4.3% operate more than ten vehicles [3].

The development of market relations within the public road transport system, along with the privatization of state-owned transport enterprises and the acquisition of vehicles through personal funds or leasing, has led to the emergence of over 30,000 passenger carriers. These carriers have obtained licenses to operate in the transport services sector. Currently, the bus service network consists of approximately 19,000 city, suburban, and international routes, with around 50,000 buses in operation [4]. Prior to the transition to a market economy, Ukraine's public transport system was served by between 45,000 and 54,000 buses across 449 specialized state-owned transport enterprises. It is important to note that passenger road transport operates as a technologically centralized system. Each vehicle must be under dispatch control and management. Experiences from both domestic and foreign transport companies indicate that the highest levels of efficiency, quality, and safety are achieved when vehicles receive regular technical maintenance, drivers receive medical assessments, and buses are monitored consistently. The transportation management system receives timely information regarding each bus's location, allowing for timely interventions in the transportation process as needed [5].

Motor transport is the most resource-intensive and hazardous mode of ground transportation for both people and the environment. It consumes over 6% of petroleum fuel, utilizes 7% of labor resources, and is responsible for more than 96% of traffic accidents. Estimates suggest that road transport contributes to 40–50% of environmental pollution, with this figure rising to 60–70% in large cities and exceeding 85% in megacities. Additionally, at least 25% of this pollution can be attributed to the technical condition of vehicles and the operational activities of road transport companies [6].

2 Life Cycle Model

The primary systemic issues facing road transport at this stage are [7]:

- loss of administrative control over road transport management due to complete privatization;
- decrease in the volume of transport services provided.
- unprofitability of passenger transport on public bus routes;
- significant aging of the fleet and lack of effective replacement mechanisms;

- mismatch between the composition of the cargo and passenger fleets and the demand for their services;
- unsatisfactory road transport safety levels and considerable economic burden on the environment.

Today, the term "life cycle" is utilized in natural sciences, technical fields, and the humanities. There are numerous definitions of the life cycle in various subject dictionaries, which share similar main characteristics [3, 7, 8].

A well-known model of the complete life cycle of an individual object describes the sequence of all phases and stages of its existence, from conception and appearance (referred to as "birth") to disappearance (referred to as "death"). The main characteristics of the full life cycle model include: - The structure (composition of stages) of the life cycle model remains consistent regardless of the specific object being described. This means that the model is invariant to changes in the type of object, industry, etc., making it universal and widely applicable. - At any given moment, there is only the process of activity of specific agents creating specific results. This initial, real process is comprehensively represented by the flows of life cycle models. A new technique in road transport involves developing information systems that provide organizational and functional support for car operation processes. This is achieved through information integration, which encompasses both the stages of a car's life cycle and systems for technical control (monitoring and diagnosing the condition of vehicles) [9] (Fig. 1).

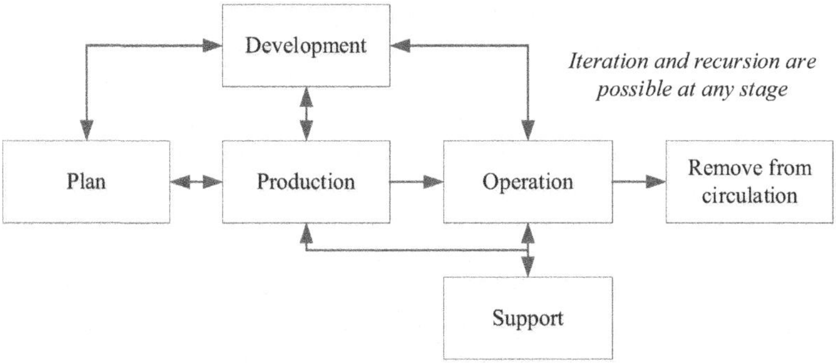

Fig. 1. Life cycle model.

2.1 IPV (Impact Protection Vehicle) / CALS (Continuous Acquisition and Life Circle Support)/ PLM (Product Lifecycle Management) - Technology Processes

The emergence of "fully responsible systems" in transportation, such as the FADEC (Full Authority Digital Electronic Control) system [4, 10], aims to create an integrated framework that encompasses on-board systems for managing the operational processes of nodes and aggregates, control and diagnostic systems, as well as organizational and functional support systems. This modern approach incorporates IPV/CALS/PLM technologies and is applied throughout all stages of a product's life cycle, including design, production, and the use of contemporary information technologies. [3, 4, 11].

In the field of transportation companies, the integrated information environment of IPV, CALS, and PDM technologies is currently in the early stages of implementation. At present, it primarily consists of a collection of networked electronic information systems that function as distributed storage. This system is heterogeneous, utilizing various technologies. A notable example of implementation in motor transport is the Torque program, which serves as the foundation for the "automotive" FADEC concept. This represents the initial step towards the FRACAS system and, consequently, the IPV, CALS, and PLM technologies. These technologies are designed to receive and display diagnostic information from the onboard self-diagnosis system. Currently, the Torque program is capable of displaying essential parameters of the engine,

environmental factors, and other systems, components, and aggregates. It can also display and decode error codes, erase errors from the electronic control unit, and automatically transmit controlled parameter values, as detected by sensors, into an integrated electronic information metaspace [5, 12].

On the basis of research carried out at KHNADU, the transport and information system of the virtual enterprise "KHNADU TESA" and the following information software complexes (IPK) "Virtual Mechanic "NADI-12"", IPK "Service Fuel Eco "NTU-HADI-12"", information software complex "MonDiaFor "HADI-15"", information software complex ""IdenMonDiaOperCon"HNADU-16"", which have the ability to work as part of intelligent transport systems (ITS) [5, 6].

The foundation for developing technology to monitor the technical condition of vehicles is based on a comprehensive approach that examines the system involving the "car - driver - operating conditions - infrastructure." This approach considers the systemic interactions among the key components of monitoring: the vehicle, the driver, the onboard information system, the operating conditions (including road, transport, atmospheric, climatic conditions, as well as environmental and work culture) [5, 14, 15], and the transport and highway infrastructure.

For example, we will present the algorithm and some experimental results from the information software complex "Service Fuel Eco NTU-HADI-12." This software includes a separate module for the "Calculation of Ecological Parameters" (Fig. 2), in which calculations are performed using the "Service Fuel Eco NTU-HADI-12" method. The module determines the consumption of fuel and lubricants, as well as the emissions and waste generated [6, 16].

Average fuel consumption for cars includes the following:

- fuel consumption during transport work, measured in liters per 100 km (or cubic meters per 100 km for compressed natural gas);
- during maintenance and repairs, the fuel consumption is 5% of the amount used during transport work, measured in liters. Additional details regarding lubricant consumption can be found in Table 1.

Table 1. Consumption of lubricants in the information software complex "Service Fuel Eco "NTU-HADI-12""

Automotive transport category	Consumption of lubricants, liters/100 L (liters/100 m^3, compressed natural gas) (in relation to fuel consumption)			
	motor oil	transmission oil	special oils	plastic lubricants
M1	0,5…1,8	0,02…0,15	0,01…0,05	0,05…0,1
M2	0,5…1,8	0,02…0,15	0,01…0,05	0,05…0,1
M3	1,5…2,8	0,1…0,4	0,05…0,1	0,1…0,3
N1	0,5…1,8	0,02…0,15	0,01…0,05	0,05…0,1
N2	1,5…2,8	0,1…0,4	0,05…0,1	0,1…0,3
N3	1,5…2,8	0,1…0,4	0,05…0,1	0,1…0,3

The mass emissions of harmful substances are calculated using the following formula: tons (t):

$$M_i = \sum_{m=1}^{m} g_{icj} G_j K_{\text{T}ikj} k_j \tag{1}$$

where g_{icj} – is the average specific emission of the i-th harmful substance from the unit mass of the j-th type of fuel, kg/tons;

Q_j – fuel consumption of the j-th car for a certain period, tons;

T_{ikj} – coefficient that takes into account the influence of the technical condition of the k-th type car, which consumes the j-th type of fuel, on the amount of specific emissions;

k_j – coefficient of bringing the car to the ecological class.

The value of the correction coefficient T_{ikj} is chosen depending on: the harmful substance (i), the emission of which is calculated; type of car (k) and type of fuel consumed (j), using the data in Table 2.

Average specific emissions g_{icj} are selected from Table 3. The value of the average specific emissions of harmful substances is given for the case when the car does not have exhaust gas neutralization systems (Euro-0) [18–20]. For ecological classes Euro-1 - Euro-6, coefficients of conversion k_j to EURO standards are used (Table 4).

In road transport, common types of waste include batteries, tires, and filters. The quantity of spent batteries is calculated based on the number of batteries installed in a vehicle, as reported by the road transport company, along with their service life and weight, measured in units per year:

$$N = \sum_{m=1}^{m} N_{\text{aBT}.i} \cdot n_i / T_i \tag{2}$$

where $N_{\text{ABT}.i}$ – number of cars equipped with batteries of the i-th type, units;

n_i – number of batteries on the car, units;

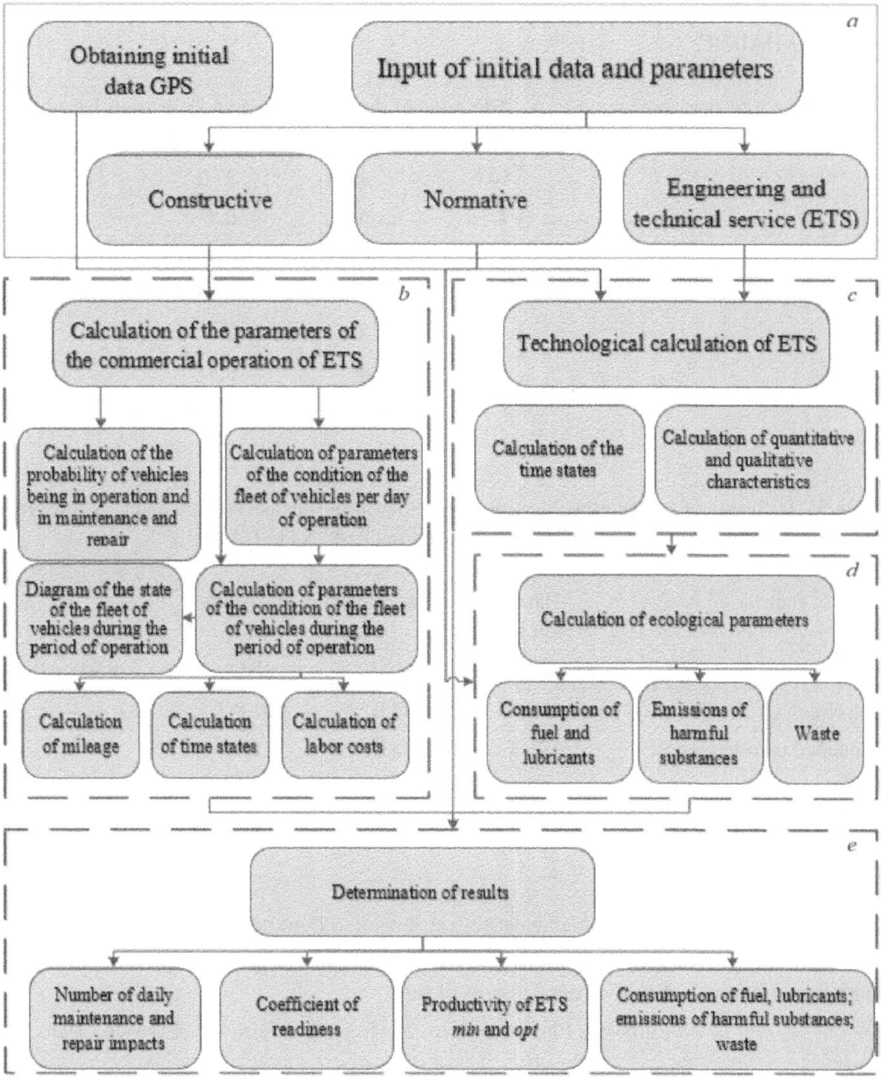

Fig. 2. Scheme of the structural information program complex "Service Fuel Eco "NTU-HADI-12"".

T_i – service life of the battery of the i-th brand, one year.

The weight of spent batteries is equal to, ton/year:

$$M = \sum_{m=1}^{m} N_i \cdot m_i \cdot 10^{-3} \tag{3}$$

where N_i – number of spent batteries of the i-th brand, units/year;

m_i – is the weight of the i-th type battery with electrolyte, kg.

Table 2. The value of the T_{ikj} coefficients in the information software complex "Service Fuel Eco "NTU-HADI-12""

Automotive transport category	The value of the T_{ikj} coefficient for various harmful substances			
	CO	C_mH_n	NO_x	PM
M1	1,5	1,5	0,9	-
M2	1,7	1,8	0,9	-
M3	1,5	1,4	0,95	1,8
N1	1,7	1,8	0,9	-
N2	1,7	1,8	0,9	-
N3	1,5	1,4	0,95	1,8

Table 3. The value of average specific emissions of harmful substances from a unit of consumed fuel in the information software complex "Service Fuel Eco "NTU-HADI-12""

Type of fuel	Harmful substances specific emissions kg/ tons of fuel			
	gCO	gCH	gNO_x	gPM
Gasoline	196,5	37,0	21,8	–
Liquefied petroleum gas	196,5	37,0	21,8	–
Diesel	36,0	6,2	31,5	3,85
Compressed natural gas	87,5	22,4	27,6	–

Table 4. Coefficients of conversion to Euro norms k_j

Harmful substances	Environmental classes of cars						
	Euro -0	Euro -1	Euro -2	Euro -3	Euro -4	Euro -5	Euro -6
CO	1	0,4	0,32	0,17	0,12	0,12	0,1
NO_x	1	0,55	0,49	0,34	0,24	0,13	0,1
C_nH_m	1	0,46	0,46	0,28	0,2	0,2	0,15
PM	1	0,51	0,21	0,14	0,03	0,03	0,2

The weight of used tires with a metal cord and with a fabric cord is determined by the formula, ton/year:

$$M = \sum_{m=1}^{m} (L_i \cdot m_i \cdot N_i \cdot n_i) / \left(L_{ni} \cdot 10^{-3}\right) \quad (4)$$

where N_i – number of cars of the i-th brand, units;

n_i – number of tires installed on a car of the i-th brand, units;

m_i – weight of one worn tire of this type, kg;

L_i – average annual mileage of a car of the i-th brand, thousand km/year;

$L_{n\,i}$ – rate of mileage of the car of the i-th brand before changing the tires, thousand km.

The weight of spent filters formed during the operation of vehicles is determined by the formula, tons/year:

$$M = \sum_{m=1}^{m} (L_i \cdot m_i \cdot N_i \cdot n_i) / \left(L_{ni} \cdot 10^{-3}\right) \tag{5}$$

where N_i – number of cars of the i-th brand, units;

n_i – the number of filters installed on vehicles of the i-th brand, units;

m_i – weight of one filter on a vehicle of the i-th brand, kg;

L_i – average annual mileage of a car of the i-th brand, thousand km/year;

$L_{n\,i}$ – rate of mileage of the car of the i-th brand before replacing the filters, thousand km.

An example of displaying information on the working windows "Calculations of environmental indicators = > Costs" is a wide range of parameters (Fig. 3):

- vehicle registration number;
- fleet of vehicles;
- fuel consumption for transportation, in liters per 100 km;
- fuel consumption during maintenance and repair, in liters;
- motor oil consumption, liters/100 L (liters/100 m^3, compressed natural gas);
- transmission oil consumption, liters/100 L (liters/100 m^3, compressed natural gas);
- consumption of special oil, liters/100 L (liters/100 m^3, compressed natural gas);
- consumption of plastic lubricants, liters/100 L (liters/100 m^3, compressed natural gas).

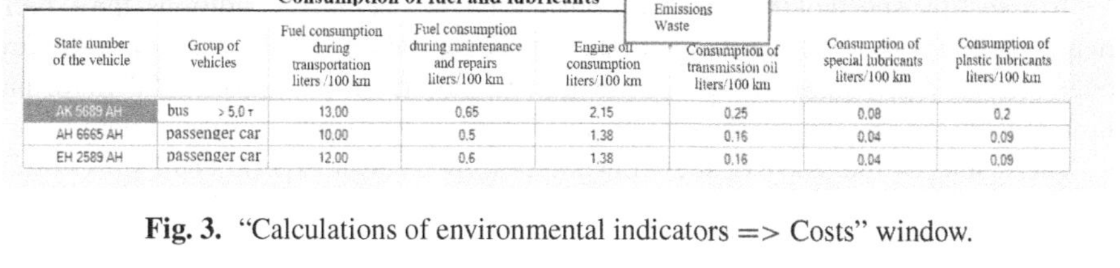

State number of the vehicle	Group of vehicles	Fuel consumption during transportation liters /100 km	Fuel consumption during maintenance and repairs liters/100 km	Engine oil consumption liters/100 km	Consumption of transmission oil liters/100 km	Consumption of special lubricants liters/100 km	Consumption of plastic lubricants liters/100 km
AK 5689 AH	bus >5,0 т	13,00	0,65	2,15	0,25	0,08	0,2
AH 6665 AH	passenger car	10,00	0,5	1,38	0,16	0,04	0,09
EH 2589 AH	passenger car	12,00	0,6	1,38	0,16	0,04	0,09

Fig. 3. "Calculations of environmental indicators => Costs" window.

To calculate emissions of harmful substances, select the window: "Calculations of environmental indicators => Emissions""in the main menu (Fig. 4).

The information of the "Emissions" window can be sorted by state registration numbers of cars and by the date of receipt (Fig. 4).

Emission parameters include the following information:

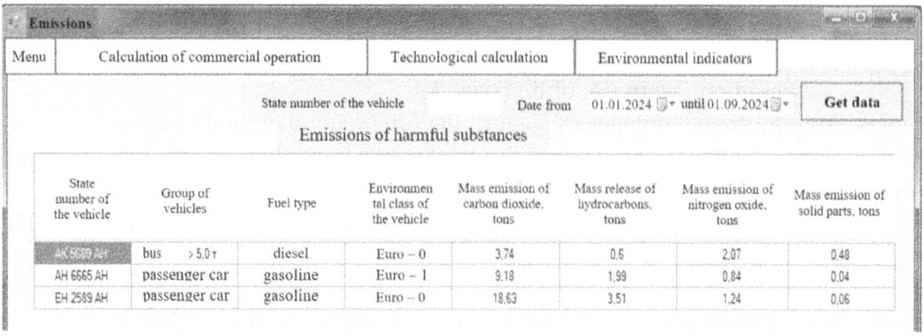

Fig. 4. "Calculations of environmental indicators => Emissions" window.

- state number of the vehicle;
- group of vehicles;
- fuel type;
- environmental classification of the vehicle;
- mass emission of carbon dioxide, tons;
- mass emission of hydrocarbons, tons;
- mass emission of nitrogen oxides, tons;
- mass emission of solid particles, tons.

To calculate waste, select the window in the main menu: "Calculations of environmental indicators => Waste" (Fig. 5).

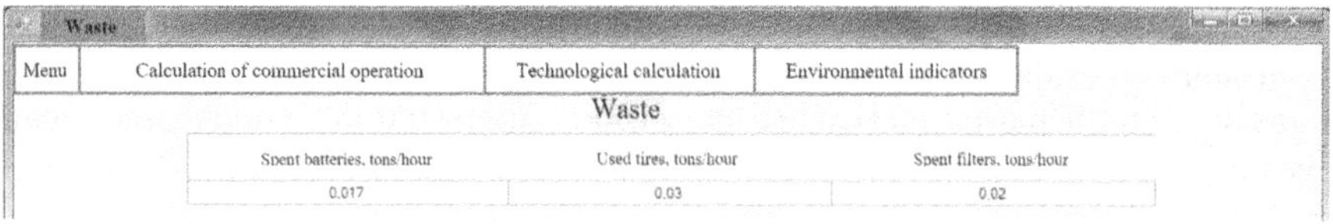

Fig. 5. "Calculations of environmental indicators => Waste" window.

To assess the speed of the car while considering the operating conditions, the experimental section of the road was divided based on the formation of geozones along its entire length. Initially, geozones within cities, where the maximum speed limit is 50 km/h, and those outside the city, with a maximum speed limit of 130 km/h, were identified.

The results of the study of fuel consumption by a car at experimental sites within the distance of the route - using the information software complex "Service Fuel Eco "NTU-HADI-12"" are shown in Figs. 6 and 7, and changes in fuel consumption by a car during movement according to the results of processing the report are shown in Fig. 8.

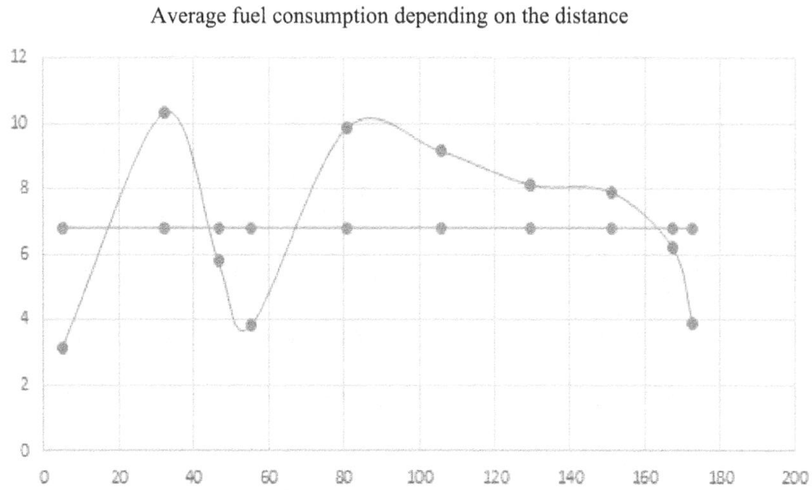

Fig. 6. Study results on car fuel consumption during test journeys: *a* - depending on the distance of the journey

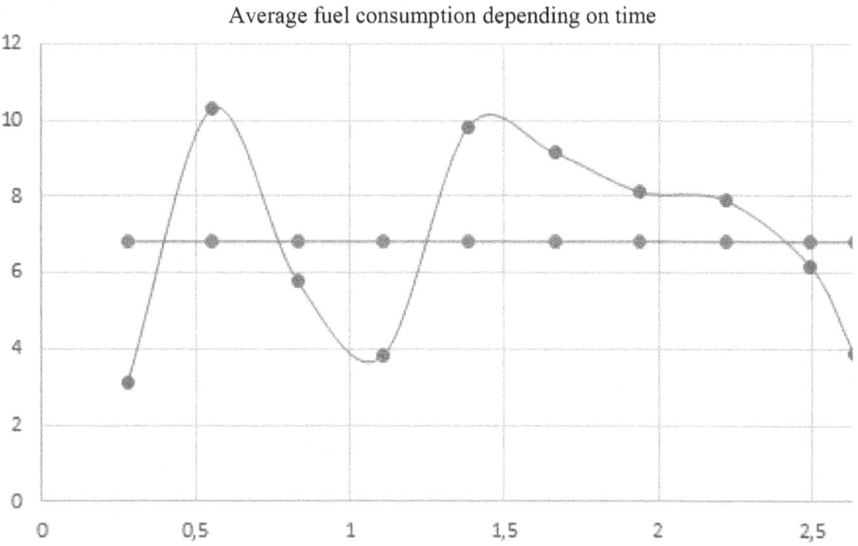

b

Fig. 7. Study results on car fuel consumption over different journey distances: *b* - the clarity of the movement depends on the timing.

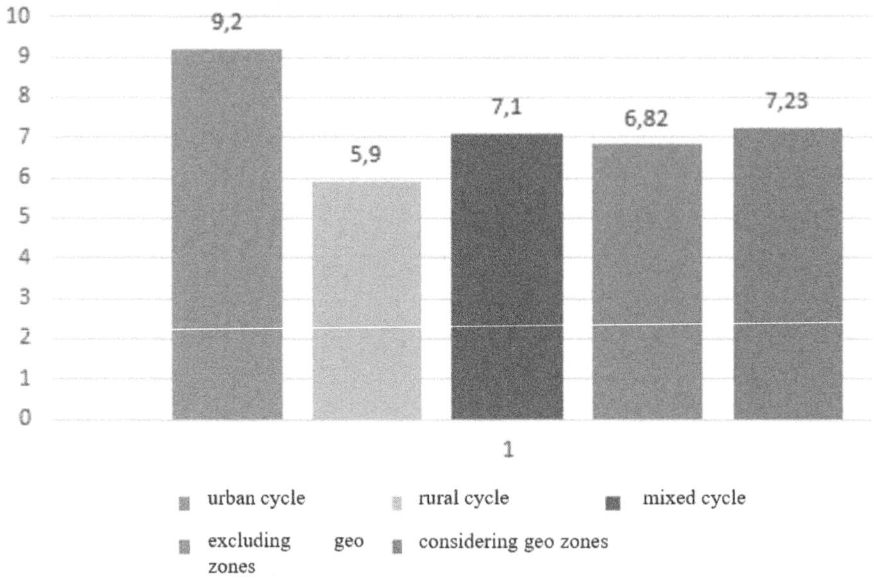

Fig. 8. The results of assessing changes in a car's fuel consumption during movement based on processed report data.

3 Conclusions

The authors developed a model based on the complete life cycle of vehicles and examined options for creating information systems that would support the processes involved in car operation. They focused on integrating information from various stages of a vehicle's life cycle and the systems used for technical control, including monitoring and diagnosing vehicle conditions. This approach led to the development of a technology for tracking the technical condition of vehicles, framed within a broader study of the interactions among the "car, driver, operating conditions, and infrastructure for car operation". The experimental section of the road was divided into geozones along its entire length. The geozones were categorized based on their location: those within city limits, where the maximum speed limit is 50 km/h, and those outside the city, where the maximum speed limit is 130 km/h. The study analyzed car fuel consumption along a designated route using the "Service Fuel Eco NTU-HADI-12" software system.

References

1. State Statistics Service of Ukraine: Transport. Official site www.ukrstat.gov.ua: website. https://www.ukrstat.gov.ua/operativ/menu/menu_u/tr.htm
2. Volkov, V., Vnukova, N., Taran, I., Pozdnyakova, O., Volkova, T.: Influence of diesel vehicles on the biosphere. Naukovyi Visnyk Natsionalnoho Hirnychoho Universytetu **5**, 94–99. (2021). https://doi.org/10.33271/nvngu/20215/094
3. Volkov, V.P. et al.: Monitoring the Technical Condition of a Car in its Life Cycle: A Textbook. KhNADU, Kharkiv (2017)

4. Volkov, V.P. et al.: Intelligent Systems for Monitoring the Technical Condition of Vehicles: a textbook. KhNADU, Kharkiv (2019)
5. Volkov, V., Gritsuk, I., Taran, I., Volkova, T., Kuzhel V., Semenov, A., Voznyak, O.: Information Systems for Vehicles Technical Condition Monitoring. Data-Centric Business and Applications Lecture Notes on Data Engineering and Communications Technologies 61–96 (2024). https://doi.org/10.1007/978-3-031-54012-7_4
6. Dmytrychenko, M.F. et al.: Software of Transport Monitoring Systems: Monograph. Kyiv: NTU, 204 p. (2016)
7. Zagursky, O.M. Teslenko, E.A.: Length of life cycle assessment of vehicle. Mach. Energ. **10**(1), 61–66 (2019), http://nbuv.gov.ua/UJRN/machener_2019_10_1_10
8. Thoruk, E., Kucher, O.: Determination of the optimal period of use of a motor vehicle as a parameter of the rolling stock renewal system. Interuniversity collection “Scientific Notes” **55**, 412–417 (2016)
9. Abirami, S., Pethuraj, M., Uthayakumar, M., Chitra, P.: A systematic survey on big data and artificial intelligence algorithms for intelligent transportation system. Case Stud. Transp. Policy **17**, 101247 (2024). https://doi.org/10.1016/j.cstp.2024.101247
10. Zhou, J., Shi, Y.: Risk control analysis of digital financial engineering based on 6G physical information system. Wireless Pers. Commun. (2024). https://doi.org/10.1007/s11277-024-11240-x
11. Wang, G., Zhao, L., Hao, Y.: Design of Active Safety Warning System for Hazardous Chemical Transportation Vehicle Information Technology and Intelligent Transportation Systems **455** (2017). ISBN: 978-3-319-38769-7
12. Na-Young, A., Donghoon, L., Seong-Jun O.: Proceedings of the Future Technologies Conference (FTC) 2018 Series: Advances in Intelligent Systems and Computing **881**, 158. (2019). https://doi.org/10.1007/978-3-030-02683-7_13
13. Santarek, K., Trojanowska, J., Kujawińska, A., Pavlenko, I., Husar, J.: Modern Technologies Supporting the Development of the Automotive Industry. Advances in Manufacturing IV. MANUFACTURING 2024. Lecture Notes in Mechanical Engineering. Springer, Cham (2024). https://doi.org/10.1007/978-3-031-56444-4_25
14. Onyshchuk, V.P., Kuznetsov, R.M., Kozachuk, I.S.: Intelligent telematics transport systems. Modern technologies in mechanical engineering and transport. Sci. J. - Lutsk: Lutsk NTU **2**(6), 110–114 (2016)
15. Hrytsuk, I.V.: Features of the structure of the information software complex for monitoring, diagnosing and predicting the technical condition of the vehicle in the conditions of ITS. Modern technologies in automotive engineering and transport: international scientific and technical conference, 15–16 October 2015: a collection of materials. Kharkiv: KHNADU, 123–125 (2015)
16. Savostin-Kosiak D., Sakhno V.: Study of fuel consumption of city buses with diesel by means of mathematical modeling. Wybrane Zagadnienia: monografia pod redakcja naukowa Kazimierza Lejdy. Series Transport, 9, 95–103 (2017)
17. Template: Real-time operating systems, https://en.wikipedia.org/wiki/Template:Realtime_operating_systems
18. Kuric, I., Mateichyk, V., et al.: The peculiarities of monitoring road vehicle performance and environmental impact. MATEC Web Conf. **244**, 143364 (2018). https://doi.org/10.1051/matecconf/201824403003
19. Volodarets, M., et al.: Optimization of vehicle operating conditions by using simulation modeling software. SAE Tech. Paper (2019). https://doi.org/10.4271/2019-01-0099
20. Hrynkiv, A.: Using of forecasting techniques to manage the technical condition of aggregates and systems of vehicles. Tech. Agric. Ind. Mach. Build. Autom. **29**, 25–32 (2016)

Using Logistics Principles to Improve the Efficiency of Freight Forwarding Services for Customers

Svitlana Shyriaieva(✉) and Iryna Diachenko

National Transport University, Mykhaila Omelianovycha - Pavlenka Str. 1, Kyiv 01010, Ukraine
s.shyriaieva@ntu.edu.ua

Abstract. The article is devoted to the problems of freight forwarding activities in the context of modern challenges, such as risk management, unpredictability of external factors and instability of transportation costs. The study examines the main factors that influence customers' choice of long-term cooperation with freight forwarders, including cargo delivery time, readiness for tariff changes, and others. Particular attention is paid to the use of logistics principles to improve the efficiency of transport forwarding services (TFS) for customers. The author proposes a feasibility study algorithm based on the use of modern technologies and logistics approaches. The algorithm consists of four sequential stages, each involving a comprehensive analysis, evaluation, and implementation of measures aimed at improving or adjusting the current level of freight forwarding services provided to clients. The results of research show that the use of logistics principles contributes to increased economic efficiency and quality of freight forwarding services to customers, which, in turn, ensures competitiveness in the market.

Keywords: Freight Forwarder · Customer · Transportation and Freight Forwarding Services · Logistics Principles

1 Introduction

1.1 Definition of the Problem

Freight forwarding plays a key role in ensuring timely and safe delivery of goods from shippers to consignees [6]. However, modern freight forwarders face a number of challenges, such as insufficient security in the transport of goods, unpredictability of external factors, fluctuations in fuel prices, and others. These problems significantly complicate the organization of effective freight forwarding services for customers [7].

1.2 Relevance of the Research

The relevance of the research is due to the need to improve the efficiency of freight forwarding services to customers in a volatile market, increasing customer requirements for service quality and the impact of external factors on the processes of providing freight

O. Slavinska et al. (Eds.): ITSESQC 2024, LNNS 1335, pp. 332–338, 2025.
https://doi.org/10.1007/978-3-031-87376-8_29

forwarding services to customers [8]. The research and implementation of logistics principles aimed at optimizing the processes of providing freight forwarding services to customers are an important elements in of the industry's development, which helps to reduce costs and improve the level of freight forwarding services to customers [9].

1.3 The Purpose and Objectives of the Research

The purpose of the research is to identify the main problems of interaction between freight forwarders and customers, and to develop an effective algorithm for freight forwarding services for customers, taking into account modern logistics approaches. To achieve this goal, the following tasks need to be solved:

1. Analyze the main problems that arise in the interaction of freight forwarders and customers.
2. Identify the factors that influence the choice of a client for long-term cooperation with a freight forwarder.
3. Develop an algorithm of freight forwarding services for customers.
4. Identify the principles of logistics to improve the efficiency of transport and forwarding services for customers.

1.4 Review of Previous Research

Previous research in the freight forwarding industry focus on the problems of the optimizing of logistics processes and supply chain management [2]. A considerable number of studies are devoted to the issues of transport logistics, transportation safety, risk management and adaptation to external conditions [3]. Some authors focus on the importance of using information technology to improve the efficiency of freight forwarding services [10]. Indeed, numerous authors emphasize the necessity of implementing Blockchain and AI in logistics [11].

Despite a significant number of research in the freight forwarding industry, the issues of using logistics principles to improve the efficiency of transport and freight forwarding services for customers.

2 Summary of the Main Research Material

According to the Law of Ukraine "On Freight Forwarding Activities" dated July 1 2004, No. 1955-IV, freight forwarding activities are defined as business activities related to the provision of freight forwarding services for the organization and provision of transportation of export, import, transit, or other cargoes; freight forwarding service is defined as work directly related to the organization and provision of transportation of export, import, transit or other cargoes under a freight forwarding agreement.

The freight forwarding service is a work directly related to the organization and provision of transportation for export, import, transit, or other cargo under a freight forwarding agreement [1].

The modern role of the freight forwarder and their interaction with customers face a number of challenges and problems that can affect the efficiency and quality of freight forwarding services. Some of the main problems in this area are shown in Table 1.

Table 1. Problems of interaction between the freight forwarders and client.

№	Title of the Problem	Brief Description of the Problems
1	Security and Risks	The constant need to manage risks associated with transportation, including theft, accidents, and loss of cargo
2	Unpredictability of External Factors (Force Majeure)	Force majeure circumstances, such as acts of war, weather conditions, natural disasters or political crises, can delay delivery
3	Unstable Fuel Prices	Fluctuations in fuel prices affect the cost of transportation, potentially leading to unforeseen expenses
4	Short Delivery Times	Pressure from customers for fast delivery, which requires clear organization and minimization of delays
5	Lack of Qualified Personnel	The lack of professionals in the industry can affect the quality of freight forwarding services and the efficiency of order fulfillment

Currently, to solve the above-mentioned problems of customers' transportation and freight forwarding services, a logistics approach is becoming increasingly popular as it is a modern approach to organizing various business processes, including freight forwarding [15].

Logistics of freight forwarding services is an activity in the field of customer feasibility studies that covers the entire range of operations and freight forwarding services for delivery of goods from shippers to consignees [16].

The logistics approach requires the coordination in time and space of the processes of transportation, storage and distribution of finished products up to their delivery to end consumers, and also requires the provision of the required level of logistics service. At the same time, modern transportation practice in terms of logistics service can be implemented with the help of the 'seven rules' of logistics and thus achieve maximum economic efficiency. Below are the 'seven rules' of logistics [13]:

1. Right product;
2. Right quality;
3. Right quantity;
4. Right time;
5. Right place;
6. Right customer;
7. Right cost.

The client chooses the forwarder with whom he will cooperate. This choice depends on the quality of the forwarding services provided by the forwarder. Therefore, the main task of the freight forwarder in the freight forwarding industry in the market conditions is to ensure a high level of freight forwarding services to customers.

During the analysis of freight forwarders' activities, the factors that influence the client's choice for long-term cooperation with freight forwarders were identified (see Fig. 1) [12]:

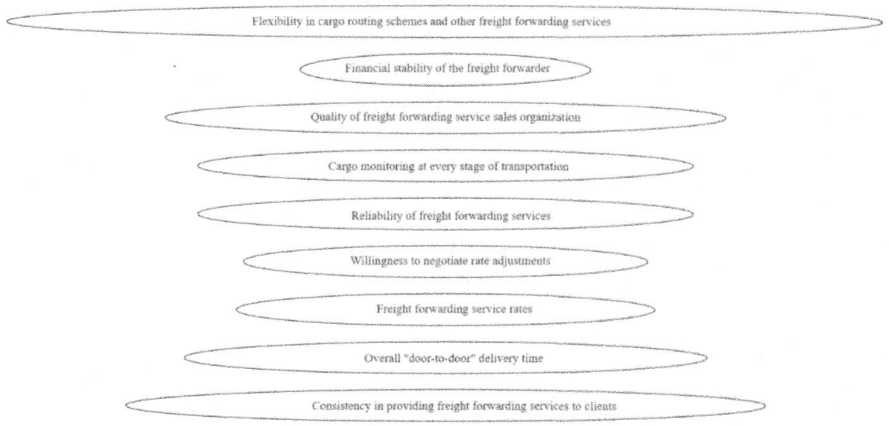

Fig. 1. List of factors influencing the client's choice to work with a freight forwarder.

Below is a brief description of each of the factors:

- Flexibility in freight routing schemes and other freight forwarding services – the ability of the freight forwarder to quickly adapt routes and delivery methods depending on customer requirements and changes in transport conditions.
- Financial stability of the forwarding agent is an indicator of the company's reliability, its ability to fulfill its obligations and ensure operations regardless of economic instability.
- The quality of the organization of sales of freight forwarding services is the efficiency and professionalism of the approach to working with clients, from consultations to paperwork.
- Monitoring of cargo at each stage of transportation – the forwarder's ability to track the movement of cargo in real time, ensuring control over the transport of goods.
- Reliability of freight forwarding services – the ability to fulfill obligations in compliance with the terms and conditions of the contract, minimizing the risk of damage or loss of cargo.
- Willingness to negotiate tariff changes – the forwarding agent's flexibility in setting pricing policy, with the ability to agree on terms that are mutually beneficial for both parties.
- Tariffs for freight forwarding services – the competitiveness of price for services provided by the forwarding agent to its customers, taking into account the cost and quality of freight forwarding services.
- Generalized door-to-door delivery time – the time required to transport goods from the shipper to the consignee, taking into account all stages of the transportation.
- Sustainability of freight forwarding services to customers– the stability and continuity of the freight forwarder's operations regardless of external factors.

- Building of an algorithm for freight forwarding customer service will significantly increase the visibility of processes and the mechanisms of interaction between participants in freight forwarding activities [14].

The algorithm consists of four consecutive stages, each of which contains a set of analysis, assessments and measures related to improving or changing the exact level of freight forwarding customer service (see Fig. 2).

The algorithm we have developed will help to structure the work of freight forwarding company and improve customer interaction.

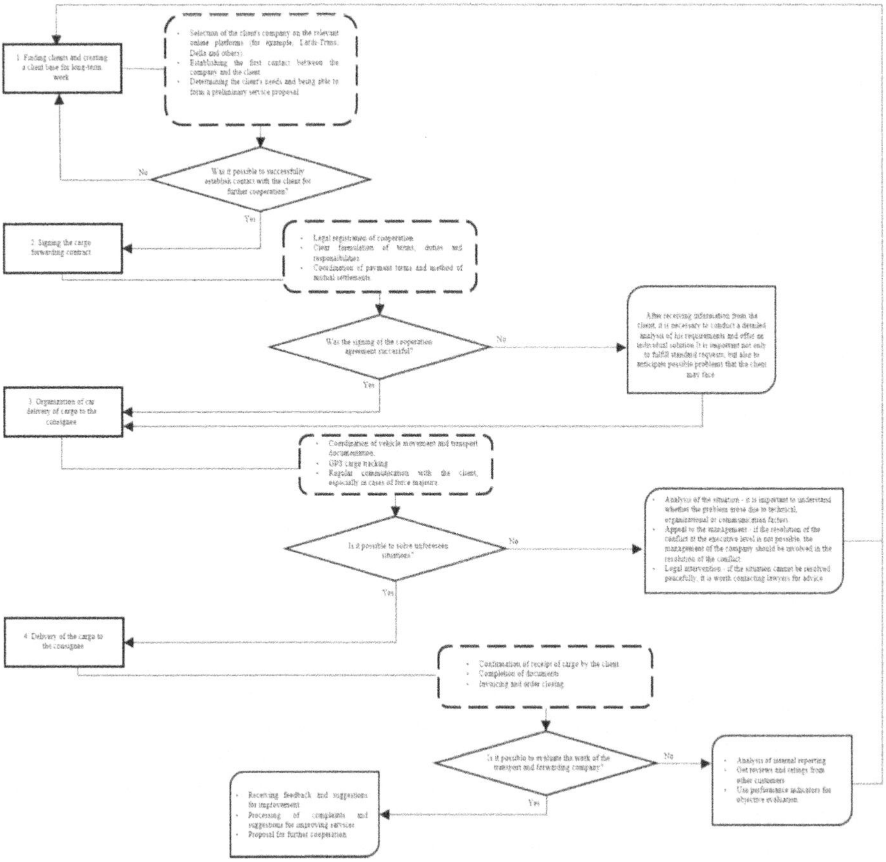

Fig. 2. Flowchart of the algorithm for freight forwarding customer service.

The principles of logistics in freight forwarding activities are the basic rules and approaches that ensure the effective organization of processes of providing freight forwarding services to customers.

The study of the principles of logistics and the need to apply them to improve the efficiency of transportation and freight forwarding services for customers was developed in Table 2, with the general characteristics of these principles [4, 5].

Table 2. Logistics principles in freight forwarding services for customers.

№	Principles	General Description
1	Safety Principle	It includes the following measures such as: rational organization of traffic; cargo safety; ensuring technical requirements for vehicle; analysis of the places of occurrence; selection of the structure of the information network for timely receipt of information; professional development of employees in the freight forwarding indusrty; medical control for drivers, etc.
2	Optimality Principle	It includes the client's choice of freight forwarding services based on criteria such as safety, time, costs, profit, quality, etc.
3	Efficiency Principle	This includes compliance with the following conditions: • Maximizing use of the cargo capacity of vehicles; • Increasing the mileage utilization rate, ensuring loading for return or next trip; Optimizing of the total costs of transportation and freight forwarding services
4	Principle of Active and Passive Adaptation to Operating Conditions	This approach allows you to adapt the provision of freight forwarding services to actual operating conditions
5	Integration Principle	It includes the integration of participants in freight forwarding services for customers
6	Reliability Principle	It involves the need to synchronise all stages of cargo transportation, as well as the creation of production and reserve stocks, as well as providing of freight forwarding in full at the specified time

Adherence to these principles will ensure the improvement of the quality, reliability, and competitiveness of transportation and freight forwarding services.

3 Conclusion

The analysis of problems in the interaction of freight forwarders with customers has shown that the main obstacles are insufficient communication, unsatisfactory quality of service, late delivery of goods and a lack of flexibility in solving problem situations. The factors influencing a client's choice for long-term cooperation with a freight forwarder are identified, including: reliability, safety, competitive tariffs, quality of service and readiness of the freight forwarder to adapt to the individual needs of the client. A customer feasibility study algorithm has been developed. The use of logistics principles will increase the efficiency of freight forwarding services and increase customer satisfaction.

For further research, it is recommended to perform modeling of the freight forwarder's relationship with the client.

References

1. Law of Ukraine "On Freight Forwarding Activities" (July 1, 2004), No. 1955-IV, https://zakon.rada.gov.ua/laws/show/1955-15#Text
2. Gattorna, J., Ellis, D.: The impact on supply chain networks of shifting demand and supply dynamics. In: Global logistics and supply chain strategies for the 2020s: vital skills for the next generation, pp. 29–48. Springer International Publishing, Cham (2020)
3. Boute, R.N., Udenio, M.: AI in logistics and supply chain management. In: Global logistics and supply chain strategies for the 2020s, pp. 49–65. Springer International Publishing, Cham (2020)
4. Reznik, N.P., Kharchevnikova, L.S., Puhachova, V.R.: Problems and prospects of the logistics sector development in Ukraine. Ukrainian J. Appl. Econ. Tech. **6**(4), 30–35 (2021)
5. Lobo, M., Pinho, T.: Lean tools applied in transport and logistics services. Frecuencia Continua **5**(1) (2019). http://portal.amelica.org/ameli/journal/167/1671501014/html/
6. Cui, Y., Gaur, V.: Supply chain transparency using blockchain: benefits, challenges, and examples. J. Bus. Logist. **42**(3), 307–326 (2021)
7. Wagner, S.M., Kurpjuweit, S.: Startups in the logistics sector: value propositions and potential impact. Int. J. Phys. Distrib. Logist. Manag. **50**(5), 421–439 (2020)
8. Gutierrez, A., Boukrami, E., Lumsden, R.: Technological, organizational and environmental factors influencing managers' decision to adopt cloud computing in logistics. J. Enterp. Inf. Manag. **32**(6), 788–807 (2019)
9. Kayikci, Y., Stix, V., Pamucar, D.: Digital logistics ecosystems: application of AI and digital twins for logistics and supply chain sustainability. J. Clean. Prod. **313**, 127927 (2022)
10. Balouei Jamkhaneh, H., Dolgui, A., Ivanov, D.: Industry 4.0-enabled smart supply chain and logistics: systematic literature review and research agenda. Comput. Ind. Eng. **161**, 107540 (2022)
11. Lumsden, R., Hackius, N.: Blockchain and AI in logistics: challenges and opportunities. In: Digitalization in supply chain management and logistics: smart and digital solutions for an industry 4.0 environment, pp. 35–58 (2019)
12. Popa, C., Goia, I.: The analysis of freight for forwarding services using the business process modelling tools. Scie. Bull. Naval Acad. **25**, 110–116 (2022)
13. Melnyk, L.G.: Enterprise economics: a textbook for higher education. University Book, Sumy (2012)
14. Kostyrko, A.H., Zhivchenko, K.V., Demchyshyn, V.D.: The state and prospects of domestic transport logistics. Eastern Eur. Econ. Bus. Manag. **6**(23), 303–308 (2019)
15. Rudkivskyi, O.A., Honhalo, Y.: Problems and ways of developing the enterprise logistics system. Market Infrastruct. **30**, 218–224 (2019)
16. Abdul Rahman, N.S.F., Karim, N.H., Md Hanafiah, R., Abdul Hamid, S., Mohammed, A.: Decision analysis of warehouse productivity performance indicators to enhance logistics operational efficiency. Int. J. Prod. Perfor. Manag. **72**(4), 962–985 (2023)

Dynamic Routing in Urban Transport Logistics in Wartime Conditions

Viktor Danchuk and Oleksandr Hutarevych(✉)

National Transport University, Kyiv, Ukraine
vdanchuk@ukr.net, oleksandr.hutarevych@gmail.com

Abstract. Research on the development of routing methods in urban transport logistics gains particular significance during wartime, when there is a high probability of reduced transport network efficiency due to possible destruction of critical infrastructure and the physical vulnerability of intelligent transport system (ITS) components. The results of the application of dynamic routing method for freight delivery by vehicles in wartime conditions, using ITS based on cloud technologies, are presented. Within the framework of the proposed ITS model, the static and dynamic state of the urban road network is monitored based on the use of consolidated data from all sources available at the moment of operating (Geographic Information Systems (GIS), traffic sensors, mobile IoT devices, etc.). An architectural implementation of such ITS is proposed, which is based on the use of Amazon Web Services cloud technologies and allows for a high level of cybersecurity, reliability and performance. AI methods with high computational efficiency are used for route optimization. To test the proposed method, simulation studies of dynamic routing were conducted within the framework of an asymmetric dynamic traveling salesman problem, using sudden changes in Kyiv urban road network operating conditions (air raid alert) as an example. Route optimization with dynamic updates at each freight delivery point was carried out using a modified ant colony algorithm based on GIS traffic data in urban road network sections. The results of the simulation studies demonstrate the potential of the proposed dynamic routing method using a cloud-based ITS in urban transport logistics under wartime conditions.

Keywords: Dynamic Route Optimization · AI Discrete Optimization Methods · GIS · IoT · Cloud Technologies

1 Introduction

Solving urban transportation logistics tasks under sudden, rapidly changing, and unpredictable external environmental influences on the state of the urban road network (URN) is highly relevant from both fundamental and applied perspectives. A key role in addressing this challenge is dynamic routing, which, within the framework of intelligent transport systems (ITS), aims to optimize routes in real-time and minimize negative environmental impacts. This requires the use of timely data collection and processing systems for traffic information, along with appropriate AI-based methods for discrete route optimization based on the analysis of this data.

O. Slavinska et al. (Eds.): ITSESQC 2024, LNNS 1335, pp. 339–349, 2025.
https://doi.org/10.1007/978-3-031-87376-8_30

Today, a wide range of modern, innovative technologies are available for obtaining and processing relevant data, such as GPS, VANET systems, the Internet of Things (IoT) with traffic sensors, big data (BD) processing, blockchain (BC), and others (see, for example, [1–8]). Typically, in solving logistics tasks, data collection occurs within the framework of V2V and V2I technologies. This data is then transmitted to control centers, where it is processed, discrete route optimization problems are solved. Finally, the corresponding information about the results obtained is provided to the user. AI methods used for discrete route optimization include swarm intelligence (SI) techniques, such as ant colony optimization (ACO), particle swarm optimization (PSO), artificial bee colony algorithms (ABC), genetic algorithms (GA), the evolutionary simulated annealing (ESA) method, as well as their modifications and hybrid approaches that combine these AI optimization techniques with artificial neural networks, machine learning, and deep learning methods [9–15].

Research on developing routing methods in the context of wartime conditions has become particularly relevant. Due to the destruction of roads, bridges, and other infrastructure elements, it often becomes impossible to use traditional or previously determined optimal routes. Additionally, for safety reasons, parts of the road infrastructure may be temporarily unavailable. For example, in Kyiv, two bridges over the Dnipro River become inaccessible to traffic during air raids. Such situation necessitates the integration of actual data about road infrastructure state and the prompt i determination of alternative optimal routes based on this data.

Furthermore, martial law imposes limitations on the use of resources such as fuel, vehicles, and personnel. Optimal routing enables the efficient use of available resources, minimizing costs.

Another important factor in the context of wartime is the high probability of unstable operation of primary data collection systems due to destruction or power outages. Relying on data from a single source (e.g., IoT traffic sensor data) may not be sufficient to form a complete picture of the URN's condition, as access to such sources could be lost. Therefore, it is necessary to develop a system that monitors the URN's state by consolidating data from all available sources, such as GIS, VANET, MANET systems, and others.

A critical aspect to consider implementing ITS during wartime is the vulnerability of the physical infrastructure hosting the system. In the event of a power outage or the destruction of a data center, the system may fail, and data could be partially or completely lost. In this regard, cloud technologies play a key role in ensuring the resilience and security of ITS. With the right precautions, these technologies enable the creation of reliable, distributed systems that are resistant to the risks associated with wartime.

In this context, the goal of this work is to develop a method for dynamic routing in urban transportation logistics under wartime conditions using the proposed ITS, which is based on cloud technologies.

2 Research Methodology

Given the high relevance of solving routing problems under wartime conditions, an ITS architecture designed to solve vehicle routing problems (VRP) using information from various data sources about the current state of the urban road network (URN) is proposed. The design of an architecture model for proposed system is illustrated in Fig. 1.

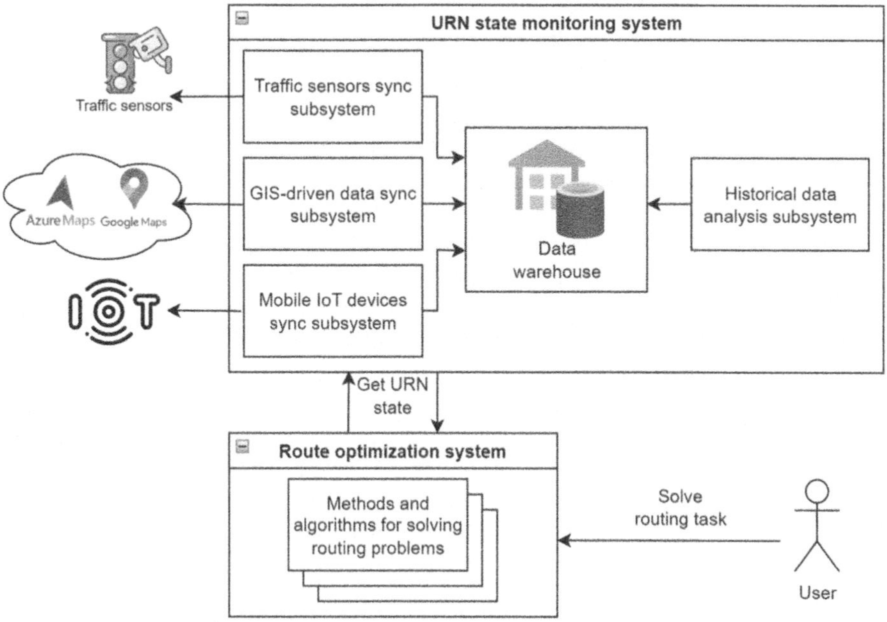

Fig. 1. Design of an ITS architecture model for solving dynamic routing problems in wartime conditions

This system consists of two partially independent components: the URN state monitoring system and the route optimization system. The URN state monitoring system consolidates information from various available sources to form a comprehensive overview of the current state of the URN, which is then used to solve routing problems. The functionality of the route optimization system is based on various methods and algorithms for solving different types of VRP.

The URN state monitoring system includes several subsystems that consolidate data from multiple sources, standardizing it for use in solving VRP tasks. These subsystems for data processing and storage, as illustrated in Fig. 1, include:

- Traffic sensors sync subsystem collects data from traffic sensors located on vari-ous sections of the URN. These sensors measure traffic flow characteristics, ena-bling the assessment of road congestion, which is critical for route planning.
- GIS-driven data sync subsystem extracts URN data from various GIS platforms, such as Google Maps [16], Azure Maps (formerly Bing Maps) [17], and others. These platforms provide up-to-date maps of urban streets, highways, and other transport routes

by combining satellite images, aerial photography, and street-level data collection. They offer information on road hierarchy (e.g., highways, major roads, local streets), traffic flow, condition of roads, and real-time congestion.

- Mobile IoT devices sync subsystem receives data from various mobile IoT devices located along URN sections, embedded in vehicles, drones, and other sources. Modern vehicles, for example, can supply a wide range of useful data, including GPS coordinates, average traffic speed, current location, and fuel consumption. Due to the variety of mobile IoT devices, this subsystem can be further divided in-to smaller subsystems focused on specific device types.

In addition to described subsystems for collecting current URN state data, the proposed URN state monitoring system architecture is designed to be open for creation of additional integration subsystems for other data sources.

The URN state monitoring system also includes a data warehouse for storing historical URN data obtained from the described subsystems (see Fig. 1). Modern cloud data warehouses allow the efficient storage and analysis of large volumes of data, using technologies such as Big Data [18].

To analyze historical data, the URN state monitoring system includes a separate historical data analysis subsystem. This subsystem aims to predict the state of the URN based on historical data analysis (see Fig. 1). Such predictions are highly useful for route evaluation and planning. Specifically, analyzing historical data helps identify long-term traffic flow patterns and congestion trends, which are crucial for urban road planning. This data enables planners to optimize road network designs by identifying areas that require expansion, traffic signal optimization, and new infrastructure development. Historical data also helps urban planners anticipate future demand, design roads to handle projected traffic volumes, and reduce bottlenecks by improving intersections or adding lanes where necessary.

The route optimization system includes various AI methods and models for solving various types of routing tasks. Users send a request to the route optimization system, including all necessary data for routing task solving. The route optimization system receives current URN state data via the URN state monitoring system, solves the user's task using the most appropriate AI method, considering the current URN state, and sends the user the optimal solution.

It is important to note that the route optimization system must include methods for solving not only static VRP tasks but also dynamic VRP (DVRP) tasks, which are particularly relevant in the context of frequent changes in URN conditions due to significant traffic changes, accidents, and other factors. In wartime, the relevance of DVRP solutions increases due to possible road infrastructure destruction, air raid alarms, evacuations, etc.

To ensure a high level of reliability and performance in solving vehicle routing tasks, the use of cloud technologies is proposed. Figure 2 shows a diagram of the architecture model of the proposed system, based on Amazon Web Services (AWS) cloud technologies [19].

The architecture diagram suggests hosting the system's computational resources on a virtual private cloud (VPC). To ensure system resilience and effective protection against cyberattacks, which pose a high risk in wartime, various AWS services are used.

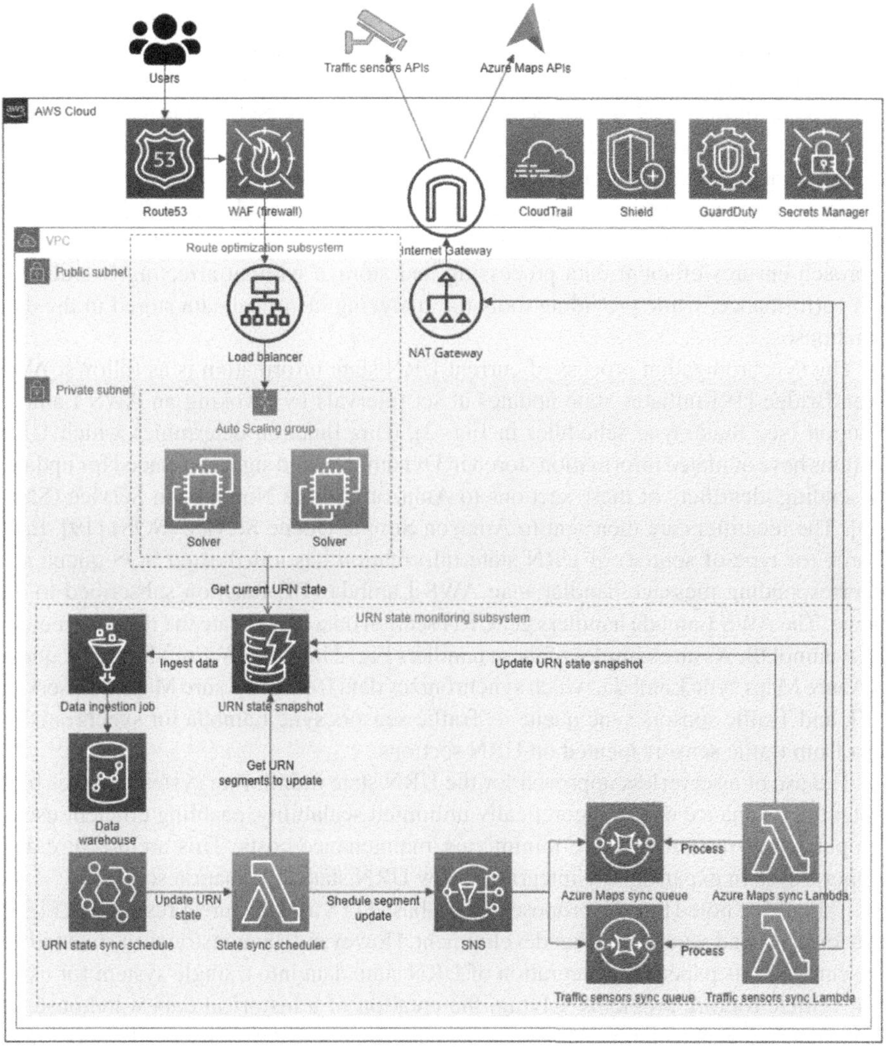

Fig. 2. Diagram of model for ITS cloud-based architecture using AWS

AWS CloudTrail [19] monitors resources to detect and track suspicious activities, while AWS Shield [19] protects against DDoS attacks. AWS GuardDuty [19] enhances cybersecurity by leveraging machine learning, anomaly detection, and threat intelligence to identify potential threats. AWS Secrets Manager [19] securely manages sensitive system information. User traffic passes through AWS Web Application Firewall (WAF) [19], which protects applications from common security threats by filtering and monitoring user requests.

The route optimization system uses Application Load Balancer (ALB) [19] and Auto Scaling Group for Amazon EC2 [19] servers, which handle the computation of optimal routes in response to user requests. This approach ensures efficient use of computational resources and scalability according to the system's current load.

The URN state monitoring system has a serverless architecture. For storing the current state of the URN, a high-performance serverless NoSQL database DynamoDB [19] is used, where each row represents an object containing the current status of a URN section from all available sources.

For storing and analyzing historical data, the system includes a modern cloud data warehouse, Amazon Redshift [19]. AWS Glue [19] is used for ETL processes. Such approach ensures efficient data processing and storage without affecting overall system performance, while providing tools for analyzing historical data stored in the data warehouse.

The synchronization process of current URN state information is as follows: AWS EventBridge [19] initiates state updates at set intervals by invoking an AWS Lambda function (see State sync scheduler in Fig. 2). This function determines which URN sections have outdated information stored in DynamoDB and signals the need for updates by sending identifiers of these sections to Amazon Simple Notification Service (SNS) [19]. The identifiers are then sent to Amazon Simple Queue Service (SQS) [19]. Each source (or type of source) of URN state information has a dedicated SQS queue and a corresponding message handler – an AWS Lambda [19] function subscribed to the queue. The AWS Lambda handlers get URN section data and update the relevant records in DynamoDB. As an examples of these handlers Fig. 2 includes Azure Maps sync queue + Azure Maps sync Lambda, which synchronizes data from the Azure Maps GIS service [17], and Traffic sensors sync queue + Traffic sensors sync Lambda for synchronizing data from traffic sensors located on URN sections.

The use of a serverless approach for the URN state monitoring system ensures high system performance due to theoretically unlimited scalability, enabling efficient use of computational resources while minimizing maintenance costs. This architecture also supports system expansion by integrating new URN state information sources.

It should be noted that the proposed cloud-based ITS architecture, presented in Fig. 2, is incomplete and requires further development. However, it demonstrates the main ideas presented in this paper: the integration of URN state data into a single system for using it in vehicle routing problems solving, the creation of a historical data warehouse for identifying relevant patterns and forecasting URN state, and the use of cloud technologies to ensure high system performance and resilience in wartime conditions.

3 Results and Discussion

In order to test the proposed method for dynamic routing in urban transport logistics using an ITS based on cloud technologies (see Sect. 2), simulation studies were conducted. These studies focused on the process of freight delivery across the urban road network (URN) of Kyiv. The dynamic asymmetric traveling salesman problem (DTSP) was considered as a bidirectional weighted graph. The nodes of the graph correspond to delivery points, and the weights of the graph's edges are given by relative discrete values according to the optimization criterion (e.g., distance between nodes, average time or speed to traverse an edge, fuel consumption, transportation cost, etc.). In this simulation, the vehicle delivers freight from a depot to 12 delivery points and returns to the depot. The delivery points correspond to branches of the popular Ukrainian postal operator "Nova Poshta" A graphical representation of the freight delivery points is shown in Fig. 3. The numbering of the delivery points in Fig. 3 corresponds to the branch numbers of "Nova Poshta".

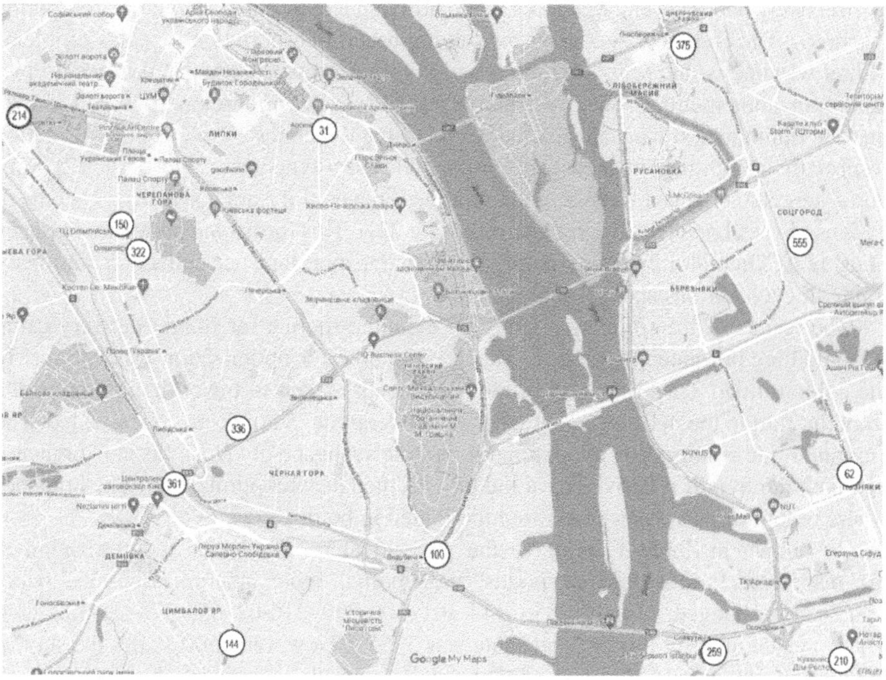

Fig. 3. Locations of freight delivery destinations on the map of Kyiv

Simulation conditions:

- The starting point (depot) is branch №214 (Saksahanskoho Street, 102).
- The freight delivery route is circular, with sequential deliveries.
- The optimization criterion is the minimization of the total time spent delivering freight to the specified points and returning to the depot.
- Reoptimization of the route occurs upon the vehicle's arrival at each delivery point.
- As a result of a prior enemy drone attack, some elements of the city's energy infrastructure were damaged, leading to a loss of traffic data from road sensors located across the URN. Under the proposed ITS (see Figs. 1 and 2), dynamic routing is based on GIS traffic data for URN sections.
- After 1,500 s from the start of the vehicle's journey, an air raid alarm is announced in Kyiv, halting traffic on the Pivdennyi and Darnytskyi bridges, while all other bridges remain operational. This results in a significant increase in travel time between delivery points located on different sides of the Dnipro River.

To perform route optimization and reoptimization based on traffic changes across the URN sections, a modified ant colony optimization algorithm, ACOmod [20], was utilized. In the proposed ITS (see Figs. 1 and 2), the ability to automatically update the weights of the graph's edges has been implemented, with the optimal configuration of the partially traversed route being saved before each update [20]. Since DTSP is classified as an NP-hard problem, the method used to solve it must deliver high performance to effectively handle even large tasks while utilizing a reasonable amount of computing resources. The choice of ACOmod for solving urban transportation logistics problems during wartime is due to its versatility and high performance compared to most other AI optimization methods (see, for example, [11, 12, 14]). Furthermore, it is well-known that the optimization mechanisms in ACO and the dynamics of traffic flow formation (especially in high-density areas) are similar and have a synergistic nature [21, 22]. Similar synergistic effects are observed in nonlinear nonequilibrium dissipative systems of various physical natures (see, for example, [23, 24]). This further supports the adequacy of using ACOmod for dynamic routing in urban transport logistics under non-stationary traffic flow conditions across URN sections.

The results of the simulation study on dynamic routing for freight delivery using ACOmod are presented in Table 1. The average time for optimization was 936 ms. In the column titled Travel time, the time (in seconds) taken to travel the optimal route from the start to the delivery point, where route reoptimization occurs, is provided. The Optimal route column lists the optimal route (the sequence of "Nova Poshta" branches to which the freight is delivered) at the time of the latest reoptimization. The branches where reoptimization takes place are highlighted in bold.

As shown in Table 1, prior to the air raid alert, even after reoptimization at the respective freight delivery points, the optimal route determined at the depot remained unchanged: 214–150-336–144-361–100-259–210-62–555-375–31-322–214. The expected time for this optimal route was 5855 s. However, 1500 s after the start of the journey, an air raid alert was announced. As a result, significant changes occurred in the configuration of the URN, specifically the closure of the Pivdennyi and Darnytskyi bridges in Kyiv (see Fig. 1). At the "Nova Poshta" branch №100, where the vehicle arrived after the air raid was declared, a reoptimization procedure was carried out, leading to a significant change in the optimal route: 214–150-336–144-361–100-375–555-62–259-210–31-322–214 (see Table 1). The recalculated route time after the air raid alarm was 6326 s. If the vehicle had continued on the originally determined route from branch №100, the delivery would have taken 6517 s, which is 191 s longer than the newly calculated route.

Thus, the results of the simulation studies indicate the potential of the proposed dynamic routing method using a cloud-based ITS in wartime conditions.

Table 1. DTSP solving results in conditions of wartime using ACOmod

Travel time (s)	Optimal route
0	**214**–150-336–144-361–100-259–210-62–555-375–31-322–214
335	214–**150**-336–144-361–100-259–210-62–555-375–31-322–214
761	214–150-**336**–144-361–100-259–210-62–555-375–31-322–214
1115	214–150-336–**144**-361–100-259–210-62–555-375–31-322–214
1387	214–150-336–144-**361**–100-259–210-62–555-375–31-322–214
AIR RAID ALARM!!!	
1846	214–150-336–144-361–**100**-375–555-62–259-210–31-322–214
2561	214–150-336–144-361–100-**375**–555-62–259-210–31-322–214
3033	214–150-336–144-361–100-375–**555**-62–259-210–31-322–214
3610	214–150-336–144-361–100-375–555-**62**–259-210–31-322–214
4139	214–150-336–144-361–100-375–555-62–**259**-210–31-322–214
4406	214–150-336–144-361–100-375–555-62–259-**210**–31-322–214
5476	214–150-336–144-361–100-375–555-62–259-210–**31**-322–214
6060	214–150-336–144-361–100-375–555-62–259-210–31-**322**–214
6326	214–150-336–144-361–100-375–555-62–259-210–31-322–**214**

4 Conclusion

Research on developing routing methods in urban transport logistics is particularly important in wartime conditions. The destruction of transport infrastructure, changes in URN operating conditions, resource limitations, power outages, and the physical vulnerability of ITS components significantly increase the possibility of decreased ITS efficiency in supporting transport and logistics operations.

The application of a dynamic routing method for freight delivery by vehicles in wartime conditions, utilizing a cloud-based ITS, is presented. Within the framework of the proposed ITS model, the static and dynamic state of the URN is monitored based on the use of consolidated data from all sources available at the moment of operating (GIS, IoT data from traffic sensors, VANET, MANET systems, etc.). The proposed architectural implementation of ITS, built using Amazon Web Services cloud technologies, provides a high level of cybersecurity, reliability, and system performance. However, the proposed architectural implementation is not yet perfect and requires additional improvements, such as distribution across different availability zones or even regions to enhance system availability and fault tolerance. AI methods with high computational efficiency are used for route reoptimization.

In order to test the proposed method, simulation studies of dynamic routing were conducted within the framework of an asymmetric DTSP, using sudden changes in Kyiv URN operating conditions (air raid alert) as an example. Route optimization, with dynamic updates at each freight delivery point, was performed using a modified ant colony algorithm based on GIS traffic data for URN sections.

The results of the simulation studies demonstrate the potential of the proposed dynamic routing method using cloud-based ITS in urban transport logistics under wartime conditions.

References

1. Rothkrantz, L.: Hybrid dynamic route planners. In: CompSysTech'18: 19th International Conference on Computer Systems and Technologies, Ruse Bulgaria. 12–19 (2018). https://doi.org/10.1145/3274005.3274012
2. Tsoukas, V., et al.: Development of a dynamically adaptable routing system for data analytics insights in logistic services. Analytics. **2**, 328–345 (2023). https://doi.org/10.3390/analytics2020018
3. Lyu, Z., et al.: Freight operations modelling for urban delivery and pickup with flexible routing: cluster transport modelling incorporating discrete-event simulation and GIS. Infrastructures. **6**, 180 (2021). https://doi.org/10.3390/infrastructures6120180
4. Badole, M.H., Thakare, A.D.: An optimized framework for VANET routing: A multi-objective hybrid model for data synchronization with digital twin. Int. J. Intell. Netw. **4**, 272–282 (2023). https://doi.org/10.1016/j.ijin.2023.10.001
5. Saoud, B., et al.: Artificial intelligence, internet of things and 6G methodologies in the context of Vehicular Ad-hoc Networks (VANETs): Survey. ICT Express. **10**, 959–980 (2024). https://doi.org/10.1016/j.icte.2024.05.008
6. Mouhcine, E., Mansouri, K., Mohamed, Y.: Solving traffic routing system using VANet strategy combined with a distributed swarm intelligence optimization. J. Comput. Sci. **14**, 1499–1511 (2018). https://doi.org/10.3844/jcssp.2018.1499.1511
7. Park, J., et al.: Intelligent trip modeling for the prediction of an origin-destination traveling speed profile. IEEE Trans. Intell. Transp. Syst. **15**, 1039–1053 (2014). https://doi.org/10.1109/tits.2013.2294934
8. Chai, H., et al.: Dynamic traffic routing in a network with adaptive signal control. Transportation Research Part C: Emerging Technologies. **85**, 64–85 (2017). https://doi.org/10.1016/j.trc.2017.08.017
9. Yavuz, M.N., Özen, H. Calibration of microscopic traffic simulation of urban road network including mini-roundabouts and unsignalized intersection using open-source simulation tool. Scientific Journal of Silesian University of Technology. Series Transport. **122**, 305–318 (2024). https://doi.org/10.20858/sjsutst.2024.122.17
10. Russo, F., Comi, A.: Sustainable urban delivery: the learning process of path costs enhanced by information and communication technologies. Sustainability. **13**, 13103 (2021). https://doi.org/10.3390/su132313103
11. Danchuk, V., et al.: The optimization of cargo delivery processes with dynamic route updates in smart logistics. East.-Eur. J. Enterp. Technol. **2**, 64–73 (2023). https://doi.org/10.15587/1729-4061.2023.277583
12. Zhang, N.: Smart logistics path for cyber-physical systems with internet of things. IEEE Access. **6**, 70808–70819 (2018). https://doi.org/10.1109/access.2018.2879966
13. Ng, K.K.H., et al.: A multiple colonies artificial bee colony algorithm for a capacitated vehicle routing problem and re-routing strategies under time-dependent traffic congestion. Comput. Ind. Eng. **109**, 151–168 (2017). https://doi.org/10.1016/j.cie.2017.05.004
14. Zajkani, M.A., Baghdorani, R.R., Haeri, M.: Model predictive based approach to solve DVRP with traffic congestion. IFAC-PapersOnLine. **54**, 163–167 (2021). https://doi.org/10.1016/j.ifacol.2021.12.028

15. Liu, H., et al.: DAACO: adaptive dynamic quantity of ant ACO algorithm to solve the traveling salesman problem. Complex & Intelligent Systems. **9**, 4317–4330 (2023). https://doi.org/10.1007/s40747-022-00949-6
16. Google Maps Platform Documentation, https://developers.google.com/maps/documentation
17. Azure Maps - Geospatial Mapping APIs, https://azure.microsoft.com/en-us/products/azure-maps
18. Nambiar, A., Mundra, D.: An overview of data warehouse and data lake in modern enterprise data management. Big Data Cogn. Comput. **6**, 132 (2022). https://doi.org/10.3390/bdcc6040132
19. Cloud Computing Services - Amazon Web Services (AWS), https://aws.amazon.com/
20. Danchuk, V., Hutarevych, O.: Method of adaptive route optimization for solving dynamic traveling salesman problem using modified ant colony algorithm. Visnyk National Transport University. **58**, 55–66 (2024). https://doi.org/10.33744/2308-6645-2024-1-58-055-066
21. Dorigo, M., Di Caro, G.: The ant colony optimization meta-heuristic. In: Corne, D., Dorigo, M., Glover, F. (eds.) New Idea in Optimization. McGrow-Hill, London (1999)
22. Kerner, B.S.: Introduction to Modern Traffic Flow Theory and Control. Springer, Berlin (2009)
23. Puchkovska, G.O., et al.: Dynamics of molecules and phase transitions in the crystals of pure and binary mixtures of n-paraffins. J. Mol. Struct. **614**, 159–166 (2002). https://doi.org/10.1016/s0022-2860(02)00237-5
24. Puchkovska, G.O., et al.: Resonance dynamical intermolecular interaction in the crystals of pure and binary mixture n-paraffins. J. Mol. Struct. **708**, 39–45 (2004). https://doi.org/10.1016/j.molstruc.2004.02.010

System-Conceptual Methods of Managing Technologically Sustainable Development in Motor Transport

Ramazan Khabutdinov(✉)

National Transport University, Mykhaila Omelianovycha-Pavlenka Str, Kyiv 01010, Ukraine
Habutd1@gmail.com

Abstract. The article proposes a new resource-consolidated approach to the conceptual and cognitive intellectualization of the main functions of the Motor Transport (MT) and the Motor Transport System (MTS) in relation to their sustainable development. The article examines the principles, isomorphic models, and strategies of systemic and conceptual management of technologically sustainable development in MT in relation to the tasks of conceptual improvement of the emergent and integrative functions of the MTS. The strategies are based on the scientific concept of high-tech and resource-saving reproduction of motor transport processes and services. It is shown, that the existing techno-empirical methods of organization and economics of the motor transport process do not provide conceptual management of technological development in MT, due to their use of the theoretical scheme of virtual transportation of goods and passengers. In connection with that naturalized, anti paradoxically harmonized models of MT's technical and technological resources and transport operations are proposed. Transport-operational and system-oriented regularities of the interactive-adaptive functioning of cars as technical and technological resources of MT subsystem and the MTS have been obtained. A new material-production, technological-innovation and system paradigm of knowledge about MT as a sphere of material, but paradoxical reproduction of motor transport processes is adopted. The patterns of complex functioning of technical-technological resources were obtained in algebraic form. It is shown that these regularities reflect a single physical and technical mechanism for the implementation of motor transport processes, as well as functions important for the sustainable development of MT and MTS.

Keywords: Principles · Concept · Resources · Technical and Technological Innovations · Motor Transport · Functions · Motor Transport Technology and System · Energy and Resource Saving · Isomorphism · Strategies · Sustainable Development

1 Statement of the Problem

On the Motor Transport (MT), the solution to the scientific and practical problem of systemic and conceptual management of its technologically sustainable development, based on the fixings of the MT [3–6] is relevant. There are three unique systemic fixings of MT

O. Slavinska et al. (Eds.): ITSESQC 2024, LNNS 1335, pp. 350–362, 2025.
https://doi.org/10.1007/978-3-031-87376-8_31

for the country: **1)** material, but paradoxically incorrect creation of motor transport processes (MTP) and services (MTSV) and their technological and innovative reproduction; **2)** subsystem-active provision of the first fixing in mass transport processes; **3)** implementation of transport service and self-service of economic entities and the population of the country. The existing methodology of the economic and organizational analysis of MTP provides only the third fixing of the MT [1, 2]. At the same time, the first and second fixing of the MT are neglected and axiomatically denied because this methodology is based on a simplified, service-commercial, non-modernizing and non-system paradigm of knowledge (SSCNN-paradigm) about MT as a field of techno-empirical transport service [1]. According to the existing simplified paradigm of knowledge, the methodology of economic and organizational analysis of MT is based on an opaque (black-box) theoretical scheme of virtual transportation of goods and passengers by virtual mobile rolling stock in an immaterial environment [1, 2]. Therefore, it opposes of sustainable development of MT. In addition, the existing methods of organizing motor transportations, due to the indicated shortcomings, led to the spontaneous formation of a creeping, fragmentary, contradictory and eclectic technical and technological empiricism at the MT. Such empiricism leads to the implementation in practice of a non-competitive method of counter-cost justification of operationally profitable and low-tech motor transport processes (MTP) and services (MTSV). In European countries, much attention is paid to the methodology of sustainable development of Motor Transport (MT) and the Motor Transport System (MTS) [4–7]. However, at the same time, there are not enough publications related to the issues of system-conceptual management of technologically sustainable development in MT.

2 The Main Part

The purpose of the article is to present the main provisions and strategies for the formation of concepts of technological and sustainable development at MT (HTRSR of MTP and MTSV), taking into account the scientific idea of operationally energy and resource saving [3–5].

Automobile transport technology (ATT) is understood as a product –focused and scientifically described set of material-creative and procedurally-detailed methods of adaptive and energy-intensive transportations of cargo and passengers in a conflict environment, as well as accident-free and resource-transforming production of high-quality material product of Motor Transport in the road motor operations.

Important for the analysis of the sustainable development of MT are the operational and technological features of natural Motor Transportations (NMT): **a)** the latter involve multifunctional cars as complex resource-technical and technological means (CRTTM) of motor transport operations (MTO), processes (MTP) and system (MTS); **b)** they are characterized by high levels of their energy and resource consumptions; **c)** the world market of cars (at the stage of their market circulation) offers a too wide range of their variable dimensions (in terms of payload and passenger capacity) and structural and technical characteristics with high dynamics of changes in their parameters over time; **d)** there is a great variety of states of structural and technical, road - traffic and terminal factors of transport and technological operation of cars. At the same time, three operational and technological requirements (adaptive productivity, trajectory-distance safety, energy efficiency) must be fulfilled in the technological processes of road transportations to ensure the successful functioning of the cars as: CRTTM of MTO, of MTP and MTS. A functional structure of ATT S(ATT) has been developed, which takes into account the specified features of MT, it is presented in the form of the following set [3]:

$$S(ATT)\,\varepsilon\,(CTT1,\ CTT2,\ CTT3.1,\ CTT3.2, CTT3.3\ ), \tag{1}$$

where CTT1 is the first service-organizational component of the forms of connection of chains of simplified transport and terminal operations according to the criteria of operational costs of rolling stock and delivery time of goods or passengers [1, 3]; CTT2 - the second - material and production component, the development of which is determined by a set of indicators (energy, energy-resource conversion technology) of the efficiency of the components of the resource-technical and technological base of MT [3,9]; CTT3.1, CTT3.2, CTT3.3– mathematical models for the analysis: of elements of CTT1, CTT2, also instructions and rules for carrying out transport and terminal operations, respectively.

The operational and technological identification of new cars as complex resource-technical and technological means (CRTTM) is implemented on the basis of the model of the modular scheme of the structural and parametric organization of the car (SPOC) Kjr. This model for the analysis of constructive-technical and product-technical innovations at MT has the following form [3]:

$$K_{jr}^{N}(t) \in \left(x_{imp}^{N},\ y_{iml}^{N}\right)_{jr} \rightarrow perfect,\ \ldots\ t = t_1,\ t_2,\ t_3 \ldots\ t_n, \tag{2}$$

where x_{imp}^{N}—is the updated subset of the p-th parameters of the m-th structural and functional module in the i-th SPOC device of the cars; N– is the upper index of the sign of SPOC innovation; y_{iml}^{N}– the updated subset of the l-th characteristics of the structure of the m-th structural and functional module of the i-th device, (i = 1, 2, 3, 4), (m = 1, 2, 3, 4), ($l = l_1, l_2, l_3 \ldots$); j – rolling stock class index (j = 1, nk, where nk –is the number of car classes); r– is the index of car varieties in the j-th class (r = 1, nв), where nв– is the number of variants of cars with implemented technical innovations).

The structure of the material and production component ATT-CTT2, which is updated, is presented in work [3]. It includes necessary elements for material production on MT: auto-technical resources $R_T(K_{jr})$ and technical-technological R_{TT} resources of MT, labor P_l and machine P_M procedures of ATT, processes of transport transformations of the car's internal energy – ET, as well as processes of ETRP - energy transformation of R_{TL} into the route physical product of MT-W_{rp}.

The basic function of MT-FA1 is formed in motor transport operations by using the resource-technical properties of the car as R_{TT}. The indicators of these properties are determined by three factors: the characteristics of car's SPOC Kjr, the above-mentioned elements of the ATT-CTT2 component, the interactive influence of the route elements of the MTS infrastructure during the adaptive functioning of the car as complex resource-technical and technological means (CRTTM) of motor transport operations (MTO), processes (MTP) and system (MTS).

The innovative development of automobile transport technology (ATT) and processes (ATP) is ensured on the basis of project identification and justification of the complex of technical and technological innovations - NVTT on MT [3]:

$$NVTT\,\varepsilon\,(NVKT,\ NVGT,\ NVTS,\ NVTL,\ NVET\ ) \rightarrow perfect, \tag{3}$$

where NVKT - structural and technical innovations in the structural and parametric organization of the new car $K_{jr}^{N}(t)$; NVGT – car's characteristics as a scientific and technical goods; NVTS – transport service innovations in the structure of CTT1 of ATT; NVTL – technological innovations in the structure of CTT2 of ATT; NVET – economic and technological innovations (resource-saving ATT and microeconomics of ATP); perfect – operator of standard-comparative improvement and development.

The functioning of the Motor Transport (MT) in accordance with its fixings as a sphere of material (but paradoxically incorrect) reproduction of motor transport processes and services (MTPS) involves the performance of a set of FA - production and economic functions [3]:

$$FA\ \varepsilon\ (FA1,\ FA2,\ FA3,\ FA4, FA5,\ FA6), \tag{4}$$

where MT functions are: technological reproduction of motor transport processes and services -FA1 (basic and existential function, on its basis all others are implemented); transport service -FA2 (service and revenue function); transport capital turnover-FA3 (this economic function is possible in two versions: in service FA3.2 and technological FA3.1); market resource provision – FA4 (resource provision of the basic function of FA1); innovative reproduction of transport processes and services - FA5 (innovative development); of technical service–FA6 (maintenance of the technical condition of cars, this function implemented only in large transport enterprises). At the same time, in the methods of economy and organization of service processes of automobile delivery of goods and passengers [1, 2] important functions of MT- FA1, FA3.1, FA4, FA5 are neglected.

Mathematical models of the analysis of two prospectively important functions of MT (FA1 and FA5), as well as models of system analysis of the technological quality of motor transport operations and processes should be based on the laws of Z_{RTT} interactive-adaptive functioning of loaded vehicles, as technical-technological $R_{TT}(K_{jr})$ resources of MT. The $R_{TT}(K_{jr})$ structure (2) technically and compositionally includes three MT material resources (technical, energy, labor) and a defined mass of cargo or passengers. Within the framework of the system approach and road- traffic subsystem, each controlled car acts as a complex resource-technical and technological R_{TT} as means of Motor transport operations, processes and system (CRTTM MTOPS) [3, 9].

Motor Transport (MT) is an active subsystem of the Motor Transport System (MTS), therefore, on the basis of operational-adaptive and interactive functioning of $R_{TT}(K_{jr})$-technical and technological resources, important functions of MT (FA1 and FA5) are formed first. Then, the integrative functions of the MTS are formed [3]. Three macro system fixings of MTS are transport and infrastructure support: **1)** massive material and technological creation of high-quality motor transport processes and services also their modernization reproduction; **2)** address-trajectory and transport-energy integration of the discontinuous economic and social space of the country by motor transport processes; **3)** transport and energy transformation of potential R_{MP} (cargo and passengers before transportation) resources of the economic and social macro-system (ESMS) into real R_{MR} (after transportation, they become economically useful), according to the logical scheme ($R_{MP} \rightarrow R_{MR}$)).

The Motor Transport System (MTS) is understood as a transport-focused, technologically composite, interactive-functional and integral set of material and production resources of MT and its three infrastructure subsystems (terminal (TMSS), road and traffic (RTSS), also service and technical (STSS)), which in the form of integrity ensures the material functioning of a higher level system, i.e. the economic and social macro system (ESMS) of the country in accordance with its threefold fixings and the characteristics of motor transport demands [3, 5–7]. Taking into account the need to ensure the fixings, a set of system functions (SF) of MTS was formed [3]:

$$SF\,\varepsilon\ (SF1,\ SF2,\ SF3,\ SF4,\ SF5,\ SF6,\ SF7,\ SF8), \tag{5}$$

SF1 –differentiated and economically necessary distribution of material resources of MT and three infrastructure subsystems (TMSS, RTSS, STSS) in the terminal- road network of the country; SF2– transport, operational and technological composition of material resources of subsystems; SF3 – function of transport and operational emergency (mass processes of interactive and adaptive functioning of technical and technological resources of MT); SF4 – function of transport and energy integration of the terminal infrastructure of the MTS and the transformation of potential resources of the ESMS R_{mp} into real R_{mr} (after transportation) based on the mass implementation of laws - Z_{RTT} interactive but-adaptive functioning of cars, such as R_{TT} and CRTTM MTOPS; SF5 – neutralization of dysfunction in subsystems and in the mechanisms of their interaction; SF6 – conceptual modernization of subsystems and system functions, SF7 – motor transport providing for the macro-system functioning (ESMS); SF8– is a function of technological development and synergetic equifinal function (the function of development of all subsystems based on the principle of ensuring energy-resource synergy during theirs interaction).

Next, the solution to the problem of establishing a complex of phenomenological and ontological regularities Z_{RTT} of the managed and complex functioning the car (as a technical and technological resources- R_{TT} of MT) in motor transport operations is presented. At the same time, they are taken into account the traction and power balance equations of locomotion of the Car under the conditions of interaction of its wheel-traction module with the rolling surface of the road in adaptively controlled traffic modes [3]. As a basis, the differential equation of the change in the kinetic energy of the Car (as R_{TT}) of the modular-generalized type, which is operationally identified as the variable set $K_{jr}(t)$ according to (2), is adopted. Based on the solution of the named system of equations, a complex of Z_{RTT} models was obtained for the analysis of the controlled functioning of technical and technological resources R_{TT} in mass motor transport operations. The complex includes models of steady and unsteady movement of the modular-generalized car (as R_{TT}). As a result of solving the above-mentioned equations, a universal model of the adaptively controlled speed of the car - $V(K_{jr}, D, t, u)$ was obtained [3]:

$$V\left(K_{jr}, D, u, t\right) = \left[V_0(u_v) +_1 \left(K_{jr},\ D, u\right)\right]{*}\exp\left(_2\left(K_{jr}, D, u\right){*}t\right) - {}_1\left(K_{jr},\ D, u_v\right), \tag{6}$$

$$t\ (t_0, t_0 + t),\ \ V\ (V_0, V_0V),\ \ 1\ (l_0, l_0 + 1),$$

where $V_0(u_v)$ – is the initial constant speed of the modular-generalized type car in motor transport operation; $A_1 = f_1(K_{jr}, D, u)$ and $A_2 = f_2(K_{jr}, D, u)$– are parameters of the model of the controlled constant and unstable speed of the car; Δ t, Δ l and Δ V – are the values of time increase, car mileage, as well as changes in its speed ($\pm\Delta$ V) in operation; u – is a set of car's locomotion control parameters, u $\in$ (u_1, u_2, u_3), where u_1, u_2, u_3 – are parameters for controlling the engine, gearbox, and brakes, respectively; t– time of motor transport operation; D – is a set of road traffic flow characteristics, D $\in$ (f_0, f_t, S_h,,N_{tf}, ρ_{tf},V_{tf}, Δ x), f_0 and f_t– coefficients of rolling resistance of undriven wheel modules and wheel-traction modules of car; $f_t = f_3(M_w, G_w, \lambda_w)$, where M_w,G_w,λ_w – are torques, axial loads and coefficients of tangential elasticity of tires of wheel-traction modules of car; N_{tf}, ρ_{tf},V_{tf},– the value of the characteristics of the road traffic flow on the given section Δ x of the road, i.e. intensity, density and speed, respectively.

Expression (6) reflects the generalized model of adaptive-discrete motion of a modular-generalized car (as R_{TT}), which includes all phases of motor transport operations (steady motion Vc = const; unsteady motion ($\pm\ \Delta$ V $\rightarrow$ var), i.e. step accelerations (+ Δ V), coasting and braking (- Δ V)). In model (6), the phases of steady motion are given by the conditions $A_1(u_V) = A_2\ (u_V) = 0$, then we have $V_0(u_V) = V_c =$ const, with $u_3 = 0$, the parameters (u_{1V}, u_{2V}) correspond to the constant car's speed of Vc. For modes of unsteady motion, we have:

$$\begin{aligned} Z_{RT}(T_H) = T_H &= [\ln(A_1(K,D,\ u_v) + Vo(u_v) + V(u)) \\ &\quad - \ln(\ A_1(K,D,\ u_v) + +Vo(u_v))]/A_2(K,D,u), \end{aligned} \tag{7}$$

$$Z_{RT}(l) = A_1(\mathrm{K},,\ \mathrm{u})T + \mathrm{V}/A_2(\mathrm{K},,\ \mathrm{u}) \tag{8}$$

$$Z_{RT}(E) = E = B_9\mathrm{V} + B_{10}V^2 + B_{11}V^3 + T(B_{12} + B_{13}\mathrm{V} + B_{14}V^2), \tag{9}$$

$$\begin{aligned} Z_{RT}(Q_H) = Q_H &= B_{16} * V + B_{17}V^2 + B_{18}V^3 \\ &\quad + T_H(B_{15} + B_{19} * V + +B_{20}V^2 + +B_{21}V^3 +\ B_{22}V^4), \end{aligned} \tag{10}$$

$$Z_{RTT} \in (Z_{RT}(l_v), Z_{RT}(E_v), Z_{RT}(Q_v), Z_{RT}(T), Z_{RT}(l), Z_{RT}(E), Z_{RT}(Q)), \tag{11}$$

where A_1 and A_2 – are the parameters of the generalized equation of the speed of movement of the car (6); T_H(V, u) – the time of change of the controlled speed of the car depending on the amount of increase (or decrease) $\pm$ V; $B_9 \ldots B_{22}$ – parameters that depend on the state of the above four sets: K_{jr}, D, M_{to} and u, as well as on the initial speed V_o of the car in the transport operation; l_n, V, T_n, E_n and Q_n – characteristics of discrete kinematics and energetics of unsteady controlled car's movement: mileage, increase (or decrease) of speed of car, time of change of speed of car, energy consumption and fuel consumption, respectively; Z_{RTT}– is a complex of transport-operational regularities of the controlled functioning of technical and technological resources R_{TT} of MT in the technological processes of MT, which occur under the influence of the properties of the elements of the road traffic and terminal subsystems of the motor transport system (MTS).

For the Vc = const modes of steady motion of the modular-generalized type car, the regularities of the movement time T_v and mileage l_v of its, as well as the energy consumption E_v and fuel consumption Q_v are derived from Eqs. (7)… (10).

Further, it is accepted that the physical and technological basis of the material realization of the production function FA1 of MT according to (4) and the important integrative functions of MTS - [SF2, SF3 and SF4] according to (5) is a complex of Z_{RTT} models (6...13) describing the controlled functioning of modularly generalized car as a technical and technological resources R_{TT} in a motor transport operations. On the basis of this premise, as well as in connection with the algebraic form of Z_{RTT} models (6...13), the logic and principle of ensuring isomorphism of models of physical and technological mechanisms of realization of production function FA1(Z_{RTT}) of MT and integrative functions of MTS (transport-emergency -SF3(Z_{RTT}) and transport-integrative - SF4 (Z_{RTT})) was formed. Next, two logical-causal (i.e., cause-and-effect) schemes for the formation of the FA1(Z_{RTT}) MT function (15) and the integrative functions of the MTS (16) in accordance with (5) were formed:

$$((R_{TT} \cap R_{In}) \cap CTT2) \rightarrow Z_{RTT} \rightarrow MMTO(Z_{RTT}) \rightarrow TPAT(Z_{RTT}) \rightarrow FA1(Z_{RTT}) \rightarrow real, \quad (12)$$

$$\text{SF2} \rightarrow \text{F1}(Z_{RTT}) \rightarrow \text{SF3}(Z_{RTT}) \rightarrow \text{SF4}(Z_{RTT}) \rightarrow \text{SF6}(Z_{RTT}) \rightarrow \text{SF8}(Z_{RTT}) \rightarrow \text{real}, \quad (13)$$

where R_{TT} and R_{IP} – are technical and technological resources of MT and resources of infrastructure subsystems of the motor transport system (MTS); MMTO – models of material motor transport operation (technological or energy equivalent); TPAT– is a model of the technological process of automobile transportation.

Logical-causal schemes (12) and (13) reflect the presence of a single physical-technological basis of the basic production function of FA1- MT and functions of MTS. In them, the function mechanisms depend on the complex of Z_{RTT} algebraic regularities, both for the basic function of FA1(Z_{RTT}) and for important integrative functions of MTS (FA3(Z_{RTT}), FA4(Z_{RTT})). Therefore, these schemes make it possible to form a methodological basis for the implementation of the principles of isomorphism of analytical models FA1(Z_{RTT}) at MT and MTS, as well as the principle of system's energy and resource synergy. According to the theory of systems the implementation of these principles makes it possible to manage the innovative development of the active subsystem (MT) and Motor Transport System (MTS) as a whole [3–7].

This allows us to record the fact that the patterns Z_{RTT} (11) of interactive and adaptive functioning of the cars (as CRTTM MTOPS - in accordance with (2)) reflect the single physical, technical and technological bases of the implementation of three levels of hierarchical functioning of the car structure (operations-processes-system). This makes it possible to establish two principles of the functional-analytical integrity of MTS: a) hierarchical (at three levels of the hierarchy), causality of the functions of MT and MTS; b) the principle of structural isomorphism of mathematical models of the named functions, which is necessary for the systematic management of sustainable innovative development at MT. The algebraic form of Eqs. (6)...(13) makes it possible to form a universal model of a generalized and testing of motor transport operation (MTO) with a flexible regime-phase structure (steady motion, acceleration, coasting, braking). Such features of the models (6...11) and the complex of generalized patterns Z_{RTT} allow to ensure the requirements of their transport-technological reliability in relation to three aspects of the cognitive potential of the proposed comprehensive theory of ATT, MTO and processes: phenomenological, ontological and system-epistemological. For the possibility of technically consistent and comparative analysis of competing variants of cars with technical innovation, an analytical models of a test transport operation was formed, consisting of n_f total number of phases of locomotion of car, which is the sum of phases of stable n_{fs} and unstable n_{us} modes of its operation:

$$T_{on}(Z_{RTT}) = \sum_{1}^{n_{fs}} t_v + \sum_{1}^{n_{us}} t_u, L(Z_{RTT}) = \sum_{1}^{n_{fs}} l_v + \sum_{1}^{n_{us}} l_u \quad (14)$$

$$E_{on}(Z_{RTT}) = \sum_{1}^{n_{fs}} E_{vi} + \sum_{1}^{n_{us}} E_{ui}, \ \ldots \ Q_{on}(Z_{RTT}) = \sum_{1}^{n_{fs}} Q_{vi} + \sum_{1}^{n_{us}} Q_{ui} \quad (15)$$

where l_{vi}, t_{vi}, E_{vi}, and Q_{vi} – are values of mileage, travel time, and fuel consumption in steady (V = const) modes of car movement in the i-th phase of the test operation; l_{ui}, t_{ui}, E_{ui} and Q_{ui} – are values of mileage, driving time and fuel consumption in non-constant (V $\neq$ const) modes of car movement in the i-th phase of the test operation.

The criteria for the production efficiency of the transport and technological functioning of the car as a resource-technical R_{TT} and complex-technological means of motor transportations (CRTTM MT) in the test operation are the coefficients of transport production and energy efficiency $CTE_{(jr)}$, of fuel efficiency $CFE_{(jr)}$ and of average operating speed $K_{v(jr)}$ of the car with the jr-th version of SPOC (modular-generalized type) in the test operation are determined [3]:

$$C_{(jr)}(Z_{RTT}) = \frac{\rho_{e(jr)}}{\rho_{eT(j)}}, \ CF_{(jr)}(Z_{RTT}) = \frac{\rho_{Q(jr)}}{\rho_{QeT(j)}}, K_{v(jr)}(Z_{RTT}) = V_{o(jr)}/V_{eT(j)} \quad (16)$$

where $\rho_{e(jr)}$, $\rho_{Q(jr)}$, $V_{(o(jr))}$ – are the values of energy return, fuel return and speed indicators of the provided car with the jr-th variant of SPOC (modular-generalized type) in the test operation; $\rho_{eT(j)}$, $\rho_{QeT(j)}$ and $V_{eT(j)}$– are the values of energy return, fuel return and speed of the reference car of the j-th class in the reference test operation of this class.

Mathematical models of all three coefficients of (16) are based on generalized patterns Z_{RTT} of adaptive and interactive functioning of car as technical and technological resources R_{TT} of MT and CRTTM of MTOPS.

Next, a complex of six strategies of conceptual management of technologically sustainable development at motor transport (MT) is presented [3].

1. The stable strategy of operationally comprehensive identification and analysis of new car as complex of resource-technical and technological R_{TT} means (CRTTM) of MTO and MTS implemented on the basis of the model (2).
2. Sustainable management strategy for the innovative development of automobile transport technology (ATT) in the two life cycles: a) of the car (LCC); b) of the operational reproduction of motor transport processes (LCOR MTP), which is provided on the basis of project identification, substantiation and implementation of a complex of technical and technological innovations – (NVTT), according to the model (3).
3. A sustainable strategy of operational management of the levels of rational productivity, conceptual and technological quality and energy efficiency of new car as CRTTM

MTOPS-is implemented on the basis of the transport energy efficiency criterion of car $TE_E\left(K_{jr}\right)$ as a resource-technical means of production taking into account its structural and parametric organization K_{jr}, which is defined as follows [3]:

$$TE_E\left(Z_{RTT},\ K_{jr}\ ,t\right) = \frac{K_{VP}\left(Z_{RTT},\ K_{jr}\ ,Pr1, Pr2\right)}{K_{EP}\left(Z_{RTT},\ K_{jr}\ ,Pr1, Pr2\right)} \cdot \frac{\gamma_s}{\left(\gamma_s + \eta_q\right)}$$
$$\to \max\ K_{jr}(t) \to Var,\ P_{r1} \to Var,\ P_{r2} \to Var \quad (17)$$

where K_{VP} and K_{EP} –are the energy coefficients of the speed of the car and its mileage for the calculated test route (determined by the method of mathematical modeling of the test processes of the energy transformation of R_{TL}- transport technological resources into the physical product W_p); P_{r1} and P_{r2} – properties of the road as a rolling surface for cars and as a traffic-regulatory communication channel for the controlled movement of cars in the flow; γ_s and η_q – are coefficients of use of the carrying capacity of the cars (or of the bus) and its equipped weight; t–stages of innovations, $t = t_1, t_2, t_3 \ldots$

4. The strategy of managing the technologically sustainable development of innovative processes of motor transportations is based on the use of energy equivalent criteria of the energy-technological efficiency of transportations [3]:

$$\begin{cases} K_w\left(K_{jr}, Z_{RTT}, , NVTT, t\right) * W_h() \to \max \\ K_{cip}\left(K_{jr}, Z_{RTT}, , NVTT, t\right) * S_w() \to \min \end{cases},\ t = t_1,\ t_2, t_3 \ldots\ t_n \quad (18)$$

where W_h and S_w – are indicators of the hourly productivity of the rolling stock and the cost of transportation for the current time t_0, which are determined according to the virtual simplified calculation scheme of the transport process and the functioning of the car as a virtual moving body [1]; K_W and K_{cip} – are the energy coefficients of the hourly productivity of cars as carriers of technical and technological resources R_{TT} of MT and the cost of innovative production of transport services [3]; F_o– is a set of operating factors of the virtual simplified calculation scheme of the transport process [1, 3]; t – is variable time.

5. The comprehensive strategy of the conceptual management of the technologically sustainable development of the micro-economics on motor transportations [1–3] is based on the target function of the high-tech (HT) maximization of future transport profit ($TF_{TP} \to$ HTmax) - through step-by-step innovative analysis and synthesis of processes of high-tech reproduction of motor transport processes HTRMTP at MT. At the same time, are taken into account the factors of the technological and innovative additional product of MP- $F_{IAP}(Z_{RTT}))$ _ and the modernization reduction of the energy resource consumption factor of the car-hour of work of the MT on the given route – $m_a(Z_{RTT})$. Taking into account models (6…14) and has the following form of TF_{TP}:

$$TF_{TP}(Z_{RTT}, t) = P_{Th}(NVNN, Z_{RTT}, t)/(W_h * T_{w0})$$

$$= F_{IAP}\left[\frac{m_m(t)}{m_a(NVTT,t)} - -m_s(NVTT,t) * F_{OTS}\right] \to HT\max \tag{19}$$

where P_{Th} – hourly motor transport profit taking into account the influence of a number of technical and technological innovations of NVTT; m_T (t) – the coefficient of the forecast increase of the current tariff T_{w0} per 1 tkm (pkm), $m_T > 1$; m_a – is an indicator of the modernization reduction of the coefficient $K_a(t)$ of the energy resource consumption of the operation of the new car in comparison with the previous version of the car with $K_a(t_0)$ on the given route, $m_a = K_a(t)/K_a(t_0)$; m_s – is an indicator of innovative reduction in the value of the energy coefficient $K_S(t)$ of the production cost of transport services when using a new car, $m_s(t) \leq 1$; F_{ots} –is the economic factor of the current organization of transport services (the ratio of the cost of transportation to the transportation tariff - 1 tkm (pkm) at $t = t_0$); $\Phi_{IAP} = (1+\beta_R(NVTT,t))$– the factor of the technologically intensive additional product of MT, $t = t_1, t_2, t_3 \ldots t_n$; β_R– is an indicator of the innovative increase in energy- resource return of the technologically transportation project, when $\beta_R > 0$, the project is implemented under the condition of ensuring the concept of processes of high-tech (HT) reproduction of motor transport processes withs smart logistics requirements.

An important feature of the model (19) is that it provides staged and conceptual management of technologically sustainable development at the MT through natural-cost analysis of the relative break-even of the new scheme of reproduction of the motor transport service during the evolution of the parameters of the car and its resource-technical properties over time. In addition, a steady increase in the conceptual quality of the resource-technological basis of motor transport (MT) is ensured in stages, as well as the design of high-tech and resource-efficient road transportations. The break-even condition of transportation in relation to the existing costs and market tariffs is verified by equating the unit values of all energy coefficients in mathematical model (19). This ensures simple fulfillment of the methodological requirements for the homogeneity and isomorphism of models for the analysis of both the operational organization of road transportations and their technological and innovative projects. The possibility of integrating engineering and economic management methods follows from the above technologically sustainable development of three fundamental product-reproducing factors of motor transport: the resource-technological base, the processes of technological transportations and the complex of functions of motor transport (4).

6. Sustainable strategy of transport-oriented improvement of the main integrative functions of motor transport system (MTS: transport-emergency - SF3 and transport-integrative functions - SF4. All motor transport operations are performed under conditions of conflicting interaction of technical and technological resources R_{TT} of MT with elements of the road and road-traffic flow, which is reflected in models (6–11). If the value of the coefficient $K\rho$ of road traffic flow density increases ($K\rho > 0.4$), then the levels of transport productivity and energy efficiency of technical and technological resources R_{TT} of MT decrease (it is a negative phenomenon of anergism of R_{TT} in road-traffic flow). This negative phenomenon of anergism occurs on motor

transportations during their adaptive and interactive functioning in a conflict environment. The indicator of such anergism of R_{TT} resources (in energy form) is the ratio $K_{RA}(Z_{RTT})$ of test values of energy $K_e(Z_{RTT})$ or fuel $K_q(Z_{RTT})$ coefficients of car mileage to its speed coefficient $K_v(Z_{RTT})$:

$$K_{RA}(Z_{RTT}) = K_{TCE} * K_E(Z_{RTT})/(K_{TCV} * K_V(Z_{RTT})) \rightarrow min, \tag{20}$$

where K_{TCE} and K_{TCV} – are indicators of operational correction of test values of energy $Ke(Z_{RTT})$ (or fuel $Kq(Z_{RTT})$) coefficients of car's mileage and it speed $K_v(Z_{RTT})$; Z_{RTT} – is a complex of transport-operational regularities of the controlled functioning of technical and technological resources R_{TT} of MT in the transport-technological processes.

For the local (within the given road section (Δx)) improvement of the emergency transport function of the MTS - SF3, the principle of minimizing the weighted average indicator of road traffic anergism R_{TT} on the road section Δx is used:

$$\sum\sum K_{RA(jr)}(K_{jr}, Z_{RTT}, \Delta x) * \alpha_{jr}(K_{jr}) \rightarrow min, \tag{21}$$

where $\alpha_{jr}(K_{jr})$– the share of cars of variants K_{jr} in the local structure of road traffic flow.

The principle of minimizing road and traffic anergism of R_{TT} during the technological improvement of transport and integrative functions of SF4 of the MTS in relation to the given -m_t number of road transportation routes of length -L_t in the provided terminal network of MTS, $t \in (1, m_t)$:

$$\sum\sum\sum K_{RA(jr)}(K_{jr}, Z_{RTT}, L_t) * \alpha_{jr}(K_{jr}) \rightarrow min, \tag{22}$$

Next, an example of a decomposition and one-parameter analysis of the influence of the maximum power of the VOLVO FH-12 + FRUEHAUF DSND-32 ATZ engine on the set of indicators of the operational and technological quality of its adaptive-interactive movement in the highway test operation (l_{to} = 4000 m) is presented, taking into account the Z_{RTT} patterns of its adaptive -interactive functioning as a complex resource-technical and technological means of motor transport operations. The graphs (Fig. 1) show that in the test operation, the modernization increase in the power of the car engine in the given range will allow to increase the comprehensive indicators of its transport-productive energy efficiency (graph 5) and fuel efficiency (graph 6) by 7% and 25%, respectively. At the same time, the patterns of changes in the values of the energy (graph 1) and fuel (graph 2)) coefficients of car mileage are different [3, 8].

The simulation results reflect the influence of the most important (in the case of technical modernization) parameter of the car from the point of view of maximizing the transport efficiency of its resource-technical properties, which are determined by the structural-parametric organization of its design. In addition, it is important in relation to the tasks of the decomposition analysis of the first three strategies of conceptual management of technologically sustainable development in MT. These results confirm the important features of the new method of decomposition analysis of sustainable development strategies. These include: variant matching, technological adequacy, as well as the possibility of systematic and quantitative analysis based on the principle of functional isomorphism of models of MT (resources, automobile transport technologies and processes, production and technological functions), as well as two integrative functions of MTS (transport and energy emergency, transport integrality).

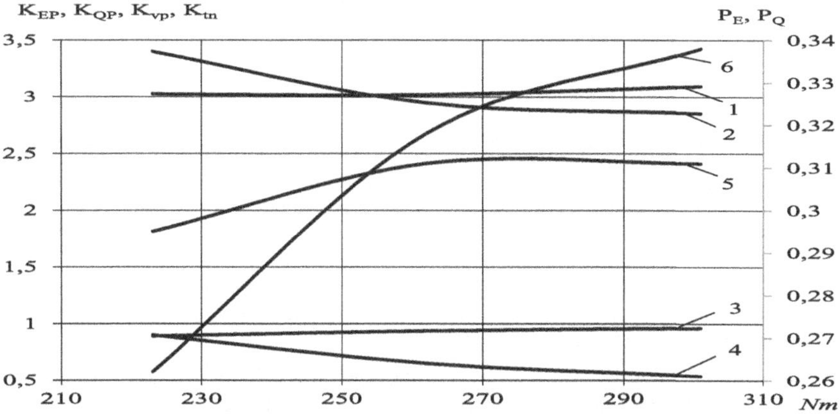

Fig. 1. Graphs of the dependence of the energy efficiency indicators of the VOLVO FH-12 + FRUEHAUF DSND-32 highway train on the maximum engine power Nm (kW) for the calculated test route [3, 8]:

$1 - K_{EP}(Z_{RTT}),\ 2 - K_{QP}(Z_{RTT}),\ 3 - K_{vp}(Z_{RTT}),\ 4 - K_{tn}(Z_{RTT}),$
$5 - P_E(Z_{RTT}),\ 6 - P_Q(Z_{RTT})$

The following designations are taken in the figure: 1 – K_{EP}- coefficient of energy mileage; 2–- K_{QP} - coefficient of fuel consumption mileage; 3– K_{vp} - average velocity coefficient; 4– K_{tn}-indicator of the share of unsteady modes of movement of the road train in the test operation; 5– P_E – complex indicator of transport productivity and energy efficiency; 6 - P_Q - complex indicator of transport productivity and fuel efficiency.

3 Conclusions

1. A new resource-consolidated approach to the conceptual-cognitive intellectualization of the main functions of the Motor Transport (MT) and the Motor Transport System (MTS) in relation to their sustainable development is proposed.
2. It was established that in motor transport processes (MTP) and in the conditions of a conflict infrastructure environment, the car functions as a complex, resource-technical and technological means R_{TT} of MT and MTS.
3. On the basis of algebraic models for the analysis of the kinematics, dynamics and energy indicators of the adaptive-discrete locomotion of the car as R_{TT}, a complex of Z_{RTT} regularities of their controlled and interactive functioning in the conflict and traffic-forced processes of road transportations was established.
4. A techno-economic concept, strategies, models and methods of management of technologically sustainable development at MT have been formulated, taking into account two integrative functions (transport-energy emergency and integrality) of the MTS.

5. The logic and principle of ensuring the isomorphism of models of physical and technological mechanisms for the realization of the production function FA1(Z_{RTT}) of MT and integrative functions of MTS (transport-emergent-CΦ3(Z_{RTT}) and transport-integrality-CΦ4(Z_{RTT})) have been formed.
6. Proposed complex of principles and methods of intellectually managing of technologically sustainable development at Motor Transport.

References

1. Vorkut A.I.: Cargo automobile transportations. Higher School, Kyiv (1986)
2. Savytska G. V.: Economic analysis of enterprise activity: Education. 3rd edn. Znannia, Kyiv (2007)
3. Khabutdinov R.A.: Theory of automobile and transport technology. Academy of Technical Sciences of Ukraine. Publisher G.M. Kushnir, Ivano-Frankivsk (2024). https://doi.org/10.63048/978-617-7926-58-9. ISBN 978–617–7926–58–9
4. Francis Vanek and Louis Thibodeau.: Sustainable Transportation Systems Engineering: Evaluation and Implementation, 1st edn. McGraw-Hill Education, New York (2014)
5. Cao, Y., Derrible, S., Le Pira, M., et al.: Advanced transport systems: the future is sustainable and technology-enabled. Nature. Sci Rep **14**, 9429 (2024). https://doi.org/10.1038/s41598-024-59438-0/
6. Critical Issues in Transportation for 2024 and Beyond. National Academies of Sciences, Engineering, and Medicine. The National Academies Press, Washington, DC (2024). https://doi.org/10.17226/27432
7. Yunpeng WANG, Xinping YAN, Guangquan LU, Chaozhong WU.: Intelligent Road Transport Systems: An Introduction to Key Technologies. Tsinghua University Press, Singapore (2022). https://doi.org/10.1007/978-981-16-5776-4
8. Khmelov, I.V., Galiona, I.I., Danylenko, I.V.: The method of analyzing the transport energy efficiency of motor vehicles taking into account changes in their design parameters. Sci. World J. **24**(1), 161–165 (2024). https://doi.org/10.30888/2663-5712.2024-24-00-069

The Role of the Maritime Transport Logistics System of Georgia in Ensuring the Stability of the Global Supply Chain

Roman Mamuladze[1,2](✉) and Giorgi Mamuladze[1,2]

[1] Batumi State Maritime Academy, Batumi, Georgia
r.mamuladze@bsma.edu.ge
[2] Batumi Shota Rustaveli State University, Batumi, Georgia

Abstract. In an era marked by increasing international trade and economic interdependence, the stability of the global supply chain has become critical. Recent disruptions, including the COVID-19 pandemic and geopolitical conflicts, underscore the urgent need for resilient trade systems. This study aims to evaluate the role of Georgia's maritime transport logistics in ensuring the stability and resilience of the global supply chain, with a particular focus on the strategic significance of the Transcaucasian 'Middle Corridor.' By assessing the current utilization and potential of Georgia's key ports, this research highlights the critical need for optimizing these assets to bolster global trade networks, especially during periods of disruption. Our analysis reveals significant underutilization of key Georgian ports, such as Batumi and Poti, where cargo turnovers fall far short of capacity. For instance, Batumi's oil terminal, designed for 15 million tons annually, handled only 2 million tons in 2022—just 13% of its capacity. This underperformance indicates systemic inefficiencies within Georgia's maritime sector. The 'Middle Corridor' presents a crucial alternative for diversifying and stabilizing the global supply chain. Enhancing the throughput and efficiency of Georgia's ports could solidify the country's role in global trade and provide a strategic buffer during supply chain disruptions. We propose forming transport clusters to optimize maritime logistics. Our findings emphasize the strategic importance of developing underutilized maritime corridors, with the goal of spurring policy reforms and leveraging Georgia's logistics capabilities as a cornerstone for global supply chain stability.

Keywords: Supply Chain · Maritime Transport · Freight Turnover · Throughput · Transport Cluster

1 Introduction

One of the defining characteristics of contemporary international economic relations is the significant increase in international trade volume and the deepening economic interdependence among nations [1, 2]. According to the World Trade Organization, the global trade volume in goods and services in 2023 reached 31 trillion US dollars, marking a 13% increase from the 2022 figure of 27.5 trillion US dollars. Notably, the volume of

O. Slavinska et al. (Eds.): ITSESQC 2024, LNNS 1335, pp. 363–371, 2025.
https://doi.org/10.1007/978-3-031-87376-8_32

world exports of goods in 2022 increased by 12% over the previous year, totaling 24.2 trillion US dollars. Trade between the European Union and China also expanded, with goods trade in 2023 growing by 31.8% compared to 2022, reaching 856.3 billion euros [3]. In 2022, China emerged as the largest importer from the EU, accounting for 626 billion euros, and the third largest exporter to the EU, contributing 230.3 billion euros [4]. Similarly, trade between the European Union and Central Asia and Transcaucasia has been rapidly increasing. In 2022, trade in goods between the European Union and Central Asian countries amounted to 47.6 billion euros, up 71% from 2021 [5]. Additionally, trade between the European Union and Azerbaijan reached 33.153 billion euros in 2022, an increase of 127% over 2021 [6].

This paper aims to critically assess the role of Georgia's maritime transport logistics, particularly within the framework of the 'Middle Corridor,' in maintaining and enhancing the stability of the global supply chain. By evaluating the capacity and efficiency of Georgia's key ports, this study seeks to demonstrate how optimizing these logistical assets can contribute to the resilience of global trade routes, offering a strategic alternative in the face of ongoing disruptions such as geopolitical conflicts and pandemics. The well-documented challenges faced by global supply chains, including those related to the Transcaucasian transport route, known as the "Middle Corridor," highlight the importance of this link in the global trade network. The analysis of trade relations between European and Asian countries reveals current trends in trade turnover and describes the features of alternative routes for goods delivery between these regions. It underscores the significance of the "Middle Corridor" in facilitating East-West trade relations. To enhance the capacity of Georgia's maritime transport system, increase the competitiveness of the Middle Corridor, and ensure the sustainability of the supply systems that traverse it, this paper proposes the formation of transport clusters, alongside national and regional transport cluster models.

2 Maritime Transport Logistics System of Georgia

The global supply chain, a complex and evolving structure, is essential for organizing investment, production, and trade worldwide. It provides significant employment and development opportunities, particularly in developing countries [7]. Effective international transport systems play a critical role in trade development; without robust connections, trade cannot prosper [8]. Investments in supply chain infrastructure profoundly impact economic development at national, regional, and local levels [9]. Participation in global value chains is linked to economic growth, and the efficiency of these chains is crucial for the smooth functioning of economies [10]. Disruptions within supply chains, such as the COVID-19 pandemic, geopolitical conflicts, and attacks on merchant ships in regions like the Red Sea, have been shown to significantly affect economic productivity and growth [11].

Despite the extensive literature on global supply chains, there is a noticeable gap in research focusing specifically on the 'Middle Corridor' and Georgia's maritime logistics system. Existing studies touch upon trends in maritime industry development in Georgia but often do not delve into the strategic importance of optimizing underutilized maritime assets [13]. This gap highlights the necessity for an in-depth analysis of

Georgia's potential role in enhancing global supply chain stability through improved maritime logistics.

3 Research Methods

This study utilizes a combination of quantitative data analysis and comparative methods to assess the operational efficiency of Georgia's maritime logistics system, with a focus on the 'Middle Corridor.' Data were collected from Georgia's port authorities covering the years 2010 to 2023, focusing on key performance indicators such as cargo throughput and capacity utilization at major ports, including Batumi and Poti.

Descriptive statistics were applied to identify performance trends and underutilization issues at these ports. The Port Capacity Utilization Rate was employed as a primary metric, calculated by dividing the port's actual cargo turnover by its designed capacity and expressing the result as a percentage [14]. This metric provides insights into the efficiency of port operations and highlights areas where improvements are necessary.

Additionally, a comparative analysis was conducted to benchmark Georgian ports against regional counterparts. This involved analyzing data from nearby ports in the Black Sea region to understand Georgia's competitive position and strategic advantages within the 'Middle Corridor.' The comparative analysis aids in identifying best practices and potential areas for collaborative development.

4 Maritime Transport Logistics System of Georgia

The Middle Corridor is a multimodal transport corridor connecting the Chinese transport system with that of Central Asia, including the Kazakhstan railway system and Aktau port. It traverses the Caspian Sea, linking to the Alyat port in Baku, Azerbaijan, and then connects to Georgia's transport system. From Georgia, it extends to the European transport system through two routes: through turkey or from the Black Sea ports of Georgia to the European Union ports in Bulgaria and Romania, continuing into the broader European transport network [12]. Thus, the Middle Corridor serves as a vital alternative to the "Northern" and "Southern" routes, connecting Europe and China. Its significance has grown notably following Russia's military aggression in Ukraine, drawing increased attention to this route.

The key components of Georgia's maritime transport infrastructure are its ports and terminals on the eastern coast of the Black Sea [13]. Main ports are Batumi Sea Port; Poti Sea Port; Poti New Marine Port (Pace Terminal); Black Sea Terminal (Kulevi Oil Terminal); Supsa Terminal and Anaklia Harbor (under construction).

In this study, we analyze the operational efficiency of Georgia's maritime logistics, specifically focusing on the strategic 'Middle Corridor.' Utilizing data from Georgia's port authorities from 2010 to 2023, we assess key performance indicators like cargo throughput and capacity utilization at major ports. Our analysis applies descriptive statistics to illustrate performance trends and pinpoint underutilization issues at Batumi and Poti ports. We also conduct a comparative analysis, benchmarking these ports against regional counterparts to understand Georgia's competitive position and strategic advantages within the 'Middle Corridor.'

An analysis of the last few years' performance reveals positive trends in cargo turnover and processing at these facilities. However, the effective utilization of existing capacity (capacity utilization) at these ports and terminals typically ranges from 60–70%. To assess the efficiency of these sea ports, we employ the Port Capacity Utilization Rate. Table 1 displays throughput and cargo turnover for individual terminals at the Batumi port, alongside one of the efficiency metrics: the degree of bandwidth utilization.

The capacity utilization rate at individual terminals of the Batumi Sea Port has been notably low over the past five years (2019–2023) [15]. Specifically, the oil terminal processed its largest volume of liquid cargo in 2022, totaling 2 million tons, but the corresponding capacity utilization rate was only 13%. In the dry cargo terminal, the highest annual cargo volume reached 1.942 million tons in 2023, achieving a bandwidth utilization rate of 72%. Additionally, the container terminal saw its peak annual throughput in 2022 with 119,471 TEU, yet the bandwidth utilization rate remained low at 34%.

Table 1. Performance and efficiency assessment indicator of Batumi port – Capacity utilization

YEAR	*OIL TERMINAL (THROUGHPUT - 15 MILLION TONS PER YEAR)*	*DRY CARGO TERMINAL (THROUGHPUT - 2.7 MILLION TONS PER YEAR)*	*CONTAINER TERMINAL (THROUGHPUT - 350,000 TEU PER YEAR)*
2014	4.5 million tons, 30%	1.203 million tons, 45%	61,980 TEU, 18%
2019	0.9 million tons, 6%	0.695 million tons, 26%	116,081 TEU, 26%
2020	1.2 million tons, 8%	0.913 million tons, 34%	103,302 TEU, 30%
2021	1.5 million tons, 10%	1.211 million tons, 45%	99,187 TEU, 28%
2022	2.0 million tons, 13%	1.742 million tons, 65%	119,471 TEU, 34%
2023	1.6 million tons, 11%	1.942 million tons, 72%	109,000 TEU, 31%

Detailed capacity and cargo turnover metrics for the Poti Sea Port, along with capacity utilization values for various cargo types, are presented in Table 2. The data from Poti seaport between 2018 and 2022 reveals varied utilization rates across different cargo types [16]. The highest annual turnover for general and bulk cargo was 6.9 million tons in 2019, with a capacity utilization rate of 41%. Liquid cargoes peaked at 0.6 million tons annually, achieving a 60% utilization rate. For containers, the maximum annual throughput was 531,735 TEU, also with a 60% utilization rate.

Table 3 presents the design capacity and cargo turnover for the Kulevi and Supsa oil terminals, and Anaklia sea port, including capacity utilization values as an effectiveness indicator. During the same period, the Kulevi oil terminal's maximum annual throughput of petroleum products was 1.9 million tons in 2021 and 2022, but the capacity utilization rate was only 19% [17]. Meanwhile, the Supsa oil terminal processed up to 4.2 million tons of petroleum products in 2020 and 2021, with a capacity utilization rate of 53%. An analysis of the performance indicators for Georgia's maritime transport logistics infrastructure suggests that the efficiency, as estimated by capacity utilization data, remains relatively low. The development of Anaklia Port is set to occur in stages, with significant

Table 2. Performance and efficiency assessment indicator of Poti Port - Capacity utilization

Year	*General/Bulk Cargo (Throughput - 20 million tons per year)*	*Liquid Cargoes (Throughput - 1 million tons per year)*	*Container Terminal (Throughput - 550,000 TEU per year)*
2018	5.9 million tons, 30%	0.5 million tons, 50%	363,936 TEU, 66%
2019	8.1 million tons, 41%	0.6 million tons, 60%	531,735 TEU, 97%
2020	6.9 million tons, 35%	0.5 million tons, 50%	387,368 TEU, 70%
2021	6.6 million tons, 33%	0.5 million tons, 50%	302,082 TEU, 55%
2022	7.2 million tons, 36%	0.5 million tons, 50%	357,623 TEU, 65%

capacity increases at each milestone. Upon completion of the first stage of construction, the port is expected to achieve a throughput capacity of 600,000 TEU, which is equivalent to handling 7.8 million tons of cargo. Once fully completed, the port's capacity will dramatically expand to accommodate a staggering 100 million tons of cargo [18] This phased development is designed to enhance Anaklia's ability to handle extensive maritime cargo volumes, supporting its role in global trade logistics.

Table 3. Performance and efficiency evaluation index of Kulevi and Supsa terminals - Bandwidth utilization

Year	*Kulevi (Black Sea) Terminal (Throughput - 10 million tons per year)*	*Supsa Marine Terminal (Throughput - 8 million tons per year)*
2018	1.4 million tons, 14%	3.7 million tons, 46%
2019	1.6 million tons, 16%	3.7 million tons, 46%
2020	1.7 million tons, 17%	4.2 million tons, 53%
2021	1.9 million tons, 19%	4.2 million tons, 53%
2022	1.9 million tons, 19%	1.0 million tons, 13%

In 2022, the maritime transport infrastructure of Georgia demonstrated significant potential for growth, with the capability to process an additional 29.5 million tons of oil products, 15.5 million tons of dry cargo, and 530,000 TEU of containers. This capacity is expected to increase to 1,000,000 TEU with the launch of the new port of Anaklia, which could significantly bolster international trade via the stable functioning of the Middle Corridor supply chain.

The formation of national and regional transport clusters is identified as a key strategy for enhancing the competitiveness of the Middle Corridor and ensuring the sustainability of supply systems. The "2023–2030 National Transport and Logistics Strategy of Georgia," developed by the Ministry of Economy and Sustainable Development of Georgia, emphasizes the promotion of clustering as a main direction for increasing logistics efficiency and competitiveness [19]. Clusters, defined as concentrations of interrelated activities within the same value chain, are pivotal in fostering competition among countries, regions, and industries [20]. They enhance productivity and efficiency through geographical concentration and collaborative ties among companies.

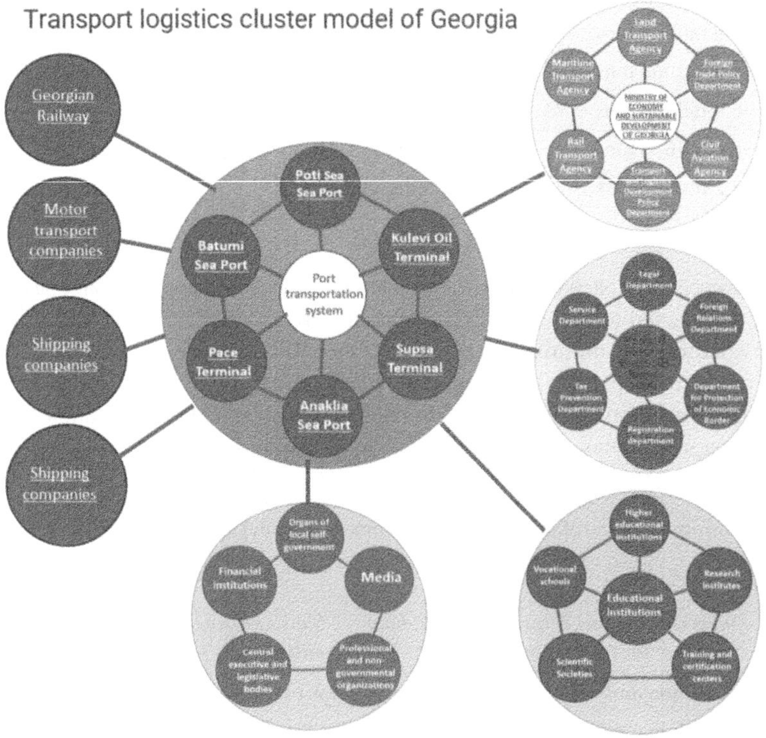

Fig. 1. Model of transport logistics cluster of Georgia

The proposed national transport logistics cluster in Georgia, as depicted in Fig. 1, would encompass major participants from the maritime system including Poti, Batumi, and Anaklia seaports, along with the Kulevi and Sufi terminals. It would also integrate elements from the land transport system such as Georgian railways, road transport companies, and warehouse logistics systems. Additional members would include shipping companies, forwarding firms, stevedoring and brokerage services, insurance companies, and other support services in the transport chain. Integral to this cluster are the structural units of the Ministry of Economy and Sustainable Development of Georgia, the Ministry of Finance, educational organizations (universities, research institutes, vocational schools, certification centers), and supporting organizations including governmental bodies, media, and financial institutions.

Looking beyond national boundaries, the next step is forming a regional transport logistics cluster [21]. This would be a cluster union based on the integration of transport logistics clusters from Georgia and other countries involved in the Middle Corridor supply chain, as shown in Fig. 2. This regional cluster would include the transport logistics clusters of Georgia, Azerbaijan, Armenia, Central Asian countries, Turkey, Ukraine, Bulgaria, Romania, and other EU countries, forming a comprehensive network that enhances the overall competitiveness and efficiency of the Middle Corridor.

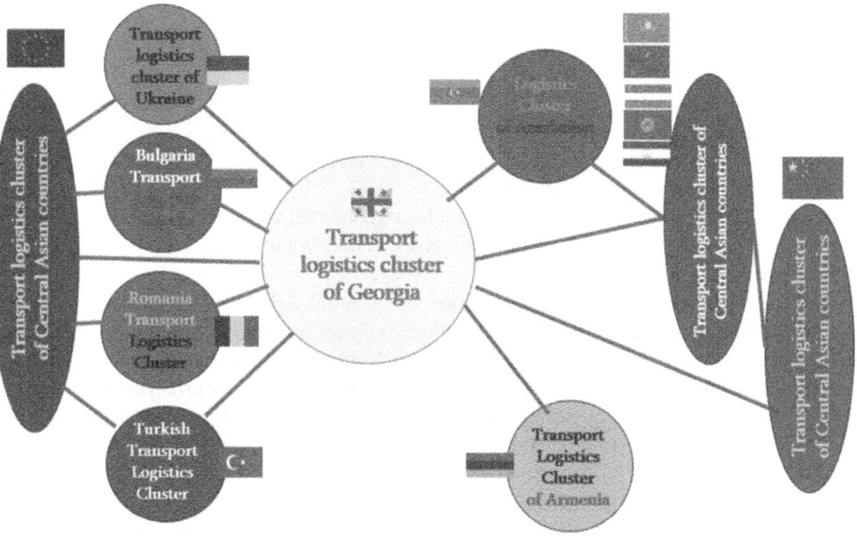

Fig. 2. Middle corridor regional logistics cluster model

5 Conclusions

The findings of this study demonstrate the evolving dynamics of international economic relations, characterized by a significant increase in international trade volume and a deepening of economic interdependence among countries. This trend underscores the pivotal role of efficient supply systems in fostering the development of national and regional economies and in maintaining the robustness of the global economic framework. Our analysis reveals that the resilience of supply systems in the face of contemporary challenges—such as geopolitical tensions and economic disruptions—is crucially dependent on the diversification of supply chains. This diversification not only enhances the stability of supply systems but also highlights the need for ongoing research into alternative communication systems that facilitate international trade.

In response to the inefficiencies identified within Georgia's maritime transport logistics, there is a clear necessity to develop transport logistics clusters. These clusters would aim to improve capacity utilization rates at Georgian ports, enhance the competitiveness of the 'Middle Corridor,' and ensure the sustainability of the supply systems that traverse this crucial juncture. The study proposes the implementation of both national and regional transport logistics cluster models. These models are designed to optimize logistics operations across Georgia and the broader region, potentially transforming the 'Middle Corridor' into a more effective and competitive route in the global supply chain network. Ultimately, this paper not only contribute to our understanding of the strategic importance of Georgia's maritime sector within the global supply chain but also suggest practical measures that can be taken to enhance this sector's contribution to global trade efficiency.

References:

1. Baldwin, R.E.: Global supply chains: why they emerged, why they matter, and where they are going (2012). https://papers.ssrn.com/abstract=2153484
2. Savoy, C.M., Ramanujam, S.R.: Diversifying supply chains: the role of development assistance and other official finance (2022)
3. EUROSTAT: China-EU - international trade in goods statistics (2024)
4. WTO: World Trade Statistical Review 2023, https://www.wto.org/english/res_e/publications_e/wtsr_2023_e.htm
5. European Commission: EU trade relations with Central Asia, https://policy.trade.ec.europa.eu/eu-trade-relationships-country-and-region/countries-and-regions/central-asia_en
6. European Commission: EU trade relations with Azerbaijan, https://policy.trade.ec.europa.eu/eu-trade-relationships-country-and-region/countries-and-regions/azerbaijan_en
7. Gereffi, G., Lee, J.: Why the world suddenly cares about global supply chains. J. Supply Chain Manag. **48**, 24–32 (2012). https://doi.org/10.1111/j.1745-493X.2012.03271.x
8. Nicita, A., Ognivtsev, V., Shirotori, M.: Global supply chains: Trade and Economic Policies for Developing Countries. Citeseer (2013)
9. Closs, D.J., Bolumole, Y.A.: Transportation's role in economic development and regional supply chain hubs. Transp. J. **54**, 33–54 (2015). https://doi.org/10.5325/transportationj.54.1.0033
10. Fagerberg, J., Lundvall, B.-Å., Srholec, M.: Global value chains, national innovation systems and economic development. Eur. J. Dev. Res. **30**, 533–556 (2018). https://doi.org/10.1057/s41287-018-0147-2
11. Goel, R.K., Saunoris, J.W.: Supply chain reliability and international economic growth: impacts of disruptions like Covid-19. CESifo, https://www.cesifo.org/en/publications/2020/working-paper/supply-chain-reliability-and-international-economic-growth-impacts
12. World Bank: The middle trade and transport corridor: policies and investments to triple freight volumes and halve travel time by 2030, https://www.worldbank.org/en/region/eca/publication/middle-trade-and-transport-corridor
13. Mamuladze, R., Gabaidze, M.: Some trends in the development of maritime industry in the world and in Georgia. Georg. Marit.E Sci. J. **1**, 131–139 (2021). https://doi.org/10.48614/gmsj120212831
14. Lee, C.-Y., Song, D.-P.: Ocean container transport in global supply chains: Overview and research opportunities. Transp. Res. Part B: Methodol. **95**, 442–474 (2017). https://doi.org/10.1016/j.trb.2016.05.001
15. Batumi Sea Port, https://batumiport.com/port/index#indexes
16. Maritime Transport Agency of Georgia: Port Infrastructure Development. Newsletter, https://www.mta.gov.ge/uploads/other/NEWSLETTER_-_MTA_1.pdf, (2023)
17. Kulevi Terminal, https://www.kulevioilterminal.com
18. Anaklia Development Agency: Project on the Public and Private Collaboration for the Anaklia New Deep Sea Port Development (2023), https://www.economy.ge/uploads/files/2017/anaklia_deep_sea_port_development_agency/soi_georgian_version_20_02_2023_.pdf
19. Ministry of Economy and Sustainable Development of Georgia: National Strategy of Transport and Logistics of Georgia for 2023–2030, https://www.economy.ge/uploads/files/2017/transport/2023/strategy/transportisa_da_logistikis_strategia_2023_2030.pdf

20. Garanti, Z., Berzina, A.: regional cluster initiatives as a driving force for regional development. European Integration Studies. 0, (2013). https://doi.org/10.5755/j01.eis.0.7.3677
21. The Geography of Transport Systems. Routledge (2024)

Smart Logistics in the Organization of Air Transportation

Olena Harazha(✉) and Dmytro Shevchuk

National Aviation University, Lubomyra Huzara Avenue, 1, Kyiv 03058, Ukraine
oasissun@ukr.net

Abstract. The article is devoted to smart logistics in the organization of air transportation of cargo and passengers in modern economic, political, and social development conditions under the influence of the digital economy and globalization processes. The theoretical provisions regarding the formation of smart logistics, which includes the latest technological and innovative processes and replaces traditional methods of managing chain freight transportation with more efficient, more complex network systems connected to the global logistics system, are considered. The main stages of the development and implementation of smart logistics in the organization of air transportation are highlighted, including: 1) the initial stage; 2) analytical stage; 3) intellectual stage. It was found that during the period of 2020–2021, the transportation of passengers increased, and cargo decreased. It has been proven that the state faces an acute problem of resuming the operation of airports and providing them with qualified staff to ensure flight safety. It was found that the closed airspace led to the loss of qualified personnel, the suspension of the development of logistics services, and the introduction of advanced smart logistics technologies. The advantages of automated processes are revealed: reduction of the cost of specific operations; exclusion of errors by the employee; reduction of delivery time; increase in orders; maximum accuracy; preventing the influence of external factors; information support.

Keywords: Smart Logistics · Air Transportation · Internet of Things (IoT) · Artificial Intelligence (AI) · Blockchain · Cargo Tracking · Route Optimization · Logistics as a Service (LAAS)

1 Introduction

Modern high-tech changes affect all spheres of life, including the global logistics infrastructure, prompting significant transformations and optimization of air passenger and cargo transportation. The integration processes that have covered the national economies of many states are formed under the influence of globalization challenges, which prompts the creation of complex supply chains in the field of air transport organization. As a result, the effectiveness of traditional methods of air transportation management decreases. In particular, the new requirements for the organization of transportation are transparency, speed, accuracy and availability. To solve them, smart logistics was created, which contains a collection of the latest technologies and approaches to the management and

O. Slavinska et al. (Eds.): ITSESQC 2024, LNNS 1335, pp. 372–384, 2025.
https://doi.org/10.1007/978-3-031-87376-8_33

organization of air passenger and cargo transportation. Therefore, they include artificial intelligence, databases, process automation, blockchain, etc. It is common knowledge that due to the implementation of these technologies, the organization and control of air transportation is changing significantly. Indeed, monitoring the location and condition of the shipped goods in real time is possible based on the use of special applications of aviation companies. Obviously, database management makes it possible to choose the optimal flight route, predict departure and arrival delays, reduce the cost of fuel and lubricants and maintenance, reduce the cost of tickets, facilitates the access of future passengers and customers to transparent and comprehensive information about the air carrier and its services in special applications. At the same time, to train the machine on the basis of large databases on the organization of air transportation and create on its basis artificial intelligence with special professional skills, it is possible to obtain accurate forecasts regarding flight delays based on changes in the weather forecast, the location of the flight and many other factors. Therefore, the elimination of errors and the human factor occurs on the basis of the application of artificial intelligence with the professional basis of conducting logistics processes during the automation of air transportation management and organization. Thus, air transport becomes not only the fastest and most expensive means of delivery, but also optimized by reducing costs for fuel, maintenance, personnel and more. At the same time, the issue of reducing the time for managing and processing transportation data based on specialized programs for tracking cargo in the supplier-carrier-client chain with minimizing the risks of delays remains quite relevant, and smart logistics is becoming an important key factor in the organization of air transportation. In the conditions of a globalized economy, there is a constant increase in the volume of air transportation and their routes, which leads to the complication of managing logistics operations. It is known that the efficiency of air transportation is characterized by timely delivery, uniform load on airports, the absence of disruptions in supply chains, and only under these conditions the efficient functioning of the economy and profitable business processes take place. Solving these problems is possible on the basis of smart logistics, which uses a combination of databases and automated programs to manage growing volumes of transportation. However, its implementation in the real management process remains quite problematic. Airlines, airports, customers and suppliers take part in the transportation process. But the regulation of interaction between them is lost due to the closedness of information, which negatively affects the time period of delivery, spending resources and funds. Therefore, smart logistics provides each of the parties with transparent and up-to-date data, but the issues of information protection and interoperability of systems require additional research. Therefore, one of the main components of the global supply system is the organization of air transportation, which creates timely and fast delivery of goods along long routes. At the same time, modern supply chains are becoming longer, more complex and dynamic, and traditional methods of transportation do not meet the needs of today, which was the problem of conducting this study. In order to successfully solve this problem, there is a need to conduct a more in-depth study on the application of new innovative methods, which are smart logistics.

2 Analysis of Recent Research and Publications

L. Luning, A. Sohaib, A. Wileman, S. Perinpanayagam suggest that the aerospace industry, including its manufacturing base, is one such active proponent of digital twins, with unprecedented interest in their individual design, development and imple-mentation in more broad operational and critical functions. They have exposed the intricacies of digital twin technologies to the aerospace community in particular and others in general to eliminate errors that affect their effective implementation for safety-critical systems [1]. Scientists Ch. Venkatesh, P. Sayooj, Ch. Vinayak, G. Vishnu believe that the introduction of artificial intelligence (AI) can reduce ineffi-ciencies and inconsistencies due to manual intervention in workloads in various areas and focused on load scheduling to reduce human dependency and error, ground handling with an autopilot vehicle that can operate in all weather conditions, se-quence of loading devices (ULD) based on priority, operating control unit to move the ULD in the cargo deck, fix the ULD when it is stuck during operation and implement predictive maintenance based on artificial intelligence for electrical and mechanical components of Cargo and making design decisions based on artificial intelligence in cargo LRUs [2]. R. Sahoo, B. Bhowmick, M. Tiwari point out that by analyzing the potential of using Blockchain technologies in smart air cargo transportation, a contri-bution is made to the field of logistics management. They proposed a distributed, multi-layered blockchain structure for efficient and safe air cargo transportation, which helps to transform aviation logistics operations in national and international trade [3]. K. Rashi, Th. Amit points to the impact of the Internet of Things (IoT) on the aviation industry, transforming air travel into an intelligent process. They empha-size that IoT helps in intelligent analysis and execution of services for better customer service and revenue generation [4]. M. Narayanan, K. Rakesh, J. Jayasundaram, K. Ranjith, J. Aarthy, C. Catherene point out that aviation big data plays a crucial role in the growth of intelligent aviation industry and the establishment of aviation big data platform and information system, and and the use of methods of multi-level network correlation analysis will significantly increase the safety and efficiency of aircraft. They propose the implementation of a national aviation big data platform and information system, as well as global cooperation in the field of aviation big data and the introduction of aviation big data technology [5]. GP Elifcan explores state-of-the-art applications for transportation planning, including baggage handling ser-vices, routing, safety and reliability, and standards including environmental effects, docking and navigation, object detection and protection, communication and inte-gration.It explains the terminology, using a practical decision support system based on Analytic Hierarchy Process (AHP) and Fuzzy Inference System (FIS). It offers the architecture of autonomous robotic units, baggage handling systems, and analysis of obstacle photography obtained from cameras. The scientist advises airport planners on sound policies and improved operations [6]. K. Kolasińska-Morawska, Ł. Sułkow-ski, P. Buła, M. Brzozowska, P. Morawski analyze the impact of the application of a smart idea in the field of sustainable development on changes in customer service logistics at the "last mile" stage, which involves reaching as many recipients as possi-ble, but has a negative impact on the natural environment , which certainly contra-dicts the concept of sustainable development, including trends that are visible in the logistics market, customers, and in the law [7]. Cai, Zihao reveal machine learning as an advanced method of modern computer modeling, which has shown good suitabil-ity and promising application in air traffic management. They propose methods and modeling processes for machine learning in air traffic management, followed by the current status of machine learning research in three areas, namely air traffic flows, air traffic services and airspace management, respectively, and finally, challenges and future prospects are indicated development of the application of machine learning in air traffic control [8]. A. Alameeri, M. Ajmal, M. Hussain, R. Nelo developed frame-works for identification, categorization and prioritization of practical sustainable development in airlines based on Analytical Hierarchy Process (AHP). They proved that public administration, as well as management, has the highest priority of innova-tion in airlines established in the United Arab Emirates (UAE) within the framework of practical sustainable development. They found the three most important sub-criteria of sustainability in these airlines are policy compliance, marketing innovation and innovation and development [9]. J. Thums, L. Künzel, M. Klumpp, M. Bard-mann, C. Ruiner point out that air freight logistics has always been at the forefront of technology development and international cooperation. They argue that autopilots used to be considered a convenient addition, today digital systems are emerging that fundamentally change the role of workers and collaboration processes at all levels of management, not least through safety and security regulations to maintain quality and efficiency [10]. Although scientists have conducted many thorough studies in the logistics field of the transport industry, the study of smart logistics in the organization of air transport requires a more thorough and in-depth disclosure of many aspects of the implementation of digital technologies in air transport systems, in particular, the problem of synergy of digital technologies in air cargo systems requires further discus-sion the current level of their development.

3 The Purpose of the Work

To analyze the historical and modern components of smart logistics, to reveal their im-pact on the aviation industry and prospects for recovery and development in the post-conflict period.

4 Research Methodology

The methodological basis is general scientific methods, namely: a dialectical ap-proach to processes and phenomena, a systemic approach, which involves consider-ing logistics processes in the form of a system that is influenced by internal and ex-ternal factors. In particular, methods of analysis and synthesis, abstraction - for writ-ing conclusions, statistical method were used to reveal them. Smart logistics is the methodological basis of digital economy research. The rapid development of the latest technologies based on virtual networks and services creates optimal manage-ment of virtual material and information resources of both a separate enterprise and the aviation industry as a whole. The smart logistics method allows the use of theoret-ical and methodological tools and practical methods of studying virtual and real logistics systems and activities. It describes specific elements of modern digitalization, which include the electronic circulation of documents for the transportation of pas-sengers and cargo by airlines, management of own electronic offices, regulation of logistic business processes. The smart approach has a set of criteria in the system of building clear rules that establish levels of responsibility in various processes of man-aging logistics operations.

5 Presentation of the Main Research Material

In the modern world, intelligence has become a popular term, which translated in-to English means "smart" and involves the use of the latest technologies, innovations that facilitate various aspects of people's lives and bring society to a new level of development - smart. The latest digital technologies have caused the transformation of the logistics sector of the aviation industry economy and the redistribution of sup-ply chains. Therefore, under the influence of digital technologies, international coop-eration developed, which led to the adoption of universal international agreements, manufacturers got the opportunity to bring the supply chains of raw materials and points of shipment of finished products closer to customers. At the same time, the digital economy is replacing the traditional economy, transforming modern life and business, developing various sectors of the transport system, where its implementa-tion is well-founded and fast. The annual dynamism of the development and im-provement of digital technologies allows us to classify them by the time of their emergence and implementation in the management and organization of air transpor-tation:

1. Initial stage: emergence of the first digital technologies and gradual automation of logistics processes (1990s - 2000s). During this period, the first automated programs for the organization of cargo transfers by air transport (Cargo Management Systems, CMS) are being implemented. For the first time, electronic air waybills (e-air waybill) are being used in order to facilitate cargo handling and eliminate paper media. For the first time, barcodes and databases are used to track the movement of goods on the delivery route. Subsequently, there is a mass application of GPS technologies for monitoring cargo transportation in real time. At the same time, there is a telecommunications technology that allows collecting data on the condition of the cargo. At the same time, information exchange between participants in the logistics process (carri-ers, customs, airport, airline) improves;
2. Analytical stage: introduction of the Internet of Things and conducting analytical actions (2000s-2010s). At this time, IoT technology is used for the first time to monitor the storage and transportation of cargo according to the main parameters defined by the Aviation Regulations (pressure, packaging, humidity, dust, temperature, etc.). Continuous updating of cargo data, its condition, location and display in real time is carried out using smart sensors attached to packages or containers. For the first time, cloud platforms are used to collect, analyze and use data, setting the level of produc-tivity based on cloud platforms in combination with IoT. At the same time, the de-velopment of analytics based on the processing of large databases (Big Data) is taking place for predicting demand and supply in the aviation services market, optimizing transportation routes involving other types of transport and complicating routes, improving the use of resource potential due to full traffic on the route, reducing costs for fuel and lubricants. Study of peak periods of cargo and passenger transportation, optimization of the organization of transportation of goods based on the implementation of current machine learning schemes. Therefore, there is a joining of global

platforms for the purpose of joint management of logistics operations and timely exchange of data between participants of air transportation. At the same time, the need for information protection is increasing and it is proposed to use blockchain technologies for the purpose of transparency, availability and relevance of cargo transportation at all points of the route in real time and confidence in its authenticity during movement and delivery. Thus, decentralized data accounting reduces the risks of loss of cargo data, forgeries or other criminal actions related to it. An attempt to implement "smart contracts" for automatic confirmation of the achievement of agreements in concluded agreements;

3. Intellectual stage. Comprehensive use of artificial intelligence, smart logistics as a service (Logistics as a Service, LaaS) (2010s - present time and future). At this stage, artificial intelligence is involved in planning transportation routes, predicting delays, and effectively distributing the resource potential of enterprises. At the same time, to automate many operations in warehouses, robots are being introduced that perform unloading and loading of goods. Therefore, important management decisions are made on the basis of options developed by artificial intelligence, which processed information about weather conditions, terrain, seasonal demand and supply in the service market, etc. In particular, the concept of smart logistics as a service (LaaS) is emerging, which offers a transition from traditional methods of management and organization of air transportation to multifaceted digitalization and global cloud platforms, which will allow effective management of logistics operations. Logistics acquires the importance of a service that will satisfy the needs of business model participants with instant connection to the global logistics system. Modern realities encourage the use of autonomous aircraft and drones for the delivery of goods to remote areas and various places.

Aviation transport is an important element of logistics service for passengers and cargo. Therefore, air transport is associated with the transportation of passengers and cargo by various airlines on the basis of charter flights or regular flights to various countries of the world. It is obvious that four directions of smart logistics service (passengers, cargo, planes, airports) are distinguished as a service area in the aviation industry.

Smart logistics activity in the field of passenger service. It provides for the use of the latest technologies and methods in order to improve the quality and efficiency of the use of the air fleet. This includes intelligent transport systems (air traffic management, flight monitoring, route optimization, flight safety). They make it possible to reduce delays and clarify flight schedules. Airlines and passengers can use updated and up-to-date information on flight status, delays and schedule changes. Mobile applications display airline services and promote the product among passengers (booking tickets, online check-in, receiving flight notifications, discounts for regular customers). Airlines present their best qualities in the service market for passenger service through a marketing strategy through the use of social media, discount programs and promotional approaches. Therefore, social media for online sales is gaining special popularity nowadays. Airlines maintain pages in social networks, and passengers have smartphones, so every potential client has the opportunity to learn about the latest news of the airline, promotions, bonuses. Therefore, communication between the client and the airline is personalized, communities and special applications are organized, which allow answering any questions and requests of the consumer in any place and at any time. Popular programs for SMS correspondence are WhatsApp, iMessage, Telegram, Viber or WeChat. Constant information notification about events in the airline company create its brand (description of the crew, aircraft, flight attendants; familiarization with the control cabin during the flight; conducting online broadcasts; stories of the crew members about interesting stories from the flight experience by the crew members in the blog or video channels). For an unforgettable impression, virtual reality methods are used in aviation marketing, during which the interior of the plane and the place of arrival are shown. Also, an important element of air travel promotion is the loyalty program (free flight, free internet, the possibility of using the highest class, exquisite meals and drinks). Therefore, airlines create positive feedback from customers (word of mouth advertising) and loyalty of regular consumers, public relations. At the same time, modern airlines widely use artificial intelligence to analyze consumer behavior and identify their needs, use it for answers in chats (chatbots) of social networks. The latest functions of the unique technologies "Smart airplane armchairs" [13] allow you to move the passenger from the waiting room to the plane. At the same time, Ukrainian airlines develop marketing strategies, but slowly implement them [14, 15]. At the same time, Ukrainian airspace is being opened for international carriers with a high level of service. Therefore, the further development of Ukrainian airlines is possible under the condition of using the experience of optimal and high-quality trends of international successful airlines in order to capture market share (Fig. 1).

Smart logistics activity in the service sector cargo. Intelligent transport systems are used to manage traffic flows, optimize routes based on traffic management, navigation systems and traffic monitoring. Real-time data helps track cargo location, assess delivery status, route changes and issues. At the same time, the collection and analysis of cargo storage and transportation data provides the Internet of Things, which provides information on temperature, humidity, etc. for various types of cargo. Loading, unloading and warehousing processes based on the automation and robotization of cargo handling processes reduce the risk of human errors, costs, and at the same time increase cargo turnover and work efficiency. The integration of logistics systems involves close interaction between airlines, customs, warehouses, etc., which reduces cargo handling time and increases their efficiency. In Ukraine, in 2024, there is an increase in the need for cross-border freight transportation due to the globalization of business and the growth of e-commerce [16].

Smart logistics activity in the field of aircraft maintenance. Application of sensors and devices for monitoring the state of the aircraft in real time (control of engine elements, hydraulic systems, temperature, humidity). Prediction algorithms based on big data analysis make it possible to organize aircraft maintenance, taking into account the state of nodes and systems, which allows to reduce breakdowns and downtime. At the same time, the automation of maintenance processes with the help of robots and automated tools makes it possible to improve the accuracy and speed of work, eliminate human errors, and increase efficiency. The Internet of Things and big data help to manage inventory in terms of the required number of parts and materials, reducing surpluses and shortages. Intelligent resource planning solutions provide the ability to find the optimal solution for aircraft maintenance, availability of technical personnel, supplies and equipment, thereby reducing aircraft downtime and increasing resource utilization. Compliance with safety and high quality standards is provided by quality management systems. Therefore, mobile applications provide technical personnel with comfortable access to instructions, documentation, checklists, quickly report on completed work and update in real time. At the same time, the training of technical personnel and the organization of virtual inspections is carried out with the help of virtual and augmented reality, which increases the quality of professional training and reduces the time for training. Therefore, the large-scale invasion led to the closure of Ukrainian airspace for civil aviation. Negative consequences arose in the industry as a whole and its infrastructure, airports. Resumption of air transportation during the conflict period is not possible in Ukraine. A combination of factors (security situation, infrastructure repair and availability of specialists) will affect the restoration of air connections. Air transport labor resources (pilots, technicians) are in short supply, which will negatively affect the recovery of the industry [17].

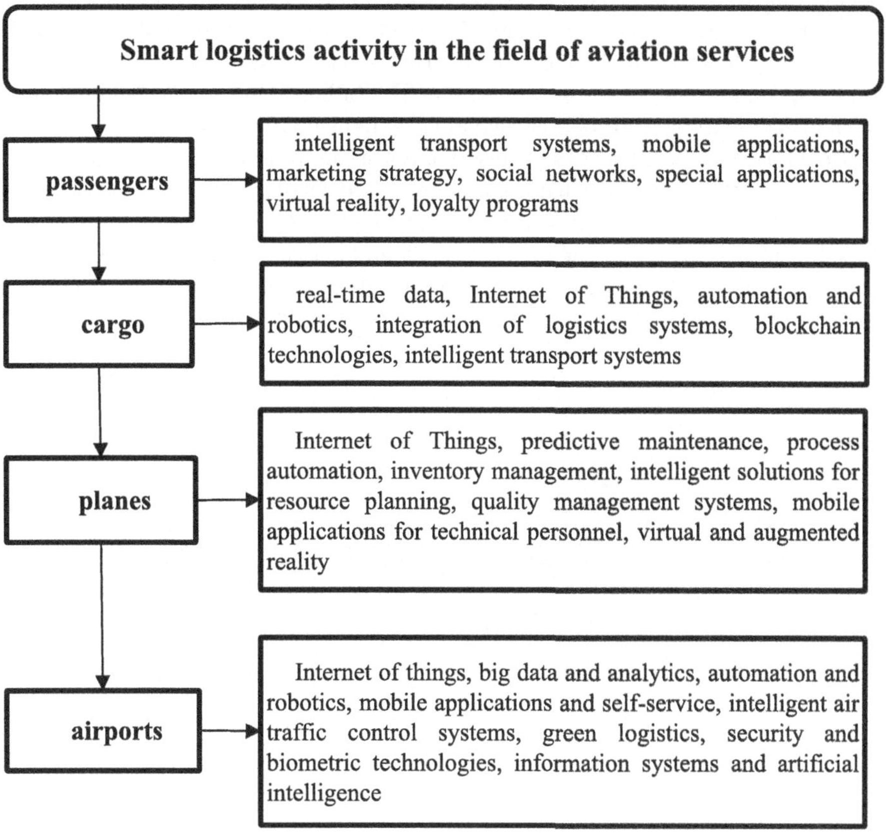

Fig. 1. Smart logistics activity in the field of aviation services

Smart logistics activities in the field of airport services. Baggage management, integration of various systems, devices, monitoring of the technical condition of equipment, control of temperature regime, humidity is based on the Internet of Things. Flight schedule optimization, terminal load planning, improved resource management, service demand forecasting is based on big data and analytics. Registration of passengers, baggage handling based on robots and automated systems takes place with the help of automation and robotics. Flight information, ticket booking, online check-in, flight status notifications, electronic boarding passes are provided by mobile applications and self-service. At the same time, coordination between aircraft, flight safety and delays, control and monitoring of airspace takes place through air traffic control. Therefore, recognition of faces, fingerprints, eye scanning provides an opportunity to conduct security control based on biometric technologies. Data analysis, resource management and flight planning are carried out with the help of artificial intelligence and information systems. The decline in export-import activity led to a decrease in the volume of freight transportation both in value and in physical volume, but the key factor in these changes was the war and, accordingly, the loss of logistics infrastructure as a result of military operations [18]. Therefore, there were 35 airfields in Ukraine. During the conflict, 19 airfields were damaged, including 12 civil airfields and 7 dual-purpose airfields. At the same time, in monetary terms, it amounted to 57.3 billion UAH or 2.04 billion dollars, which includes airports, airfields and aviation equipment [19].

Therefore, the specified stages reflect the transition from traditional logistics to its new level - smart logistics, which involves the replacement of paper media with computer technologies, full automation of loading and unloading operations, and the introduction of intelligent systems for managing the transportation of passengers and goods. During the period of 2020–2021, the transportation of passengers increased, and cargo decreased. The number of passengers on regular flights was 68.4–74.9% (Fig. 2).

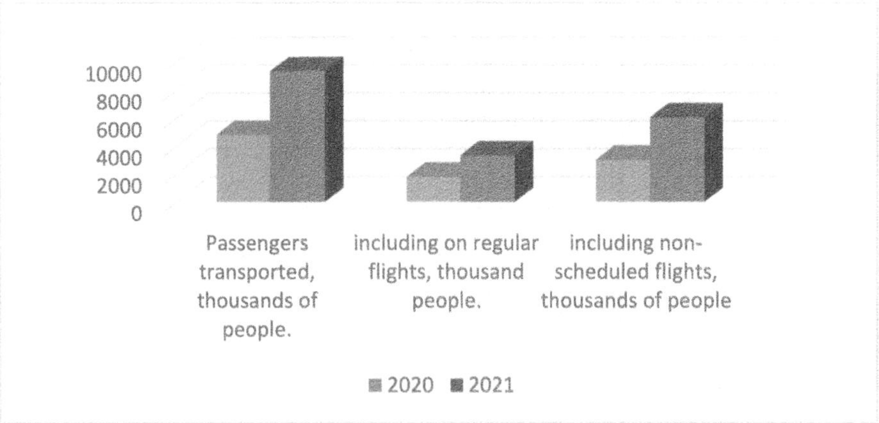

Fig. 2. Transportation of passengers of Ukraine as of 2020–2021 [11]

A comparable amount is also made up of indirect losses of the industry from hostilities, since from the moment the beginning of the war, in fact, the work of the entire civil aviation industry was stopped; income from passenger traffic stopped completely. According to preliminary estimates, the total indirect losses of the aviation industry for a period of 21 months since the beginning of the war is estimated at 154.7 billion UAH (5.3 billion dollars), of which 13.6 billion UAH (0.46 billion dollars) are losses of airports, 125.2 billion UAH (4.28 billion dollars) - losses of airlines, - 7.4 billion UAH (0.25 billion dollars) - the state-owned air traffic service enterprise of Ukraine (Ukrairorukh), UAH 8.4 billion ($0.29 billion) - losses of other entities of commercial activity on airport territories. Due to the complete shutdown of the industry, to the above estimates of indirect losses the amount for the maintenance of personnel from the beginning of the war and until June 1, 2022, i.e. for the period, is included March - May, based on the average monthly amount of labor costs [19]. Currently, Ukraine faces an acute problem of resuming the operation of airports and providing them with qualified staff to ensure flight safety. The closure of the airspace led to the loss of qualified personnel, the suspension of the development of logistics services, the introduction of advanced technologies of smart logistics.

The introduction of automation has a significant impact on the logistics processes in the organization of air transportation during the passage of goods from the point of departure to the point of arrival. During the pandemic, deficiencies in the supply system were revealed and prerequisites were created for accelerating the automation of contactless delivery processes. Airlines and airports are investing heavily in automation to improve their competitiveness. Labor-intensive processes must pass without errors, but the human factor is not excluded, so it is better to entrust such operations to a machine. Autonomous management technologies based on machine learning and artificial intelligence are widely used to eliminate shortcomings in work, with time savings and increased profits, increased volumes of cargo processing and attracting new customers. The advantages of automated processes are the reduction of the cost of specific operations (a person completes a task in a few hours, and a machine in minutes); exclusion of errors by the employee (the person accurately performs operations); reduction of delivery time (possibility to calculate machine work time and determine the delivery date); increasing the volume of orders (the machine can handle a large amount of cargo processing around the clock); maximum accuracy (elimination of errors during cargo processing, accumulation of information in one place); prevention of the influence of external factors (a machine can work around the clock, unlike a person, who is affected by many external factors); information support (collection of information about warehouses, cargo, customers, etc. takes place in one digital environment). The use of information and communication technologies at enterprises according to the type of economic activity "Transport, warehousing, postal and courier activities" showed that access to the Internet increased by 0.7% in 2021–2022, in 2021 they had 21.4% of enterprises have their website, the reduction in the use of cloud computing services was 0.2% [12]. The use of smart logistics has many obstacles. Among them, it is possible to single out a low level of personnel training, maintenance and lack of modern telecommunication technologies, an imperfect regulatory and legal framework.

Air transport logistics is the most promising supply chain. Using the main direction of cargo delivery efficiency, in which minimum costs provide maximum profit to the carrier, air transportation becomes the main link of the logistics supply chain, which will meet the requirements of high-quality service and fast delivery of cargo to the destination. Therefore, the delivery of cargo by airlines is much more expensive than by other modes of transport. However, the preservation of the cargo and its reliable transportation is ensured. This gives rise to a set of advantages of the delivery of goods by air transport: fast delivery of goods, especially over long distances; transportation range, which involves the movement of goods over long distances; ensuring cargo storage, while its damage becomes minimal, and perishable goods are delivered to their destination in a short period of time; continuous monitoring of cargo transportation over long distances regardless of its location on the delivery route.

In today's world, many different companies use the services of public and private air carriers. The peculiarities of transportation by air transport are such obstacles as: passing customs control, which involves the long duration of the procedure for issuing customs documents (waybills, information, etc.); special requirements for cargo, which include restrictions on weight, dimensions, declaration of goods in accordance with the legislation between the countries of the buyer and the seller; the formation of a logistics system, which involves laying the route, reducing transportation costs and eliminating problems; the cost is more expensive when transported by plane, because it is necessary to load it as much as possible, while small cargoes can be transported in a large volume, therefore, the transportation is postponed according to the full loading of the ship. The cargo capacity of the aircraft is limited, and the cost of cargo transportation takes into account the ratio of weight to a certain volume. Therefore, the cost of air transportation consists of air freight - the cost of transportation; fuel, airport, security allowances for transportation; the cost of airport activity (loading, unloading, document processing). The delivery infrastructure is of great importance in the logistics chain, because the aircraft must take off and land at a specially equipped point, and there is no possibility of delivery to the customer's door.

6 Research Conclusions and Prospects for Further Research in This Direction

The modern digital economy is built on the latest technologies, which are rapidly developing virtual material and information flows, which must be managed by smart logistics, which updates a large amount of information using cloud computing. Scientific studies of smart logistics in the organization of air transportation reflect virtual logistics processes on various cargo transportation routes. But a set of issues require further research: justification for the creation of analytical technologies for business evaluation to avoid delays in the delivery of goods in real time; conclusion of smart contracts in blockchain technology, management of "big data" from various locations of cargo tracking along the route. Smart logistics requires qualified personnel with knowledge in many areas, investment in the latest technologies to obtain high profits in the future, but their development exceeds the pace of their introduction into production and this becomes a problem for small and medium-sized airlines. Therefore, digital technologies are replacing outdated logistics methods through the use of a networked logistics supply system instead of a traditional chain one. They become open and accessible to all air transportation participants in the digital supply network and avoid delivery delays and various problems in the transportation of cargo and passengers.

References

1. Luning, L., Sohaib, A., Wileman, A., Perinpanayagam, S.: Digital twin in the aerospace industry: a gentle introduction. IEEE Access. 1 (2021)
2. Venkatesh, Ch., Sayooj, P., Vinayak, Ch., Vishnu, G.: Artificial Intelligence in Air Cargo System (2022)
3. Sahoo, R., Bhowmick, B., Tiwari, M.: Smart integration of blockchain in air cargo handling for profit maximization. IFIP International Conference on Advances in Production Management Systems (APMS), 107–114 (2021)
4. Rashi, K., Amit, T.: A brief review on IoT, its applications, challenges & future aspects in aviation industry. Curr. Sci. **12**, 2250–1770 (2022)
5. Narayanan, M., Rakesh, K., Jayasundaram, J., Ranjith, K., Aarthy, J., Catherene, C.: Big data analytics and an intelligent aviation information management system. Turkish Journal of Computer and Mathematics Education (TURCOMAT). **12**, 4328–4340 (2021)
6. Elifcan, G.: Smart Airport: Evaluation of performance standards and technologies for a smart logistics zone, transportation research record: Journal of the Transportation Research Board. 2675, 036119812110197 (2021)
7. Kolasińska-Morawska, K., Sułkowski, Ł, Buła, P., Brzozowska, M., Morawski, P.: Smart logistics—sustainable technological innovations in customer service at the last-mile stage: the polish perspective. Energies **15**, 6395 (2022)
8. Cai, Z.: Overview of research of machine learning in air traffic management. Appl. Comput. Eng. **8**, 103–112 (2023)
9. Alameeri, A., Ajmal, M., Hussain, M., Nelo, R.: Sustainability practices in the aviation sector: A study of UAE-based airlines. Int. J. Sustain. Soc. **9**, 119 (2017)
10. Thums, J., Künzel, L., Klumpp, M., Bardmann, M., Ruiner, C.: Future air transportation and digital work at airports – Review and developments, Transportation Research Interdisciplinary Perspectives. 19 (2023)
11. State Aviation Service of Ukraine: Information on the results of the activities of enterprises in the aviation industry, https://avia.gov.ua/pro-nas/statistika/periodychna-informatsiya
12. Statistical yearbook of Ukraine: Statistical collection, https://ukrstat.gov.ua/druk/publicat/kat_u/publ1_u.htm
13. International airlines "WestJet", https://www.westjet.com
14. Gorokhova, T., Fomenko, Y.: Peculiarities of managing logistics processes in air transport: basic tools and modern trends in the development of marketing in the industry. Econ. Soc. **15**, 266–271 (2018)
15. Yakymov V.: Peculiarities of logistics service in the aviation industry in the conditions of foreign trade, Seventy-second economic and legal discussions, series: Social sciences and humanities, NAU. 1–2 (2023)
16. Logistics 2024: New opportunities and challenges of cargo air transportation to Ukraine, Kompromis. Info, pp. 1–2, (2024)
17. Suvorova, I., Ignatusha, M.: Prospects for the development of the aviation industry in post-conflict Ukraine. State and regions, series: Economy and entrepreneurship. **1**(131), 62–66 (2024)

18. Moroz, S., Levchenko, S.: The logistics industry of Ukraine: the economic importance of the industry, losses and prospects for development in the conditions of war. Dev. Serv. Ind. Manag. **3**(11), 75–82 (2023)
19. Project of the Recovery Plan of Ukraine. The National Council for the Recovery of Ukraine from the Consequences of the War. 86 (2023)

Digital Control Tower Model for Public Transport City Network

Stanislav Popov(✉) and Inna Vyhovska

National Transport University, Kyiv 01010, Ukraine
popov@ntu.edu.ua

Abstract. Research explores the concept of a Digital Control Tower (DCT), defining it as a centralized, data-driven system that integrates real-time information from various sources to monitor, control, and optimize operations. The study delves into different kinds and functions of DCTs, particularly focusing on their applications in transportation systems. Specifically, we explore their role in managing public transport systems, such as city bus routes and networks, where DCTs ensure efficient route management, dynamic scheduling, and real-time decision-making to enhance service reliability. We also investigate the mathematical approaches used to solve the complexities of managing public bus networks. Optimization algorithms, including linear programming, dynamic programming, and machine learning models, are employed to minimize operational costs and improve service performance. Reinforcement learning algorithms are applied to dynamically adjust bus routes and schedules in real-time, while predictive maintenance models are used to reduce vehicle downtime and enhance fleet management. This paper describes a mathematical model for a city public bus transport network that integrates real-time data and predictive analytics. The proposed model demonstrates how a DCT can optimize public transportation services by efficiently solving challenges such as traffic congestion, route planning, and resource allocation. Finally, we suggest potential areas for further research, including advanced AI techniques and the integration of autonomous vehicles into city transport networks, to further improve the efficiency and sustainability of public transportation systems.

Keywords: Digital Control Tower · Public Bus Transport Network · Fleet ManAgement · Urban Mobility · Transportation Systems Optimization · Smart City Transportation

1 Introduction

A Digital Control Tower provides real-time visibility, control, and decision-making across complex systems, such as public transport networks [1]. The integration of technologies such as artificial intelligence (AI), machine learning (ML), the Internet of Things (IoT), and big data analytics into Digital Control Towers allows for effective monitoring, optimization, and management of bus operations in real-time. This is crucial as public transport systems are increasingly complex and must respond to various challenges.

O. Slavinska et al. (Eds.): ITSESQC 2024, LNNS 1335, pp. 385–394, 2025.
https://doi.org/10.1007/978-3-031-87376-8_34

Managing public transport efficiently, particularly city bus networks, has become more challenging due to increasing urbanization, congestion, and fluctuating passenger demand [2–4]. Traditional approaches, which often rely on static scheduling and reactive responses, frequently fail to adapt to the dynamic nature of modern urban traffic and passenger flow [5]. Existing studies have highlighted the limitations of such conventional methods in achieving operational efficiency [6, 7]. Additionally, previous research has demonstrated the effectiveness of integrating predictive analytics and real-time data to enhance service quality and reduce delays [8]. However, comprehensive solutions that combine real-time data analysis, predictive modeling, and automated decision-making remain underexplored.

This study builds upon these insights and proposes a novel Digital Control Tower model specifically designed for city bus networks. The primary objective of this model is to enhance operational efficiency by leveraging data-driven insights to optimize bus routes and schedules in real time [9]. Key functions of the proposed Digital Control Tower include real-time visibility, predictive analytics, automation and optimization, collaboration, and risk management. By incorporating these features, the model aims to provide a holistic approach that overcomes the challenges associated with traditional methods and supports more resilient and adaptive public transportation systems.

This research contributes to the field by detailing a methodology for the integration of AI and data analytics into public transport management. The proposed model aims to improve the reliability and effectiveness of city bus operations, fostering a sustainable and user-friendly urban transport environment. Furthermore, this paper discusses the potential benefits of the Digital Control Tower in terms of reducing operational costs, minimizing delays, and enhancing passenger satisfaction.

2 Research Methodology

The research methodology for developing and implementing the proposed Digital Control Tower model for public bus network optimization involves several key phases. The proposed model integrates various technological components such as real-time data streams, predictive analytics, and AI-driven decision-making algorithms. To ensure the effectiveness of these elements, parameters were selected based on a combination of empirical data analysis and prior studies that highlight best practices in transport management [10]. The primary parameters considered include traffic density, average bus speed, passenger load, and schedule adherence.

Each parameter was chosen to reflect critical factors influencing public bus performance. Traffic density data were sourced from city traffic management systems, while passenger load information was obtained through onboard sensors. These data points were processed using ML algorithms to enable adaptive scheduling and route adjustments. The assumptions underpinning this model include steady data flow from sensors and consistent communication between the control tower and buses, ensuring minimal delays in data processing.

To validate the model, historical data from previous bus routes and schedules were used alongside real-time data inputs. Validation involved testing the model under different traffic and demand scenarios to determine its ability to respond adaptively. A comparative analysis was conducted with existing static scheduling methods to evaluate improvements in performance metrics such as on-time arrival rates and passenger satisfaction [11].

The model's validation process included cross-referencing results with independent data sources to ensure reliability. Additionally, a sensitivity analysis was performed to identify how variations in key parameters affected model outputs. This approach helped to establish the model's robustness and confirmed that it could maintain high performance across diverse conditions.

The methodology is based on several assumptions, including consistent data transmission and the accuracy of predictive models under varying conditions. It is assumed that the data collection systems in place can provide real-time updates without significant latency. These assumptions are crucial for the model's practical implementation and were outlined to ensure that readers understand potential constraints.

3 Results and Discussion

In modern public transport systems, the effective integration of real-time data is essential for optimizing operations. The Digital Control Tower collects real-time data from multiple sources, such as GPS, IoT sensors, and smartphone-based ticketing systems, to monitor various aspects of public bus transport, including bus locations, traffic conditions, and passenger demand. This data-driven approach ensures that the transport system remains adaptable and responsive to dynamic urban environments. His real-time data integration allows transport operators to monitor the current status of buses and anticipate issues such as traffic congestion or bus delays. The control tower can then make proactive adjustments to routes and schedules, reducing delays and improving the overall efficiency of the bus network. By combining real-time data collection with predictive modeling, the system ensures that passenger demand is met while maintaining optimal bus utilization. As shown in previous research [12], integrating real-time data into public transport systems significantly enhances decision-making and overall system performance.

Real-Time Data Integration. Digital Control Tower collects real-time data from various sources, including GPS, IoT sensors, and smartphone-based ticketing systems, to dynamically monitor the position of buses, traffic conditions, and passenger demand. Mathematically, let $x(t)$ represent the location of a bus at time t, derived from GPS data, and let Tt represent traffic conditions at time t, influenced by factors like road congestion $C(t)$ and real-time incidents. Passenger demand Dt is collected through ticketing systems or mobile apps, and this data is used to optimize the flow of buses across the route [13]. This real-time integration allows operators to monitor current conditions, anticipate traffic congestion, and make proactive adjustments to minimize delays, as supported by Singh et al. [14].

Optimization Algorithms. The Digital Control Tower employs optimization algorithms, such as linear programming, dynamic programming, and stochastic modeling, to minimize operational costs Cop and maximize service efficiency. For example, the total operational cost Cop can be modeled as:

$$Z = min\left[C_{op} = \sum (FuelCost\,(t) + LaborCost\,(t) + MaintenanceCost\,(t))\right], \quad (1)$$

subject to constraints on travel time, passenger demand, and fleet size [15].

AI-Powered Decision Support. AI algorithms, particularly reinforcement learning, assist in real-time decision-making. The Digital Control Tower can reroute buses dynamically based on the current traffic flow T_t or adjust the number of deployed buses N_{bus} according to peak demand periods. A Markov Decision Process can be applied where the system learns the optimal policy $\pi_{(s)}$, mapping states s (e.g., traffic conditions, demand levels) to actions a (e.g., rerouting, dispatching extra buses). The objective is to maximize cumulative reward *R(t)*, which represents service reliability or punctuality [16].

Predictive Maintenance. Predictive maintenance algorithms are used to minimize vehicle downtime. Given operational data *V(t)*, such as vehicle usage and engine health, machine learning models predict the likelihood of breakdowns $P_b(t)$, allowing for timely maintenance. Maintenance is triggered when the probability of failure exceeds a certain threshold τ:

$$M_s = argminP_b(t),\ ifP_b(t) > \tau, \tag{2}$$

as described predictive maintenance models [13]. This predictive approach reduces the risk of unexpected breakdowns and ensures higher fleet availability.

The model of a Public Bus Transport Network Control Tower can be similarly framed using optimization and control theory, with the goal of minimizing operational costs while maximizing efficiency and service reliability in managing public bus operations, including scheduling, routing, and fleet management.

Basic outline of a mathematical framework for a Public Bus Transport Network Control Tower:

3.1 Variables and Parameters

Decision Variables.

S_t: Bus schedule at time t.

Rt: Route assignment for buses at time t.

$N_{bus,t}$: Number of buses deployed on route Rt at time t.

T_t: Travel time on route R_t at time t.

D_t: Passenger demand at time $_t$.

M_t: Maintenance performed at time t.

Parameters.

C_f: Fuel cost per unit distance traveled.

C_m: Maintenance cost per bus.

C_l: Labor cost per bus per unit time.

C_s: Shortage cost for unmet passenger demand.

P_{bus}: Maximum passenger capacity of each bus.

T_{max}: Maximum allowable travel time.

N_{max}: Maximum number of buses available.

3.2 Objective Function

Using the approach of minimizing time-related factors in public bus transport systems to enhance operational efficiency, improve passenger satisfaction, and optimize resource usage. By reducing travel time, wait time, and delays, the system becomes more responsive, cost-effective, and environmentally sustainable.

The objective function minimizes the total operational costs of the public bus transport system, including fuel, labor, maintenance, and unmet passenger demand.

$$Minimize\, Z_{costs} = \sum (C_f \cdot T_t + C_l \cdot N_{bus,t} + C_m \cdot M_t + C_s \cdot (D_t - P_{bus} \cdot N_{bus,t})^+) \tag{3}$$

where:

$(D_t - P_{bus} \cdot N_{bus,t})^+$ represents unmet passenger demand, i.e., the number of passengers who cannot be accommodated due to insufficient buses or capacity.

T is the total planning horizon.

For time optimization objective function minimizes the total travel and operational time, including travel time on routes, bus scheduling time, maintenance time, and delays caused by unmet passenger demand.

$$\begin{aligned} Minimize\, Z_{time} = & \sum (C_{Tt} \cdot T_t + C_{wait} \cdot W_t + C_{\text{maint}} \cdot M_t \\ & + C_{delay} \cdot (D_t - P_{bus} \cdot N_{bus,t})^+) \end{aligned} \tag{4}$$

where:

C_{Tt}: Weight or cost associated with travel time T_t, representing the total travel time for buses on route Rt at time t.

C_{wait}: Cost associated with passenger wait time W_t (e.g., due to buses not meeting demand or delays).

C_{maint}: Maintenance time cost for performing maintenance on buses at time M_t.

C_{delay}: Penalty cost for unmet demand or delays due to bus capacity shortfall.

$(D_t - P_{bus} \cdot N_{bus,t})^+$: Represents unmet demand or delays in passenger service, i.e., the difference between demand Dt and the capacity provided by $N_{bus,t}$ buses at time t.

The approach of minimizing time factors in public bus transport systems ensures more efficient operations, better resource utilization, improved passenger satisfaction, and a reduced environmental impact. it directly contributes to the overall effectiveness and sustainability of the public transport network, making it a valuable strategy in both short-term operational management and long-term urban mobility planning.

3.3 Constraints

The following constraints are crucial for ensuring the system functions effectively:

Bus Schedule Balance: Ensures that the number of buses on the route is enough to meet the demand, within the limits of bus capacity.

$$P_{bus} \cdot N_{bus,t} \geq D_t, \forall t \tag{5}$$

This constraint ensures that the number of buses operating on a route is sufficient to meet the demand at any given time. It prevents underutilization of buses or overloading, helping maintain service quality and efficiency. The system must deploy enough buses based on real-time passenger demand while considering bus capacity limits.

Fleet Capacity: The total number of buses deployed should not exceed the maximum available fleet size.

$$N_{bus,t} \leq N_{max}, \forall t \quad (6)$$

This constraint ensures that the total number of buses in operation does not exceed the available fleet size. The method guarantees that the deployment of buses is within the physical capacity of the fleet, preventing overscheduling and ensuring that resources are used optimally.

Travel Time: The travel time on each route must not exceed the maximum allowable time, considering traffic conditions and delays

$$T_t \leq T_{max}, \forall t. \quad (7)$$

This ensures that the travel time for each route does not exceed a predefined maximum, considering factors like traffic conditions and delays. It maintains service punctuality and reliability, crucial for passenger satisfaction. Limiting travel time helps reduce delays and improve route efficiency.

Non-Negativity: Variables like the number of buses and travel times must be non-negative.

$$N_{bus,t}, \cdot T_t \geq 0, \forall t \quad (8)$$

This constraint ensures that the number of buses and travel times are non-negative, meaning buses cannot be deployed in negative numbers, and time cannot be negative. It is a basic requirement in optimization models to maintain logical and feasible solutions.

Demand Satisfaction: The control tower should strive to meet the passenger demand within available resources.

$$P_{bus} \cdot N_{bus,t} \geq D_t, \forall t \quad (9)$$

This constraint reinforces that the system must meet passenger demand using available resources. Ensuring demand satisfaction guarantees that passenger needs are prioritized and no route is left underserved, balancing capacity and demand.

Maintenance Constraints: Maintenance must be scheduled to prevent vehicle breakdowns, ensuring operational readiness of the buses.

$$M_t \geq \tau, \forall t \quad (10)$$

where τ is the threshold for predictive maintenance, triggered when vehicle condition metrics exceed a predefined limit.

This constraint ensures that buses are regularly maintained, with maintenance scheduled to avoid breakdowns. The threshold $\tau\tau$ represents the point at which maintenance must be performed, based on vehicle condition metrics. It prevents unexpected service interruptions and ensures fleet readiness, promoting operational reliability and safety.

Constraints like fleet capacity, travel time, and demand satisfaction ensure that the system is not overstressed or inefficiently deployed. Regular maintenance constraints help prevent breakdowns, improving long-term operational stability. By incorporating these constraints into the optimization model, decision-makers can balance the network's performance and service delivery, making the public transport system more effective and responsive to real-time conditions.

Extensions. Only two levels Advanced models may incorporate stochastic and dynamic elements to handle the uncertainties in demand and traffic conditions, including:

Stochastic Programming. Accounts for variability in demand, traffic, and incidents affecting the bus network.
Dynamic Programming. Optimizes real-time decisions on route adjustments and bus scheduling over time, considering future demand and conditions.
Network Optimization Models. To manage multiple routes and the interaction between bus routes and traffic congestion

Simulation Techniques. Only two levels Simulation can be used to model various scenarios, such as traffic congestion, accidents, or changes in passenger demand, to analyze the robustness of the Public Bus Control Tower optimization model. This provides insights into the system's resilience and the effectiveness of different strategies under varying conditions.

This mathematical model for a Public Bus Control Tower aims to optimize bus fleet operations by minimizing operational costs while ensuring reliable service. It uses optimization techniques such as linear programming or mixed-integer programming to solve for the optimal bus schedules, routes, and fleet management. By integrating real-time data, the Public Bus Control Tower allows dynamic adjustments to bus deployment and maintenance schedules, providing more efficient and responsive public transportation services.

The model demonstrated a significant improvement in operational efficiency. Optimization algorithms effectively reduced fuel and labor costs, with AI-based decision support improving schedule adherence during peak hours. Predictive maintenance algorithms contributed to a reduction in vehicle breakdowns, enhancing fleet availability by 10%. These findings align with previous research, underscoring the potential of Digital Control Towers to transform public bus networks [17].

Artificial intelligence (AI) and machine learning (ML) models can significantly enhance the management of a Digital Control Tower (DCT) by automating decision-making processes, improving predictive capabilities, and optimizing operations in real time. AI models are particularly effective for dynamic scheduling and routing, predictive maintenance, and demand forecasting. Reinforcement learning, for instance, can optimize bus schedules and routing strategies by analyzing real-time data on traffic and passenger demand, while supervised learning models can predict maintenance needs based on historical and real-time vehicle data. Time series forecasting models can be used to predict passenger demand based on historical trends and external factors such as weather conditions.

Several AI and ML models are suitable for DCT applications, including reinforcement learning for real-time decision-making, supervised learning for predictive maintenance, and time series models for demand forecasting. Additionally, optimization algorithms such as genetic algorithms or linear programming can improve resource allocation and route planning. The scientific foundation for these models is grounded in reinforcement learning theory, predictive analytics, and operations research, which provide the tools for optimizing complex transportation networks. Big data and sensor networks play a critical role in feeding the AI models with the necessary real-time information for decision-making.

AI improves the DCT model by enabling real-time adaptation to changing conditions, enhancing predictive accuracy for maintenance and demand forecasting, and optimizing overall resource management. However, practical challenges exist, including issues related to data quality, the complexity of integrating AI with existing infrastructure, the need for computational resources, scalability concerns, and compliance with ethical and regulatory standards. Despite these challenges, the integration of AI and ML models into DCT systems holds considerable potential for transforming public transportation networks by improving efficiency, reliability, and responsiveness.

4 Conclusion

This research demonstrates the significant potential of a Digital Control Tower in optimizing city public bus transport networks by integrating real-time data and advanced optimization techniques. The proposed Digital Control Tower model enhances operational efficiency through dynamic scheduling, real-time route adjustments, and predictive maintenance, effectively addressing key challenges such as fluctuating passenger demand, traffic congestion, and resource allocation.

By employing optimization algorithms like linear programming, dynamic programming, and machine learning models, the model successfully minimizes operational costs related to fuel consumption, labor, and maintenance. Reinforcement learning further supports real-time decision-making, enabling the control tower to dynamically adjust bus routes and schedules based on real-time traffic data and demand fluctuations. Predictive maintenance, powered by machine learning, reduces unplanned vehicle downtime and ensures better fleet availability.

The study's results indicate improved service reliability, reduced fuel costs, and higher fleet efficiency. Moreover, the integration of IoT sensors, GPS, and ticketing systems ensures that passenger demand is met with greater precision, minimizing wait times and enhancing the overall passenger experience.

In addition to these findings, the research identifies further directions for improving public bus transport networks, including the application of artificial intelligence technologies into the Digital Control Tower model. Special attention should be given to the enhancement of the model for optimizing not only bus route networks but also transport hubs, where passenger transfers between different modes of transport occur. Describing a comprehensive mathematical model that accounts for both the bus network and the overall transport network is essential for efficient transport infrastructure management.

To create an effective model for interactions between various types of public transport at hubs, parameters such as transfer time, vehicle load, arrival and departure times, and route and schedule planning must be optimized. The use of recurrent neural networks, particularly Long Short-Term Memory, is recommended for forecasting in complex transport systems. This AI model can analyze and predict traffic and passenger flow changes, considering historical data and interactions between various transport system elements. This leads to improved accuracy in predicting vehicle arrivals, passenger demand, and other critical indicators, thereby enhancing the management of transport hubs.

Further research should explore applying this model to other forms of public transport, such as railways, trams, and on-demand services, with the ultimate goal of developing a comprehensive urban mobility management system. Real-world implementation of this model, coupled with the collection and analysis of actual operational data, is crucial for validating the model's effectiveness in live environments.

References

1. Singh, S., Goel, N., Sharma, A.: Real-time data integration in urban public transport systems. J. Transp. Eng. **142**(4), 123–136 (2020). https://doi.org/10.1016/j.jtreng.2020.04.012
2. Sun, L., Zhang, Y., Li, M.: Digital control tower framework for smart city transportation systems. IEEE Trans. Intell. Transp. Syst. **68**(1), 908–922 (2023). https://doi.org/10.1109/TITS.2023.3040192
3. Zhang, X., Li, Q., Tan, S.: AI-powered real-time bus scheduling in public transport networks. IEEE Trans. Syst. Man Cybern. **60**(4), 765–779 (2022). https://doi.org/10.1109/TSMC.2022.3213427
4. Li, W., Liu, J., Tan, X.: Machine learning models for predictive maintenance in public transport systems. J. Transp. Eng. **146**(5), 04020023 (2020). https://doi.org/10.1061/(ASCE)TE.1943-5436.0000473
5. Karlaftis, M.G., Tsamboulas, D.: Efficiency measurement in public transport: are findings specification sensitive? Transp. Res. Part A: Policy Pract. **46**(2), 206–216 (2012). https://doi.org/10.1016/j.tra.2011.09.002
6. Gkiotsalitis, K., Cats, O.: Public transport planning adaptation under the COVID-19 pandemic crisis: literature review of research needs and directions. Transp. Rev. **41**(3), 374–392 (2021). https://doi.org/10.1080/01441647.2020.1857886
7. Ceder, A., Teixeira, J.C.: Public transport network design: A Bi-Level approach. Transp. Res. Part B
8. Bimpikis, K., Candogan, O., Iyengar, G.: Spatial pricing in ride-sharing networks. Oper. Res. **67**(3), 744–769 (2019). https://doi.org/10.1287/opre.2018.1832
9. Ma, Y., Wang, Y., Chen, H.: Smart urban transit systems: from integrated framework to scheduling optimization. Transp. Res. Part C: Emerg. Technol. **114**, 147–164 (2020). https://doi.org/10.1016/j.trc.2020.02.003
10. Mahmassani, H.S., Koutsopoulos, H.N.: Real-time bus operations control: review of research and practical applications. Transp. Res. Rec. **2675**(6), 662–675 (2021). https://doi.org/10.1177/03611981211992060
11. Chen, Z., Fan, W., Ma, Y.: AI-Powered bus dispatch and scheduling in public transport systems. IEEE Trans. Intell. Transp. Syst. **22**(8), 5471–5481 (2021). https://doi.org/10.1109/TITS.2020.3010120
12. Jaiswal, R., Agarwal, P., Jain, V.: Reinforcement learning-based framework for real-time transit vehicle scheduling. J. Adv. Transp. **2020**, 1–12 (2020). https://doi.org/10.1155/2020/9046789

13. Hasan, S., Mesbah, M., Ferreira, L.: Optimizing public transport priority routes in real-time using data-driven approaches. Transp. Res. Part C: Emerg. Technol. **36**, 145–158 (2013). https://doi.org/10.1016/j.trc.2013.08.006
14. Almasri, A., Mesbah, M.: AI-Based demand-responsive public transport system for urban environments. Transp. Res. Part C: Emerg. Technol. **126**, 103005 (2021). https://doi.org/10.1016/j.trc.2021.103005
15. Xie, J., Wang, F., He, Y.: Smart City bus systems: integrating IoT and AI for real-time operations. IEEE Access **7**, 117036–117047 (2019). https://doi.org/10.1109/ACCESS.2019.2936389
16. Catalano, S., Cartenì, A., Cascetta, E.: Smart mobility in smart cities: action research and decision support system development. Transp. Res. Procedia **14**, 14–23 (2016). https://doi.org/10.1016/j.trpro.2016.05.004
17. Zhao, J., Yang, H., Ran, B.: Data-Driven bus transit scheduling optimization: a smart city approach. J. Transp. Eng. **139**(8), 863–870 (2013). https://doi.org/10.1061/(ASCE)TE.1943-5436.0000567

Mathematical Models of Goods Flows Transport Providing in Marketing

Iryna Maksiutenko[1](✉), Iryna Shevchenko[2], and Victoriia Akmaldinova[2,2]

[1] Kyiv National University of Construction and Architecture, 31, Povitroflotsky Avenue, Kyiv 03037, Ukraine
irikmax@gmail.com

[2] National Aviation University, 1, Liubomyra Huzara Avenue, Kyiv 03058, Ukraine

Abstract. The research is devoted to the formation of innovative models for the implementation of goods flow process provided that various transport modes are involved. The "supplier - transport - consumer" interaction algorithm is an effective tool for logistics management to ensure goods flow optimization. It allows to minimize the transportation, storage costs and to ensure timely delivery of goods to final consumers. The process begins with collecting data on suppliers, consumers and transport, including volumes of goods, location, demand, availability of transport, as well as time requirements. The offered algorithm makes it possible to calculate transportation time and distances between different points, which allows to determine optimal routes. An important step is the estimation of transportation costs considering the distance, transportation cost per unit of goods, and the carrying capacity of vehicles. The model takes into account the processes of stock optimization, including the warehouse resources management, to avoid shortages or surpluses, as well as possible risks and unpredictable factors, such as delays or changes in demand, suggesting alternative scenarios to ensure supply flexibility. As a result of the proposed algorithm implementation, the total costs of transportation, storage and possible fines due to delays have been calculated, an optimized delivery plan has been formed, including the transport schedule, routes and realization control. This model provides an effective logistics chain management, reduces costs and improves customer service quality, at the same time making the transportation process flexible and adaptable to external conditions.

Keywords: Mathematical Model · Providing with Transport · Goods Flow

1 Introduction

At the current stage of the globalization processes development, the role of transport is significantly increasing, but the processes of passenger and cargo transportation themselves are very complicated due to their high density. For a transport company, developing an optimal route is a relevant problem that can provide it the highest profitability. At the same time, for a manufacturing company or trading enterprise, the criterion of materials, resources and goods availability in a specified place at the specified time is important.

O. Slavinska et al. (Eds.): ITSESQC 2024, LNNS 1335, pp. 395–405, 2025.
https://doi.org/10.1007/978-3-031-87376-8_35

Non-compliance with these conditions will lead to such companies' losses, while transport companies, under the conditions of correctly constructed normative and legal aspects of interaction, will fall under compensation payment for such losses. The processes of mathematical modeling currently allow to create the most favorable conditions for the interaction of all participants in the process. In addition, modern mathematical tools, such as artificial intelligence, machine learning make it possible to simplify the calculation process as much as possible and to choose possible alternative options meeting the requirements of all participants in the process.

The problem of the research is whether all participants in the goods flow process can use the same methods and models in their calculations for planning and forecasting.

The purpose of the study is to propose a model that would satisfy the needs of all participants in the goods flow process in the "supplier - transport company - consumer" chain for profit maximization.

The task of the research is to consider and analyze the existing models, to determine their advantages and disadvantages for individual participants in the goods flow process, and to offer an optimal method of ensuring profitability.

The review of the existing modern research on the supply chain and transportation optimization issues can be summarized in three main areas: by decision-making levels, by topics as well as the used optimization models. In particular, in the study of Huang et al. (2003) proposed four classification criteria that can be used for the supply chain and transportation optimization, namely: the structure of the supply chain itself, the level of decision-making, the modeling approach, in particular the model and the optimization goal, and the criteria of shared information will be used in future research.

Analyzing scientific studies related to the level of decision-making, it can be noted, that they are mainly classified according to the scope or effect of the decision to be made in terms of time (Mula et al., 2010). The reviewed studies are classified according to the levels of adoption into 3 levels: the strategic (Garg et al. 2015), (Varseia & Polyakovskiy, 2015), (Pasandideh et al. 2015), (Ardalan et al. 2016), tactical (Darvish et al. 2016), (Zanjani et al., 2016), (Bajgiran et al. 2016), (Carvalho & Nascimento, 2016) and the operational (Bilgen and Çelebi, 2013), (Shi et al., 2015). At the same time, not enough attention is paid to management issues at the operational level.

A review of previous studies has shown that in Lóránt Tavasszy's and Vasco Reis's researches emphasis is placed on cargo transportation modeling according to the geographical characteristics of the market segments. Fredriksson Henrik in his research on the application of traffic flows mathematical modeling for transport systems improvement and the multimodal transportation planning and management proposed to integrate advanced technologies (ICT and AI) in various transport modes with the use of smart urban mobility models to optimize last-mile delivery. Vasylenko I, Kozachok L., Mnif G. and Bouamama S. proposed a software implementation of a mathematical model for optimizing the interaction of urban passenger transport, passenger-railway transfer points and urban trains in the main transport hubs in the conditions of sustainable development and digitalization. Prause Hunnar and Hoffmann Thomas in their research, analyzed the impact of the smart contract system and blockchain technologies implementation on the effectiveness of green transport corridors management, as well as identified the prospects and threats of using smart contracts to ensure legal security in the management and distribution of rights to manage common resources. Liudmyla Deineko and colleagues determines in their research the importance of creating global value chains in the goods flow process by applying multiple linear regression. Junfang Cao deals with the multi-index transportation problem by establishing a multi-index mathematical model and algorithm to find a scientific transportation method for the goods to be transported, so as to save the cost and time of transportation.

The gaps in existing researches are that there is no interests interaction model of the parties directly involved in the goods flow process.

The research methodology provides for the use of transport optimization models that make it possible to minimize the costs of transporting goods from several points of departure to several destinations and are based on the solution of the transport problem. Inventory management models helping to determine the optimal level of product stocks in warehouses are also used, taking into account storage costs, transport costs and risks of product shortages. It is possible to use transport routing models in order to determine the optimal routes for transporting goods to minimize fuel costs, delivery time and other logistics costs as well. The use of network models will allow to optimize planning and management of logistics supply chains, where each stage of goods flow can be represented in the form of a graph. This allows to visualize and optimize the goods flow through various points of the logistics network.

The use of simulation modeling will make it possible to determine alternative scenarios of goods flow, allowing to assess the consequences of changes in transport or warehouse processes before their actual implementation.

Currently, there are many optimization models used in supply chain processes, namely: the Linear Programming (LP), the Mixed Integer Programming (MIP), the Multi-Objective Linear Programming (MOLP), the Multi-Objective Mixed Integer Programming (MOMIP), the Fuzzy Mathematical Programming (FMP), the Stochastic Programming (SP) and the Heuristics (HEU) (Zare-Reisabadi & Mirmohammadi, 2015), (Varseia & Polyakov-skiy, 2015), (Fahimnia et al., 2015).

The generalization of the methods and models listed above through the application of stochastic models will allow to take into account random factors of the external environment determining the change in the market situation and the occurrence of various types of risks in the process of goods flow. They will give the possiblity to make production processes more flexible and adaptable to changing conditions.

2 Results and Discussion

For the effective organization of goods flow in the "supplier - transport - consumer" system, it is advisable to propose a mathematical model based on the optimization of the interaction between these elements.

In general, the goods flow channel (marketing or market channel) is a partially ordered set of intermediaries (retail and wholesale, transport and forwarding companies) for bringing goods from a specific manufacturer to its final consumers. That is, it is a system where the product, product liability or product ownership are transferred from level to level (from intermediary to intermediary), eventually reaching the consumer.

Here we can distinguish between the "distribution channel", which in turn includes the supply channel, and the goods flow channel [13].

The main tasks of the model are to minimize transportation costs, to ensure timely delivery and satisfy consumer demand [4].

Thus, we refer to the main model components the following:

- Supplier (S) – point or several points of goods origin.
- Transport (T) – goods transportation means (trucks, railway transport, etc.).
- Consumer (C) – point or multiple destinations (customers, consumers).

Based on the goal of this model, which is to minimize total costs (Z) and consists of the following constituent elements [15]:

- costs of goods transportation (Zt),
- costs of goods storing (Zs),
- costs associated with unsatisfied demand (Zp) (for example, fines or losses due to shortage of goods).

the objective function of the model will be:

$$Z = Zt + Zs + Zp \tag{1}$$

At the same time, this model should take into account the following limitations:

- The balance of supply and demand, which involves satisfying consumers needs with the total volume of goods supplied by suppliers. This can be expressed by the following formula:

$$\sum_{i=1}^{n} S_i \geq \sum_{j=1}^{m} C_j \tag{2}$$

where S_i—the volume of goods from the i-th supplier, C_j—the need of the j-th consumer, n—the number of suppliers, m— number of consumers.

- The vehicles capacity must ensure that each vehicle with a limited carrying capacity (T_{max}) determines the maximum volume of goods that can be transported at one time:

$$T_k \leq T_{max,} \forall k \tag{3}$$

where T_k—cargo carried by the k-th vehicle.

- Time constraints provide that the goods must be delivered within the specified terms. This takes into account the delivery time (t_d), which must meet the consumers' requirements:

$$t_d \leq t_{max} \tag{4}$$

where t_{max} is the maximum allowable delivery time.

The problem of finding the optimal route (R) that will connect the supplier with consumers through the minimum possible costs and time can be characterized by the formula:

$$R = min \sum_{i=1}^{n} d_{i,j} \times c_{i,j} \tag{5}$$

where di,j—distance between supplier i and consumer j, a ci,j—costs of goods transportation in this area.

In addition, it is necessary to take into account the optimal amount of goods that should be in stock to prevent shortages, here it is advisable to use the EOQ (economic order) model minimizing the total costs of storage and replenishment of stocks:

$$\mathrm{EOQ} = \sqrt{\frac{2\mathrm{DS}}{\mathrm{H}}} \tag{6}$$

where D—annual demand, S—ordering costs, H—storage costs.

Now we will apply a stochastic approach to the known optimization models to take into account possible delays, changes in demand or supply. This will help to adapt the transportation process to unforeseen situations.

For this purpose, we shall create an optimization model, Fig. 1.

Thus, at the output, the model should provide an optimal plan for supplying goods, delivery routes, the number of goods to be transported and the level of stocks in warehouses. This will ensure minimal costs and efficient fulfillment of consumer needs [1].

This approach provides a balanced interaction between the supplier, transport and consumer, optimizing processes and reducing risks and costs.

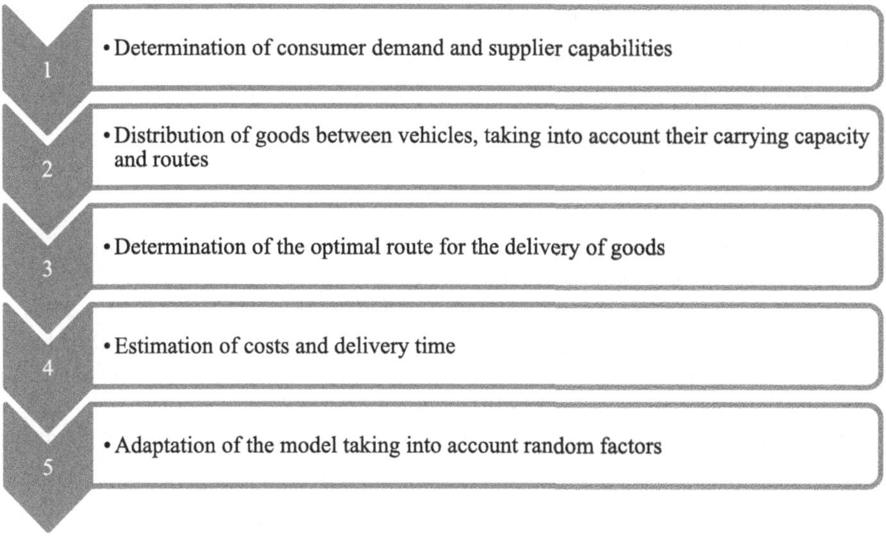

Fig. 1. Model of optimizing the interaction between the "supplier-transport-consumer" scheme participants

In order to calculate the proposed interaction model "supplier - transport - consumer" specific data are required for each stage of the goods flow process. Here are the key categories of data needed for the model to work effectively, Table 1.

Logistics indicators need special attention in the optimization process, namely: distance between points; storage costs, especially at intermediate points; restrictions on the route (by type of transport or prohibition of movement at a certain time, etc.); transportation time taking into account the characteristics of transport types and alternative routes.

The main category of indicators on which this study is focused are economic data, which take into account fixed and variable costs, inventory management costs, possible fines and penalties.

The expense part can also include the stocks characteristics, the level of which is affected by the supplier's (consumer's) limited financial resources or the limited production resources of the transport company. These shortcomings can be quickly eliminated by calculating and defining the appropriate inventory management model, but still, inventory remains a significant costs component.

In addition, it is necessary to take into account time model characteristics related to the duration of production and shipment; transportation and order processing terms.

Stochastic data from unpredictable factors can be combined into a general group with external factors, because their presence determines the emergence of new costs.

The main tasks of the limitations and metrics of this model are:

- Delivery costs - each route has its own cost, depending on the distance, type of transport, conditions of transportation, etc.
- Resource limitations—limitations on the quantity of goods from suppliers and consumer needs.
- Delivery time is an important factor for products with a short shelf life or strict time requirements.

Stochastic variable models include limited by such characteristics as:

- Demand, which may vary depending on market conditions, seasonality, etc.
- Delivery time depends on weather conditions, traffic jams and other factors.
- Expenses that may change depending on fuel prices, labor costs, etc.

The "supplier - transport - consumer" interaction algorithm is shown in Fig. 2 allocating critical costs points of each of the participants in the goods flow.

Thus, it is advisable to use a method that would combine the positive properties of time series analysis and causal methods. It is possible to use a combined method based on the use of the predictive factor-time mathematical model consisting of the predictive factor and time mathematical models [9].

At the same time, we calculate transportation costs for each individual route using the formula:

$$\text{Costs} = \text{Distance x Cost per kilometer x Product volume} \tag{7}$$

This algorithm describes the sequence of steps to optimize the process of goods flow between the supplier, transport and consumer. It ensures effective interaction of all process participants and minimizes costs.

With help of the simulation, it is possible to evaluate various scenarios of the events development: growth in demand, delays in transportation, fluctuations in the cost of transportation.

After modeling the scenarios, sensitivity analysis can be used to identify the key factors that most affect system performance.

Table 1. Initial data for calculating the optimal "supplier - transport - consumer" model (author's development)

Supplier data (S)	Consumer data (C)	Transport data (T)
Product Volume: The amount of product that suppliers can ship in a given time period	Consumer demand: the amount of goods that each consumer needs in a given period	Type of vehicle: transport characteristics (trucks, rail transport, air transport, etc.)
Location of suppliers: geographical location of each supplier for distance and route calculation	Location of consumers: geographical position of each consumer for calculating distances and optimal route	Carrying capacity: the maximum amount of goods that each vehicle can carry
Production time: the time required to manufacture products (if the supplier manufactures the goods)	Consumers' time requirements: terms by which the goods must be delivered (specified final dates for each order)	Transportation cost: transportation costs of one kilogram of goods per kilometer for each type of transport
Cost of production/supply: costs of manufacturing products or obtaining them for further transportation	Criticality of the product: some customers may require priority delivery (for example, due to the scarcity or importance of the product)	Transport speed: the average speed of movement used to calculate the delivery time

To calculate the criteria of this model, we will assume that: suppliers with a limited supply of goods; carriers have different shipping costs between suppliers and consumers; consumers with a certain level of demand for the product.

Under the conditions of the two suppliers presence (A and B), two carriers (X and Y) and two consumers (C1 and C2), it is advisable to use the following algorithm implementation code.

```
# Problem definition.
problem = pulp.LpProblem("Transport_Optimization", pulp.LpMinimize).
# Output data.
supply = {'A': 100, 'B': 120} # Stock from suppliers.
demand = {'C1': 80, 'C2': 140} # Demand from consumers.
# Shipping costs.
cost_supplier_to_carrier =
{ ('A', 'X'): 10, ('A', 'Y'): 15,
('B', 'X'): 20, ('B', 'Y'): 30}.
cost_carrier_to_consumer =
{ ('X', 'C1'): 5, ('X', 'C2'): 8,
('Y', 'C1'): 7, ('Y', 'C2'): 6}.
# Variables.
```

vars_supp_carrier = {(s, c): pulp.LpVariable(f"x_{s}_{c}", lowBound = 0, cat = 'Continuous') for s in supply for c in ['X', 'Y']}.

vars_carrier_consumer = {(c, d): pulp.LpVariable(f"y_{c}_{d}", lowBound = 0, cat = 'Continuous') for c in ['X', 'Y'] for d in demand}.

Objective function: cost minimization.

problem + = (pulp.lpSum(cost_supplier_to_carrier[s, c] * vars_supp_carrier[s, c] for s in supply for c in ['X', 'Y']) + pulp.lpSum(cost_carrier_to_consumer[c, d] * vars_carrier_consumer[c, d] for c in ['X', 'Y'] for d in demand),

"Total_Transport_Cost".

Limitations on the volume of suppliers.

for s in supply:

problem + = pulp.lpSum(vars_supp_carrier[s, c] for c in ['X', 'Y']) < = supply[s], f"Supply_limit_{s}".

Restrictions on consumer demand.

for d in demand:

problem + = pulp.lpSum(vars_carrier_consumer[c, d] for c in ['X', 'Y']) > = demand[d], f"Demand_requirement_{d}".

Balance between carriers.

for c in ['X', 'Y']:

problem + = (pulp.lpSum(vars_supp_carrier[s, c] for s in supply) = = pulp.lpSum(vars_carrier_consumer[c, d] for d in demand), f"Carrier_balance_{c}").

Solving the problem.

problem.solve().

Display results.

print("Status:", pulp.LpStatus[problem.status]).

print("Optimal Total Cost:", pulp.value(problem.objective)).

for v in problem.variables():

print(v.name, " = ", v.varValue).

At the same time:

Variables: vars_supp_carrier represent transport volumes from suppliers to carriers, and vars_carrier_consumer from carriers to consumers.

Objective function: Minimize the total cost based on the cost for each pair.

Limitation:

- Stocks with suppliers should not exceed the specified volumes.
- Consumer demand must be satisfied.
- Carrier Balance: The quantity of goods a carrier receives from a supplier must equal the quantity it supplies to consumers.

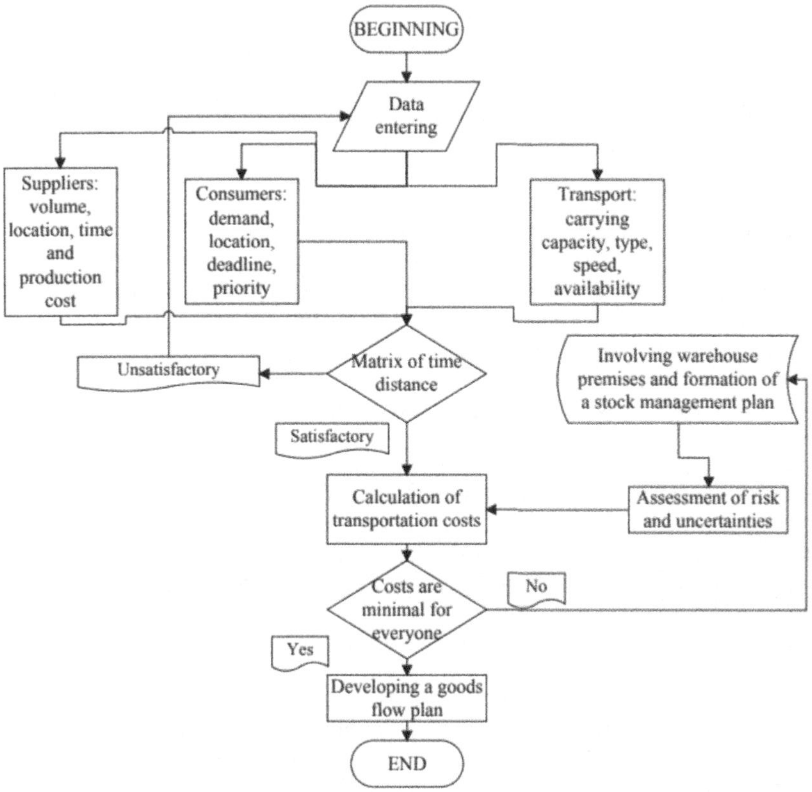

Fig. 2. Algorithm for optimizing the "supplier - transport - consumer" scheme participants interaction (author's development)

Problem solving: we use problem.solve() to find the optimal solution.

3 Main Conclusions and Possible Prospects for Further Research

Based on the importance of the goods flow process in the marketing activities of any enterprise, the optimization of product promotion processes is of great importance due to the significant costs of these operations. The use of mathematical methods and models to optimize the movement of goods allows to optimize costs and increase the competitiveness level of all enterprises interacting in the "supplier-transport-consumer" system. At the same time considering the peculiarities of transport, namely: its distribution by types, carrying capacity, availability and cost, and the need for auxiliary products transshipment operations, it is advisable to introduce an intermediary organization into the cost optimization algorithm, that has warehouse facilities and equipment at its disposal, and undertakes costs of performing these operations. The application of mathematical methods and models, as well as matrix calculations for the transport operations optimization according to the distance and time criteria, will allow all participants of the "supplier-transport-consumer" system to reduce their costs.

The requirements for computer tools when using research data modeling are best met by data analysis and technical graphics software, which differs from electronic spreadsheets and corresponding software by a wide range of data analysis and graphing tools, the ability to work with different types and large arrays of data, as well as increased productivity due to the automation of small tasks.

References

1. Ardalan, Z., Karimi, S., Naderi, B., Khamseh, A.: Supply chain networks design with multi-mode demand satisfaction policy. Comput. Ind. Eng. **96**, 108–117 (2016). https://doi.org/10.1016/j.cie.2016.03.006
2. Arifin, Y.: The impact of green supply chain management on the economic performance of SMEs, considering environmental conditions. Economic Annals-XXI. 9–10, 34–38 (2023). https://doi.org/10.21003/ea.v205-03
3. Bajgiran, O., Zanjani, M., Nourelfath, M.: The value of integrated tactical planning optimization in the lumber supply chain. Int. J. Prod. Econ. **171**(1), 22–33 (2016). https://doi.org/10.1016/j.ijpe.2015.10.021
4. Bilgen, B., Çelebi, Y.: Integrated production scheduling and distribution planning in dairy supply chain by hybrid modelling. Ann. Oper. Res. **211**(1), 55–82 (2013). https://doi.org/10.1007/s10479-013-1415-3
5. Carvalho, D., Nascimento, M.: Lagrangian heuristics for the capacitated multi-plant lot sizing problem with multiple periods and items. Comput. Oper. Res. **71**, 137–148 (2016). https://doi.org/10.1016/j.cor.2016.01.019
6. Darvish, M., Larrain, H., Coelho, L.: A dynamic multi-plant lot-sizing and distribution problem. Int. J. Prod. Res. **54**(22), 6707–6717 (2016). https://doi.org/10.1080/00207543.2016.1154623
7. Deineko, L.: An institutional mechanism for integrating domestic manufacturing into global value chains. Problems and Perspectives in Managemen. **3**, 438–451 (2019). https://doi.org/10.21511/ppm.17(3).2019.35
8. Fahimnia, B., Davarzani, H., Eshragh, A.: Planning of complex supply chains: A performance comparison of three meta-heuristic algorithms. Computers and Operations Research, in Press. 89, 241–252 (2015). https://doi.org /https://doi.org/10.1016/j.cor.2015.10.008
9. Fredriksson, H.: On the use of traffic flows for improved transportation systems: Mathematical modeling and applications. Department of Mathematics and Natural Sciences, Blekinge Institute of Technology, Karlskrona (2021)
10. Garg, K., Kannan, D., Diabat, A., Jha, P.: A multi-criteria optimization approach to manage environmental issues in closed loop supply chain network design. J. Clean. Prod. **100**, 297–314 (2015). https://doi.org/10.1016/j.jclepro.2015.02.075
11. Goncharenko, O., Tohochynskyi, O., Anishchenko, V., Sirenko, K.: Econometric modelling of innovation-driven development of Ukraine's industrial capacity. Science and Innovation. **18**(1), 20–28 (2022). https://doi.org/10.15407/scine18.01.020
12. Junfang, C.: Mathematical model and algorithm of multi-index transportation problem in the background of artificial intelligence. J. Adv. Transp. **1**, 1–11 (2022). https://doi.org/10.1155/2022/3664105
13. Kozachok, L., Kozachok, A.: Mathematical model construction of transport systems operation processes using fuzzy logic methods. Socio-Economic Research Bulletin. 3–4 (78–79), 98–106 (2021). http://vsed.oneu.edu.ua/collections/2021/78-79/pdf/98-106.pdf

14. Laiko, O., Ilchenko, S., Maslii, N., Demianchuk, M.: Scholarly research and applied approach to the interdependence of manufacture and transportation of products based on intermodality and multimodality. Science and Innovation. **1**, 86–98 (2023). https://doi.org/10.15407/scine19.01.086
15. Tavasszy, L.: Appraisal of freight projects and policies. Adv. Transp. Policy Plan. **7**, 205–233 (2021). https://doi.org/10.1016/bs.atpp.2021.01.002
16. Lysenko, O., Alexeeva, I., Dieniezhkin, M., Semenenko, O., Gamula, I.: Forecast method based on factor-time mathematical model. Interdisciplinary Studies of Complex Systems. 22, 19–31 (2023). https://doi.org/10.31392/iscs.2023.22.019
17. Mnif, G., Bouamama, S.: Multi-objective optimization methods for transportation network problems: definition, taxonomy, and annotation. Int. J. Oper. Res. Inf. Syst. **11**(1), 1–36 (2020). https://doi.org/10.4018/IJORIS.2020010101
18. Nasution, D., Mawengkang, H., Fahmi, Zarlis, M.: Mathematical model for optimization in air traffic scheduling management during the COVID-19 pandemic. East.-Eur. J. Enterp. Technol. 3 (126), 18–26 (2023). https://doi.org/10.15587/1729-4061.2023.293514
19. Pasandideh, S., Niaki, S., Asadi, K.: Optimizing a bi-objective multi-product multi-period three echelon supply chain network with warehouse reliability. Expert Syst. Appl. **42**(5), 2615–2623 (2015). https://doi.org/10.1016/j.eswa.2014.11.018
20. Prause, G., Hoffmann, T.: Innovative management of common-pool resources by smart contracts. Mark. Manag. Innov. **1**, 265–273 (2020). https://doi.org/10.21272/mmi.2020.1-22
21. Serhieieva, O.: Trade in Ukraine – today's realities. Entrepreneurship and innovation. 11–1, 113–117 (2020). https://doi.org/10.37320/2415-3583/11.17
22. Shi, L., Jiang, Y., Wang, L., Huang, D.: A novel two-stage Lagrangian decomposition approach for refinery production scheduling with operational transitions in mode switching. Chin. J. Chem. Eng. **23**(11), 1793–1800 (2015). https://doi.org/10.1016/j.cjche.2015.08.017
23. Varseia, M., Polyakovskiy, S.: Sustainable supply chain network design: A case of the wine industry in Australia. Omega **66**, 236–247 (2017). https://doi.org/10.1016/j.omega.2015.11.009
24. Vasylenko, I.: Implementation of the mathematical model of interaction of passenger transport modes for sustainable economic development. MATEC Web of Conferences. **390**, 1–6 (2024). https://doi.org/10.1051/matecconf/202439003008
25. Zanjani, M., Bajgiran, O., Nourelfath, M.: A hybrid scenario cluster decomposition algorithm for supply chain tactical planning under uncertainty. Eur. J. Oper. Res. **252**(2), 466–476 (2016). https://doi.org/10.1016/j.ejor.2016.01.048
26. Zare-Reisabadi, E., Mirmohammadi, S.: Site dependent vehicle routing problem with soft time window: Modeling and solution approach. Comput. Ind. Eng. **90**, 177–185 (2015). https://doi.org/10.1016/j.cie.2015.09.002
27. Zhang, R., Xie, P., Wang, C., Liu, G., Wan, S.: Classifying transportation mode and speed from trajectory data via deep multi-Scale learning. Comput. **162** (1) (2019). https://doi.org/10.1016/j.comnet.2019.106861

Author Index

O. Slavinska et al. (Eds.): ITSESQC 2024, LNNS 1335, pp. 407–408, 2025.
https://doi.org/10.1007/978-3-031-87376-8

The series "Lecture Notes in Networks and Systems" publishes the latest developments in Networks and Systems—quickly, informally and with high quality. Original research reported in proceedings and post-proceedings represents the core of LNNS.

Volumes published in LNNS embrace all aspects and subfields of, as well as new challenges in, Networks and Systems.

The series contains proceedings and edited volumes in systems and networks, spanning the areas of Cyber-Physical Systems, Autonomous Systems, Sensor Networks, Control Systems, Energy Systems, Automotive Systems, Biological Systems, Vehicular Networking and Connected Vehicles, Aerospace Systems, Automation, Manufacturing, Smart Grids, Nonlinear Systems, Power Systems, Robotics, Social Systems, Economic Systems and other. Of particular value to both the contributors and the readership are the short publication timeframe and the world-wide distribution and exposure which enable both a wide and rapid dissemination of research output.

The series covers the theory, applications, and perspectives on the state of the art and future developments relevant to systems and networks, decision making, control, complex processes and related areas, as embedded in the fields of interdisciplinary and applied sciences, engineering, computer science, physics, economics, social, and life sciences, as well as the paradigms and methodologies behind them.

Indexed by SCOPUS, EI Compendex, INSPEC, WTI Frankfurt eG, zbMATH, SCImago.

All books published in the series are submitted for consideration in Web of Science.

For proposals from Asia please contact Aninda Bose (aninda.bose@springer.com).

Lecture Notes in Networks and Systems 1335